The Upper Atmosphere and Solar-Terrestrial Relations

From The Times

TRANS-ATLANTIC MESSAGE

Monday, December 16, 1901 —
From Our Correspondent, St. Johns, NF, Dec. 14 —

Signor Marconi authorizes me to announce that he received on Wednesday and Thursday electrical signals at his experimental station here from the station at Poldhu, Cornwall, thus solving the problem of telegraphing across the Atlantic without wire. He has informed the Governor, Sir Cavendish Boyle, requesting him to apprise the British Cabinet of the discovery, the importance of which it is impossible to overvalue.

The Upper Atmosphere and Solar-Terrestrial Relations

An introduction to the aerospace environment

J. K. Hargreaves
Department of Environmental Sciences
University of Lancaster
Lancaster, England

formerly with

Space Environment Laboratory
National Oceanic and Atmospheric Administration
Boulder, Colorado, USA

VNR VAN NOSTRAND REINHOLD COMPANY
New York — Cincinnati — Toronto — London — Melbourne

**Published by Van Nostrand Reinhold Company Ltd.,
Molly Millars Lane, Wokingham, Berkshire, England**

*Published in 1979 by Van Nostrand Reinhold Company
A Division of Litton Educational Publishing, Inc.,
450 West 33rd Street, New York, N.Y. 10001, U.S.A.*

*Van Nostrand Reinhold Limited
1410 Birchmount Road, Scarborough, Ontario, M1P 2E7,
Canada*

*Van Nostrand Reinhold Australia Pty. Limited
17 Queen Street, Mitcham, Victoria 3132, Australia*

Library of Congress Cataloging in Publication Data

Hargreaves, John Keith, 1930—
   The upper atmosphere and solar-terrestrial relations

      Includes bibliographies and index.
      1. Atmosphere, Upper. 2. Solar radiation.
I. Title.
QC879.H28   551.5'14   78—1457
ISBN 0-442-30215-1
ISBN 0-442-30216-9 pbk.

Printed by Adlard & Son Ltd, Bartholomew Press, Dorking, Surrey

Preface

Apology, clientele and scope

Like so many other textbooks this one grew from a lecture course: a set of 25 lectures given to students of the 'Physics of the Environment' at the University of Lancaster. When I started to prepare the lectures in 1969 there was a marked lack of books in this field, and it soon became obvious that, drawing from review papers, the original literature and my own experience, I was in effect compiling my own textbook as I went along. Some good textbooks have appeared since then, but they are mainly written at a more advanced level than the one I required. Moreover, I wished to treat the subject more broadly than could be done from any one existing book. The present offering is at a level suitable for the undergraduate with a basic knowledge of classical physics — a first year physics course should be quite sufficient, though if physics courses are continuing beyond that level, so much the better.

Looking beyond the academic sphere, however, I feel that no apology should be needed. The ionosphere long ago went into operational service, and now we see that after an initial period of basic exploration the magnetospheric and near-space environments are increasingly involved in technological activities. A landmark of this progress is the development of the Space Shuttle. With up to 70 flights a year, and the capacity to carry people without full astronaut training, the Shuttle surely marks a new era in the utilization of near space. In addition, satellite techniques are finding new applications in many fields, and the procedures of communications and navigation involving the upper atmosphere become ever more sophisticated. There is a need for suitable books, both for those who study the topic as part of academic training, and for those who need a background for vocational purposes. It is hoped that this text will fill both requirements.

I have tried to match the level of presentation to the intended readership. The object is to explain the topics as simply as possible, covering the matter with reasonable breadth so that the reader who comes fresh to the subject matter will emerge with a concept of what upper-atmosphere studies are all about and why they are important, as well as gaining a useful quantity of facts and theories on the way. Of course, the main thrust of the text is to put the subject in the context of physics, since that is the only way to generate understanding; but much descriptive matter is included also, and that may interest some groups of students who are not too keen on physics but still want to know something about the subject. For all groups, advanced mathematics is avoided. At this level a basic physical insight is vastly more useful than mathematical sophistication. The book is not intended primarily for graduate students or for the specialized courses sometimes offered to third-year honours physics students in British universities. However, it may serve them as background reading to add breadth.

Structure and use of the book

The main part of the material is contained in Chapters 4–11, which describe the structure and behaviour of the upper atmosphere as now known. Chapter 3 summarizes the major observational techniques, with emphasis on the principles behind the techniques; without some knowledge of the techniques it is difficult to appreciate the development of the subject and the limits they impose. Chapter 2 discusses briefly some topics of classical physics that are basic to the succeeding chapters. It would be possible to begin with Chapters 4–11, referring back to Chapters 2 and 3 as necessary. The final chapter indicates some practical applications of upper-atmosphere and space studies – clearly a most important aspect.

To assist the reader the text has been peppered with cross-references. These should help in building up a total picture of the subject and in revision. A few passages which are more difficult, or optional, are printed in smaller type: these could be skipped at a first reading, though the serious student should take them in at some stage. Questions are given at the end of the book. It should be possible to find the answers from within the present text, though the student may have to hunt through more than one chapter to do so. The questions should help to test comprehension and to assist revision, either in private study or in tutorial sessions. For those who wish to pursue the subject in more detail – and I hope that many will – material for 'further reading' is listed at the end of each chapter. The further reading is generally at the graduate or research level.

Language

The English language as used in science varies with the nationality of the writer, and in this case may be found slightly 'mid-Atlantic'. Where British and American versions differ I have simply used the one that seemed most natural to me. I hope the result will neither offend the Americans with British pedantry, nor put off the British with Americanisms. If any passages turn out obscure as a result, I shall no doubt hear about it!

Both SI (MKS) and cgs units are used. Although SI are now preferred, many of the relevant books and papers are still in cgs and at the present time it is necessary to be able to follow both systems.

Thank-yous

Generally the material I have selected derives from a wide range of 'open literature' – advanced books, review journals, and original papers – with some use of private reports that have come my way. In Section 11.5, however, the discussion of Sun-weather relationships has drawn principally on the published work of Drs A. J. Dessler, J. W. King, W. O. Roberts, J. M. Wilcox, D. M. Willis and collaborators. I am indebted to personnel in the Institute of Telecommunication Sciences and in the Space Environment Laboratory, both in Boulder, Colorado, for discussions on various practical applications of upper-atmosphere science (Chapter 12), and to Dr T. B. Jones for comments on the manuscript.

I appreciate the kind permission of many publishers and authors to make use of their diagrams, and helpful suggestions in relation to several of them. Most of the diagrams have been slightly modified for the present purpose.

I particularly wish to thank my wife, Sylvia, for much general assistance and for constant encouragement, and Jonathan and Julia for giving up, though not without protest, so much of my time that properly belonged to them.

Contents

Symbols for Quantities

Symbols are defined as they occur in the text, but those most frequently used are summarized here for reference.

A	displacement (in a wave), anisotropy
B	magnetic induction
C	capacity
C_D	drag coefficient
c	speed of light in free space, speed of sound (in atmospheric wave theory)
D	diffusion coefficient
d	distance
E	electric field strength, energy
e	2.7183, charge on electron
F	force
F_N, F_B	ion plasma frequency, ion gyrofrequency
f	frequency
f_N, f_B	electron plasma frequency, electron gyrofrequency
G	gravitational constant
g	acceleration due to gravity
H	magnetic field strength, scale height
h	Planck's constant, altitude
I	electron content, ionization potential, integral invariant (of trapped particles), intensity (of light), magnetic dip
$\mathbf{i}$	unit vector
J	current density
j	$\sqrt{-1}$
$\mathbf{j}$	unit vector
k	Boltzmann's constant, reaction rate, wavenumber
$\mathbf{k}$	unit vector, propagation vector
L	inductance
M	molecular weight
M_E	mass of the Earth
m	mass of a particle
m_e, m_i, m_p	mass of electron, ion, proton
N	number density of charged particles
N_e, N_i	number density of electrons, ions
N_0	Avogadro's number
n	number density of neutral particles, phase refractive index
n_g	group refractive index
p	pressure, period, momentum

Q	charge
q	ion–electron production rate
R	gas constant, rigidity (of energetic particle), sunspot number
r_B	radius of gyration
r_e, r_i	gyroradius of electron, ion
s	speed of sound
T	temperature, period
T_e, T_i	temperature of electrons, ions
t	time
U, u	neutral air speed
u	group refractive index
V	volume, electrical potential
v	velocity, phase velocity (of wave)
v_A	Alfvén speed
v_e, v_i	velocity of electron, ion
$v_\parallel, v_\perp$	velocity components parallel, perpendicular, to magnetic field
W	energy, work
x, y, z	coordinates
z	atomic number, reduced height
α	pitch angle (of charged particle), recombination coefficient
β	attachment coefficient
γ	ratio of specific heats
ϵ_0	permittivity of free space
η	ionization efficiency
θ	angle
κ	absorption coefficient, dielectric constant
λ	magnetic latitude, ratio of negative ions to electrons, wavelength
λ_D	Debye length
Λ	invariant latitude
μ	magnetic moment, real part of refractive index
μ_0	permeability of free space
ν	collision frequency, frequency (of light wave)
ν_e, ν_i	collision frequency of electron, ion
π	3.14159
ρ	correlation coefficient, mass density
σ	absorption cross-section, electrical conductivity, standard deviation
τ	characteristic time, optical depth, slab thickness (of ionosphere)
χ	imaginary part of refractive index, solar zenith angle
ψ	angle
ω, Ω	angular frequency
ω_B	angular gyration frequency, Brunt–Vaisala frequency
ω_e, ω_i	angular gyrofrequency of electron, ion
ω_N	angular electron plasma frequency
ω_a	acoustic resonance frequency
Ω_N	angular ion plasma frequency

1
Nature of the Medium

1.1 Introduction

This book deals with the Earth's upper atmosphere and near-space environment. The regions to be discussed begin some 50–70 km above the Earth's surface and extend to distances which are measured in tens of earth radii (1 earth radius ~ 6400 km). Broadly, they include the terrestrial matter which is furthest from the solid earth yet is unquestionably associated with the planet Earth rather than with the Sun or any other astronomical body.

The Greeks recognized four 'elements' which are surprisingly similar to the first four states of matter known to us today. The first three states are familiar to all: solid (earth), liquid (water), and gas (air). Our particular subject introduces the fourth state: plasma (fire). The four states of matter describe different degrees of aggregation. In the solid state the atoms are fixed relative to each other but can vibrate about average positions. In a liquid they can move around but must remain within proximity of each other. In a gas there is free movement and the atoms are independent except when they collide; the atom, however, remains as an entity, the electrons being firmly bound to the nucleus. The characteristic of a plasma is that one or more electrons become detached, leaving behind an *ion* with a positive charge. However, the electrons and ions must remain loosely associated because of the electrostatic attraction between charges of opposite sign. Geophysics is the branch of physics dealing with all the states of matter that make up the Earth. Upper atmosphere geophysics is characterized by its emphasis on naturally occurring plasma.

Apart from the lightning flash, plasma phenomena are somewhat outside the everyday experience of the average terrestrial inhabitant. Yet we should remember that in the universe as a whole most of the matter is so hot or so tenuous that it exists in the plasma state. It is only in peculiar places like the Earth that matter occurs mainly as solid, liquid and neutral gas.

1.2 Essential characteristics of the upper atmosphere

We introduce here some basic properties of the upper atmosphere to set the scene for detailed discussions of the physical processes which are to follow in subsequent chapters.

1.2.1 A tenuous gas

The variations of some basic properties as a function of altitude are shown in Fig. 1.1. At 50 km the density and pressure are only about 10^{-3} (0.1%) of their values

1

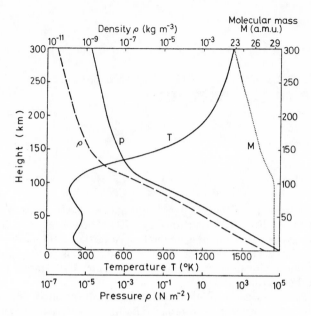

Figure 1.1 Typical variation with altitude of atmospheric pressure, density, temperature and mean molecular weight. (From H. Rishbeth and O.K. Garriott, *Introduction to Ionospheric Physics*, Academic Press, 1969).

at ground level. Therefore, if we start the upper atmosphere at 50 km we are dealing with only 0.1% of the total mass of the Earth's atmosphere. Most of our discussion will relate to heights above 100 km, and will deal with less than 10^{-6} (one part per million) of the total mass. The volume occupied is very large, however, and so the important point emerges that the gas is tenuous, at pressures which in the laboratory would be considered a good vacuum. Given only the chemical nature of the atmosphere and the nature of the emissions arriving from the Sun, all the characteristic properties of the upper atmosphere follow from the low pressure.

1.2.2 Ionization

The upper atmosphere is ionized by energetic radiations from the Sun. Once produced, the free electrons and ions tend to recombine, and a balance is established between the electron–ion production and loss. The net concentration of free electrons is greatest at heights of several hundred kilometres. Although the electron concentration at these heights amounts to only 1% of the concentration of neutral gases, it is the ionized component that introduces important phenomena to the upper atmosphere that are negligible or absent in the troposphere. Such phenomena include electric currents and fields, the reflection of radio waves, and various plasma processes. The ionized region of the upper atmosphere is the *ionosphere*.

1.2.3 Photochemistry

Once electrons and ions have been formed the ionic species may undergo chemical reactions to produce further species. The processes of ion production, loss and transformation in the presence of sunlight and other ionizing radiations are collectively known as *photochemistry*. As we shall see, the photochemistry is fairly simple at some altitudes but very complicated at others.

1.2.4 Heating

The photochemical processes release energy which heats the upper atmosphere and leads to a temperature profile such as that shown in Fig. 1.1. The decrease in temperature with height in the first few kilometres above the surface is soon halted, and eventually reversed, by these sources. Above 150—200 km the atmosphere is distinctly hot.

1.2.5 Composition

The increases in temperature at the higher levels produces an atmosphere which is stable to vertical motion and thereby inhibits the mixing of gases from different levels. In consequence, the composition changes with altitude. This may be seen in Fig. 1.1 as a decrease of mean molecular weight as the height increases above 100 km. Since the initial composition of the neutral air is a major starting point of the photochemistry, and thence of the ion production and the heating, we see immediately that many of the properties to be discussed are interlinked.

1.2.6 The geomagnetic field

As we shall see in later chapters, the magnetic field of the Earth exerts a strong influence in the upper part of the ionosphere over plasma motions, electric currents, and electrical properties in general. Further out the geomagnetic field causes the trapping of charged particles of high energy, and plays the dominant role in the more distant region known as the *magnetosphere*.

1.2.7 Solar activity

One of the characteristics of the upper atmosphere is that it depends on many variables. In addition to height, latitude, time of day and season, the level of solar activity must also be taken into account. The state of activity of the Sun varies greatly; corresponding variations appear in the upper atmosphere, which is ionized and heated by solar emissions, both electromagnetic and corpuscular. The response of the upper atmosphere to solar disturbances is complex and forms a large part of the subject of *solar—terrestrial relations*.

It will be clear from the foregoing sketch that the upper atmosphere is rather different from the troposphere with which most people are more familiar. The physics of a tenuous gas, the presence of substantial amounts of ionization, electrical conduction, energetic particles, and the controlling influences of photochemical processes and of the geomagnetic field are all aspects of the atmosphere that are not found near the ground. Transport processes are important throughout the atmosphere, and the upper regions display a dynamic variability which may be different in kind but is no less apparent than that of the troposphere. One constituent, water, that is very important in the lower atmosphere is less significant higher up.

It will also be clear from the foregoing remarks why an understanding of the upper atmosphere rests on a knowledge of classical physics.

1.3 Nomenclature of the subject

The area of science that will be treated here as 'the upper atmosphere and solar–terrestrial relations' may be encountered under various other names, and usage varies somewhat between different authors. We therefore consider the nomenclature at this point.

The upper atmosphere is a fairly general term for the higher regions, though some might use it for the neutral gas only. Note, however, that what meteorologists call 'the upper air' is within the troposphere and stratosphere and so does not come within our scope. *Upper atmosphere physics* or *upper atmosphere geophysics* refer in particular to physical processes in the upper atmosphere. *Aeronomy* is a term of more recent origin which is applied to the processes, both physical and chemical, of the ionosphere. Coming to *ionosphere* and *magnetosphere*, we find two terms with a demarcation problem, since the definition of either one includes much of the other. Most authors would use 'ionosphere' if concerned primarily with the ionization that maximizes at several hundred kilometres altitude, and 'magnetosphere' if considering the outer regions of the Earth's environment that are dominated by the geomagnetic field. One should recognize that in some circumstances the terms are virtually interchangeable. *Space physics* would strictly be the physics of space beyond the Earth's atmosphere, which would take it largely beyond our scope. In practice it is often used to mean the study of physics using space vehicles, and that certainly comes within the present topic. *Solar–terrestrial relations* is self-explanatory, but in practice tends to find much of its subject matter at the terrestrial end of the relationship and in effects at the higher altitudes.

In reading around the subject the student should keep all these italicized terms in mind.

2
Basic Physical Principles

The upper-atmosphere scientist has to be a bit of a chemist and a bit of an engineer, but more than anything he has to be a physicist. Thus the subject of this book is based in physics — mainly classical physics. It is not essential that the physics is treated at a particularly advanced level — as physics goes — though anyone who goes on to do professional work in the field will very likely have to acquire expertise in one or more branches of physics or engineering. At the introductory level what is important is a familiarity with fundamental physical concepts, such as temperature, energy, molecules, quanta, waves, heat, electric and magnetic fields, and so on. That level of familiarity is assumed. The present chapter goes one stage further, to summarize some topics that will be needed at a rather higher level and to which the student will need to refer from time to time.

Although it is clearly a good idea to learn the basics of a subject before studying the applications, the reverse process is not to be despised. It often happens that a problem encountered in the application can lead the student back to check some fundamental point and, in consequence, improve his understanding of it. (In a somewhat analogous way, the scientific exploration of the upper atmosphere from time to time stimulates new developments in basic physics.) The present chapter is intended as a link in this two-way process, as well as a summary of the relevant material.

2.1 Units, and some basic quantities

Units, it seems, are perpetually in a state of transition. Except for the persisting use of nautical miles as an altitude unit in some learned quarters, the old British Foot–Pound–Second system that many of us learned in our schooldays seems to have passed into history. The current transition from Centimetre–Gram–Second (cgs) to Metre–Kilogram–Second (MKS), also known as the International System (SI), is unfortunately not yet complete. The situation is somewhat peculiar, since many students enter university familiar only with SI, while much of the professional work in the field is still being conducted in cgs. The serious student will find that the advanced texts, the review papers, and the contributions to learned journals to which he will need to refer may be written in either cgs or SI. The only practical response is for him to become bilingual, if necessary multilingual, in units, while recognizing SI as the preferred system.

An extra complication of electromagnetism is that within the cgs system there are several possibilities, the most common being the electrostatic units (e.s.u.), the electromagnetic units (e.m.u.) and the mixed units usually called 'Gaussian'. In e.s.u. the force between two electric currents is used to define the unit of charge. In e.m.u. the force between two electric currents is used to determine the unit of current. As can be seen from Table 2.1, the ratios between the basic units in the two systems involve a factor of 3×10^{10}, the speed of light in cgs. In Gaussian units some quantities are in e.s.u. and some in e.m.u., the speed of light 'c' appearing explicitly. We shall avoid Gaussian units. When using cgs our formulae will be in e.s.u. or e.m.u. The MKS units will be 'rationalized', involving a factor of 4π more or less in certain formulae.

Table 2.1   Comparison of Units

Quantity	MKS/SI	cgs		Relation
		e.s.u.	e.m.u.	
Time	second (s)	second (s)		
Length	metre (m)	centimetre (cm)		1 m = 100 cm
Mass	kilogram (kg)	gram (g)		1 kg = 1000 g
Force	newton (N)	dyne		1 N = 10^5 dyne
Pressure	pascal (Pa) = newton/m^2	dyne/cm^2		1 Pa = 10 dyne/cm^2
Energy	joule (J)		erg	1 J = 10^7 erg
Charge	coulomb (C)	e.s.u	e.m.u.	1 C = 0.1 e.m.u. = 3×10^9 e.s.u.
Potential	volt (V)	e.s.u.	e.m.u.	1 V = 10^8 e.m.u. = 1/300 e.s.u.
Electric field strength	volt/m	e.s.u.		1 V/m = $1/(3 \times 10^4)$ e.s.u.
Electric displacement (flux density)	coulomb/m^2	e.s.u.		1 C/m^2 = 12×10^5 e.s.u.
Electric current	amp (A)	e.s.u.	e.m.u.	1 A = 0.1 e.m.u. = 3×10^9 e.s.u.
Magnetic field strength	amp/m		oersted	1 A/m = 4/1000 oersted
Magnetic induction (flux density)	tesla = weber/m^2		gauss (G)	1 Wb/m^2 = 10^4 G
Magnetic flux	weber (Wb)		maxwell (M)	1 Wb = 10^8 M
Capacity	farad (F)	e.s.u.		1 F = 9×10^{11} e.s.u.
Inductance	henry (H)		e.m.u.	1 H = 10^9 e.m.u.

Table 2.2   Some Useful Constants

		MKS/SI	cgs
Planck	h	6.624×10^{-34} J s	6.624×10^{-27} erg s
Boltzmann	k	1.381×10^{-23} J $^\circ$K^{-1}	1.381×10^{-16} erg $^\circ$K^{-1}
Gas	R	8.31 J $^\circ$K^{-1} mol^{-1}	8.31×10^7 erg $^\circ$K^{-1} mol^{-1}
Gravitational	G	6.67×10^{-11} Nm^{-2} kg^{-2}	6.67×10^{-8} cm^{-3} g^{-1} s^{-1}
Speed of light in free space	c	2.998×10^8 m s^{-1}	2.998×10^{10} cm s^{-1}
Electronic charge	e	1.602×10^{-19} C	4.803×10^{-10} e.s.u.
Electronic mass	m_e	9.11×10^{-31} kg	9.11×10^{-28} g
Proton mass	m_ρ	1.673×10^{-27} kg	1.673×10^{-24} g
Gravitational acceleration at Earth's surface	g	9.80 m s^{-2}	980 cm s^{-2}
Radius of Earth	R_E	6370 km	
Permittivity of free space	ϵ_0	8.854×10^{-12} F m^{-1}	1.0 (e.s.u.)
Permeability of free space	μ_0	1.257×10^{-6} H m^{-1} ($= 4\pi \times 10^{-7}$ H m^{-1})	1.0 (e.m.u.)

Table 2.1 compares some units of the MKS and cgs systems. To these should be added the electron-volt as a unit of energy and the gamma as a unit of magnetic flux density: two *ad hoc* units that are particularly useful in our subject. The electron-volt is commonly used to express the energy of a charged particle. As the name suggests, 1 eV is the energy gained by a particle with one electronic charge in falling through a potential difference of one volt:

$$1 \text{ eV} = 1.602 \times 10^{-12} \text{ erg} = 1.602 \times 10^{-19} \text{joule}$$

The gamma is a small unit of magnetic flux density, useful for the weak fields of the magnetosphere:

$$1\gamma = 10^{-5} \text{ gauss} = 10^{-9} \text{ tesla}$$

Table 2.2 lists some useful constants in MKS and cgs.

One minor nuisance of the MKS system is the need to include ϵ_0 and μ_0, the permittivity and permeability of free space. In cgs the magnetic induction (B) in free space is numerically equal to the magnetic field strength (H), whereas in MKS one has to write $B = \mu_0 H$. Similarly for electric displacement and electric field, $D = \epsilon_0 E$. ϵ_0 and μ_0 are related by:

$$\epsilon_0 \mu_0 = 1/c^2$$

c being the speed of light. The formulae of electromagnetism differ between MKS and cgs mainly because of ϵ_0 and μ_0 and because of the 4π due to rationalization. For example, the magnetic energy density is $B^2/8\pi$ in cgs (e.m.u.) but $B^2/2\mu_0$ in MKS (in free space). When in doubt which system an author is using, look for such clues.

2.2 Aspects of kinetic theory

2.2.1 Gas laws and continuity

The upper atmosphere being gaseous, kinetic theory, describing the bulk properties of a gas in terms of the microscopic behaviour of its constituent molecules, is fundamental.

The pressure (P), volume (V) and absolute temperature (T) of one gram-molecule of gas are related by the universal gas law

$$PV = RT \tag{2.1}$$

For N gram-molecules the relation is

$$PV = NRT \tag{2.2}$$

Here R is the gas constant, equal to $8.31 \times 10^7 \text{ erg }^{\circ}\text{K}^{-1} \text{ mol}^{-1}$. Since the density of the gas is $\rho = NM/V$ where M is the molecular weight, an alternative form is

$$P = \rho T/M \tag{2.3}$$

The gas constant is related to Boltzmann's constant $(k = 1.381 \times 10^{-16} \text{ erg }^{\circ}\text{K}^{-1})$ by $R = N_0 k$, N_0 being Avogadro's number, the number of molecules in one gram-molecule. The gas law can also be written

$$P = nkT \tag{2.4}$$

where n is the number of molecules per unit volume, often called the *number density*.

The temperature of a gas is an expression of the random motion of the molecules:

$$\overline{mv^2} = 3kT \tag{2.5}$$

where m is the mass of one molecule and $\overline{v^2}$ is the mean square velocity of the assembly of molecules. Equation 2.5 expresses the equivalence between the molecular kinetic energy $(\overline{mv^2}/2)$ and the internal energy of the gas at $kT/2$ per molecule for each degree of freedom.

Superimposed on the thermal motions there might well be a bulk motion of the gas as a whole. Continuity requires that every molecule be accounted for, and if the drift speed varies from place to place molecules will accumulate in some regions and be depleted from others. Consider an open box with dimensions x, y, z (Fig. 2.1) through which gas particles move at speed v in the x direction only. If the particle density and speed are n_1 and v_1 at face 1, and n_2 and v_2 at face 2, the number entering the box in unit time is $n_1 v_1 yz$ and the number leaving is $n_2 v_2 yz$.

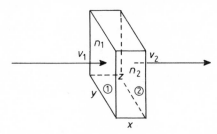

Figure 2.1   Continuity requires that the number density, n, is related to the drift speed, v, by $\partial n/\partial t = -\partial(nv)/\partial x$.

The rate of accumulation is therefore $(n_1 v_1 - n_2 v_2)yz$, and the rate of change of particle density in the box is $(n_1 v_1 - n_2 v_2)/x \to \partial(nv)/\partial x$ as the box is made infinitesimally small. Hence

$$\frac{\partial n}{\partial t} = -\frac{\partial}{\partial x}(nv) \tag{2.6}$$

In the general case, allowing movement in three dimensions

$$\frac{\partial n}{\partial t} = -\operatorname{div}(n\mathbf{v})$$

2.2.2 Thermal equilibrium

The molecules within a body of gas exchange energy by collisions and come to a state of thermal equilibrium, when the velocities are distributed according to the Maxwell–Boltzmann law:

$$N(v)\,dv = 4\pi N_T \left(\frac{m}{2\pi kT}\right)^{3/2} \exp(-mv^2/2kT)\,dv \tag{2.7}$$

where $N(v)\,dv$ is the number of molecules with speeds between v and $v + dv$, N_T is the total number of molecules, m is the molecular mass, T is the absolute temperature, and k is Boltzmann's constant. The distribution is illustrated in Fig. 2.2. If $v_{r.m.s.}$ is the root mean square speed and $v_{m.p.}$ the most probable speed (v at the peak of the distribution), then

$$v_{r.m.s.} = (3kT/m)^{1/2}$$

$$v_{m.p.} = (2kT/m)^{1/2}$$

(2.8)

In thermal equilibrium the energy is shared not only between the particles of one gas but, if the gas is a mixture, all the components are involved. Thus if two species have molecular masses m_1 and m_2, the mean square speeds $\overline{v_1^2}$ and $\overline{v_2^2}$ are related by

$$m_1\overline{v_1^2} = m_2\overline{v_2^2}$$

This is just saying that both species come to the same temperature (Equation 2.5).

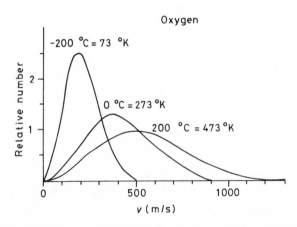

Figure 2.2 Maxwell–Boltzmann distribution of molecular speeds in oxygen at three temperatures. (After S. Borowitz and L.A. Bornstein, *Contemporary View of Elementary Physics*, copyright McGraw-Hill, 1968, used with permission of McGraw-Hill Book Company.)

The lighter gas has the higher molecular speed

$$\overline{v_2^2}/\overline{v_1^2} = m_1/m_2$$

(2.9)

2.2.3 Collisions

It is through collisions that one gas particle affects another. If there are more collisions a gas will come to equilibrium more quickly, and bulk motion of one component will be more effectively transmitted to another. The simplest definition of the collision frequency (ν) is just the number of collisions that one particle makes with others each second. The average distance that the particle travels between collisions is the *mean free path*. If the r.m.s. velocity is $v = (\sqrt{3kT/m})$, the mean free path is

$$l_f = (3kT/m)^{1/2}/v$$

Since l_f is approximately the distance between molecules in the gas, one would expect that, other things being equal, v should be directly proportional to the number density and proportional to $T^{1/2}$. In fact the collision cross-section may also be temperature dependent. As a result of the different interactions that arise, we have v varying as T for an electron colliding with neutral particles, as T^0 for an ion with neutrals (i.e. independent of T) and as $T^{-3/2}$ for an electron with an ion.

In the distant regions of the atmosphere the mean free path can be remarkably long. If it exceeds the dimensions of the containing 'box' or the typical size of the region itself, a long free path implies that thermal equilibrium cannot be assumed and the gas laws cannot be applied in their simple form.

Although it is convenient to think of collision frequency as the rate of collisions, a better definition is in terms of momentum transfer. Consider a gas mixture of two species of masses m_1 and m_2 with bulk velocities v_1 and v_2, assumed for simplicity to be in the same direction. If a particle of the first species collides head on with a particle of the second species, and if the collision is elastic, the conservation of momentum requires that the first particle gains momentum

$$2m_1m_2(v_2 - v_1)/(m_1 + m_2)$$

and the second particle loses the same amount. If there are v_{12} such collisions per unit time, and n_1 particles of the first species per unit volume, the total force (the total rate of change of momentum) on the first species due to collisions with the second is

$$F_{12} = n_1 v_{12} \cdot 2m_1m_2(v_2 - v_1)/(m_1 + m_2) \qquad (2.10)$$

The advantage of this approach is that it avoids the need to know in detail what happens at each collision. Equation 2.10 involves measurable quantities like force, mass and velocity, and thus provides a definition of v_{12}. This is the *momentum-transfer collision frequency*. Since the forces between the two species must be equal and opposite, we can write

$$F_{12} = F_{21} \qquad v_{21}/v_{12} = n_1/n_2$$

The drag on a solid body moving through the air can be derived from the same principles. If a gas contains n molecules per unit volume, the number encountered per second by an object of cross-section A moving through the gas at speed v is nAv. The change of momentum at each collision is approximately mv, and thus the drag force is

$$D = nmAv^2 = \rho v^2 A$$

where ρ is the gas density. When molecules meet the object some will bounce off but others will be deflected around it, so transferring less momentum than mv. Therefore we write

$$D = \tfrac{1}{2}\rho v^2 A C_D \qquad (2.11)$$

where C_D is a *drag coefficient* that depends on the shape of the object.

10

2.2.4 Diffusion

If there is a pressure gradient within a gas, molecules will move down the gradient to equalize the pressure. At a given moment the net velocity, superimposed on the random thermal movements, is proportional to the gradient. The drift speed may be written

$$v = -\frac{D}{n} \cdot \frac{\partial n}{\partial x} \tag{2.12}$$

where n is the particle number density, D is the diffusion coefficient, and movement in one dimension (as along a narrow column) has been assumed. (In three dimensions, $\mathbf{v} = -(D/n)\,\text{grad}\,n$.)

The rate of change of particle density at a given point is obtained by writing (Equation 2.6)

$$\frac{\partial n}{\partial t} = -\frac{\partial}{\partial x}(vn) = -\frac{\partial}{\partial x}\left(-D \cdot \frac{\partial n}{\partial x}\right) = +D\,\frac{\partial^2 n}{\partial x^2} \tag{2.13}$$

(In three dimensions, $\partial n/\partial t = D\,(\nabla^2 n)$.)

We see that the diffusion coefficient has dimensions length^2/time.

An expression for the diffusion coefficient of a gas at constant temperature can be derived by equating the pressure gradient $\partial p/\partial x$ $(= kT\,\partial n/\partial x)$ to the drag force due to collisions, $nvmv$ (Equation 2.10). Thus

$$kT\,\frac{\partial n}{\partial x} = -nvmv$$

and, by comparison with Equation 2.12,

$$D = kT/mv \tag{2.14}$$

Taking $v \propto nT^{1/2}$ (Section 2.2.3), $D \propto T^{1/2}n^{-1}$. Molecular diffusion in a neutral gas is more rapid at higher temperature and lower pressure.

The mathematical theory of diffusion has many applications besides molecular diffusion in gases, heat conduction being perhaps the best known. In the atmosphere gases can be mixed by turbulence, which differs from molecular diffusion in that the relative movement is between bubbles or cells of gas, rather than between individual molecules. The mixing due to turbulence can be treated by diffusion theory and one may define an *eddy diffusion coefficient.* There is no simple derivation for the eddy diffusion coefficient from first principles, though values may be obtained empirically. The atmospheric turbulence depends, among other things, on the vertical temperature gradient. It will be obvious that the relative importance of eddy diffusion and molecular diffusion will vary with altitude.

Other diffusion-like processes that will concern us affect the pitch angles of geomagnetically trapped particles and the motion of plasma through a magnetic field.

2.3 Some properties of a magnetoplasma

Much of the upper atmosphere is ionized, and the ionized gas is permeated by the geomagnetic field. The combination of plasma and magnetic field makes possible properties and behaviour not found in a plasma without magnetic field or in a magnetic field in a vacuum. We therefore outline here some properties of a magneto-plasma that are basic to subsequent discussions.

2.3.1 Energy density of electric and magnetic fields

The total energy of a magnetoplasma derives not only from the energy of the particles but also from the magnetic field, and from any electric field that may be present. The behaviour of the medium may depend on whether the particle energy or the field energy is the greater. By Equation 2.5, the kinetic energy of the particles is just $3nkT/2$. The formulae for the electric and magnetic energy densities are readily proved as follows.

Consider a parallel-plate capacitor of capacity C, with charge Q and potential V. The work done in charging the capacitor is $W = Q^2/2C$, since

$$W = \int V \, dq = \int_0^Q (q/C) \, dq = Q^2/2C$$

This work may be attributed to the energy in the space between the plates of the capacitor. If the plate area is A and the separation is d, the energy density is

$$w = W/Ad = Q^2/2A \, dC = \epsilon V^2/8\pi d^2 = \epsilon E^2/8\pi \qquad \text{in cgs (e.s.u.)} \qquad (2.15a)$$

(since $C = \epsilon A/4\pi d$ and $E = V/d$). In air $\epsilon = 1$, and $w = E^2/8\pi$. In MKS, the capacity $C = \epsilon A/d$. Thus

$$w = \epsilon E^2/2 \qquad\qquad\qquad (2.15b)$$

which becomes $\epsilon_0 E^2/2$ in air.

The formula for magnetic energy density is similar, and may be derived by considering a solenoid. If a solenoid of N turns, length l, area A, carries a current i, the total magnetic energy is $Li^2/2$ where L is the inductance. In cgs, $L = 4\pi N^2 A\mu/l$. Thus the energy density within the solenoid of volume Al is

$$w = 4\pi\mu N^2 A i^2/2lAl = \mu H^2/8\pi \qquad \text{in cgs (e.m.u.)} \qquad (2.16a)$$

since $H = 4\pi Ni/l$. In MKS, $L = \mu N^2 A/l$, and the magnetic field strength $H = Ni/l$; thus

$$w = \mu N^2 i^2/2l^2 = \mu H^2/2$$
$$(2.16b)$$

In air

$$w = \mu_0 H^2/2 \, (= HB/2 = B^2/2\mu_0)$$

2.3.2 Gyrofrequency

The gyrofrequency of a charged particle is its rate of gyration in a magnetic field. If the magnetic induction is **B**, the charge on the particle is e, and v is its velocity, the Lorentz force on the particle is

$$\mathbf{F} = e \cdot \mathbf{v} \times \mathbf{B}$$

12

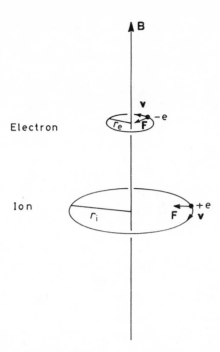

Figure 2.3 Gyrations of an electron and an ion with positive charge in a magnetic field.

acting at right angles to **v** and **B** (see Fig. 2.3). Thus the particle moves along a curved path. If there is no component of velocity in the direction of the magnetic field the path will be a circle, the Lorentz force providing the centripetal acceleration v^2/r_B, where r_B is the radius of the circle. If m is the mass of the particle, the gyro-radius is

$$r_B = mv/Be \qquad (2.17)$$

The period of revolution is $P = 2\pi r_B/v = 2\pi m/Be$. Then the angular frequency is

$$\omega_B = 2\pi/P = Be/m \qquad \text{rad/s} \qquad (2.18a)$$

and

$$f_B = Be/2\pi m \qquad \text{Hz} \qquad (2.18b)$$

r_B and ω_B depend not only on the strength of the magnetic field but also on the nature of the charged particle. If electrons and ions have the same kinetic energy, E,

$$v = (2E/m)^{1/2}$$

Then

$$r_B = (2mE)^{1/2}/Be \qquad (2.19)$$

Using subscripts i and e for ions and electrons, we see that

$$\omega_i/\omega_e = m_e/m_i \qquad r_i/r_e = (m_i/m_e)^{1/2}$$

As an example to indicate typical magnitudes, take a magnetic field of 1/2 gauss,

13

$B = 0.5 \times 10^{-4}\,\text{Wb/m}^2$, $e = 1.6 \times 10^{-19}\,\text{C}$, $m_e = 9.1 \times 10^{-31}$ kg; then

$$\omega_e = \frac{0.5 \times 10^{-4} \times 1.6 \times 10^{-19}}{9.1 \times 10^{-31}} = 8.8 \times 10^6 \text{ rad/s } (\equiv 1.4 \text{ MHz})$$

For an oxygen ion (O^+), ω_i is smaller than ω_e by a factor of 29 380, giving a gyro-frequency of only 48 Hz. For many purposes heavy ions can be considered stationary relative to electrons. If $E = 10$ keV $= 1.6 \times 10^{-15}$ J, $r_e = 6.7$ m, and $r_i(O^+) = 1.2$ km.

2.3.3 Betatron acceleration

If the magnetic induction, B, is gradually increased, the gyration frequency is increased in proportion (Equation 2.18). However, the angular momentum, given by $m v r_B = m v^2/\omega_B$, is conserved. Therefore the kinetic energy

$$E = m v^2/2 \propto \omega \propto B$$

The particle is thereby energized in proportion to the change in magnetic field. This is an important acceleration mechanism in the magnetosphere for particles moving transverse to the geomagnetic field.

2.3.4 Plasma frequency

Consider that the positive and negative charges from a slab of plasma are separated by a distance y over an area A. If the plasma contained N electrons and N ions per unit volume, the parallel-plate capacitor thereby formed (Fig. 2.4) has charge $Q = N A y e$ and capacity $C = A/4\pi y$ in cgs and $\epsilon_0 A/y$ in MKS. Working in MKS, the potential across the plates is $N e y^2/\epsilon_0$, the electric field strength in the gap is $N e y/\epsilon_0$; therefore the force on an electron is $N e^2 y/\epsilon_0$ and the restoring acceleration is

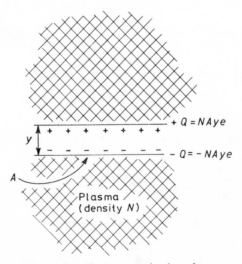

Figure 2.4   Charge separation in a plasma.

$Ne^2y/\epsilon_0 m$. Since the restoring acceleration is proportional to the separation the charged particles will undergo simple harmonic motion with angular frequency

$$\omega_N = \left(\frac{Ne^2}{\epsilon_0 m}\right)^{1/2}$$

In cgs (e.s.u.) the formula is

$$\omega_N = \left(\frac{4\pi Ne^2}{m}\right)^{1/2}$$

(2.20)

ω_N is the characteristic oscillation frequency for electrostatic perturbations in the plasma, and is an important quantity in upper atmosphere physics. It is usually expressed for numerical purposes as the frequency in Hertz, $f_N = \omega_N/2\pi$.

Since the particle mass appears in the denominator of Equation 2.20 there is a different plasma frequency for each ionic species, the frequency being lower for the heavier ones. For each, the plasma frequency is proportional to the square root of the plasma density. In general we shall use ω_N, f_N for electrons, Ω_N, F_N for ions. A convenient approximate formula for the electron plasma frequency is f_N (kHz) $\approx (81\,N)^{1/2}$ where N is in cm^{-3}.

2.3.5 Debye length

If an electron oscillating at the plasma frequency ω_N has an amplitude l, its maximum speed is $v = \omega_N l$. Suppose now that the speed is determined by the thermal motion, with $mv^2/2 = kT/2$ for motion in one dimension only. Then

$$v = (kT/m)^{1/2}$$

and

$$l = \frac{1}{\omega_N} \cdot \left(\frac{kT}{m}\right)^{1/2}$$

The amplitude of oscillation is the *Debye length*, usually written λ_D. Substituting from Equation 2.20

$$\lambda_D = \left(\frac{kT}{4\pi Ne^2}\right)^{1/2} \quad \text{in cgs (e.s.u.)}$$

$$\lambda_D = \left(\frac{\epsilon_0 kT}{Ne^2}\right)^{1/2} \quad \text{in MKS}$$

(2.21)

Numerically, $\lambda_D = 6.9\,(T/N)^{1/2}$ cm, where T is in $^\circ$K and N is in cm^{-3}.

The Debye length is the fundamental unit of length in a plasma, since it represents the distance beyond which the electrostatic field due to one particle is cancelled by the effects of particles of opposite sign. It is also known as the *Debye shielding distance*. For distances less than λ_D particles have to be considered individually, but over distances exceeding λ_D the effects of individual particles are not seen and the plasma may be treated as a fluid. The concept of a plasma requires that there shall be at least one particle in a sphere of radius λ_D. This is the *Debye sphere*.

15

2.3.6 Frozen-in field

Consider the simple dynamo in Fig. 2.5. A thin conducting annulus of width l is rotated between frictionless contacts of unit length, magnetic flux density B being applied normal to the annulus. By Faraday's law of electromagnetic induction,

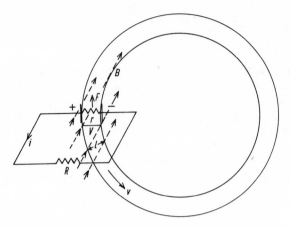

Figure 2.5   Simple dynamo, to illustrate 'freezing' between conductor and magnetic field.

the potential difference generated between the contacts is the rate of change of flux, $V = Bvl$. If the circuit is completed by a resistor R, and the resistance of the section of annulus between the contacts is r, a current

$$i = V/(R + r) = Bvl/(R + r)$$

flows in the circuit. The power dissipated is

$$P = Vi = B^2v^2l^2/(R + r) \tag{2.22}$$

Since the current flows across the magnetic field, it experiences a force $F = Bil$, acting in a direction opposing the rotation of the disc. Thus mechanical work is being done at a rate $Fv = Bilv = B^2v^2l^2/(R + r)$, equal to the electrical dissipation rate, as would be expected.

Now suppose the external circuit has zero resistance, and the conductivity of the material of the disc is σ. Then from Equation 2.22,

$$v^2 = rP/B^2l^2 = P/B\sigma l$$

If $\sigma \to \infty$, $r \to 0$. If v remains finite, $i \to \infty$, and the retarding force $\to \infty$. Clearly this is impossible. The consequence of making σ very large while keeping P finite must be to make v very small. At the limit, when the conductivity is infinite, the disc velocity is zero.

The above model illustrates an important property of the upper atmosphere: when the electrical conductivity is very large, relative motion between the plasma and the magnetic field becomes virtually impossible. The field is then said to be *frozen in*. Similarly, a magnetic field cannot enter a region already occupied by a highly conducting plasma, and it is then *frozen out*. These are significant concepts in understanding the structure and behaviour of the magnetosphere.

16

The freezing-in of magnetic field may be rigorously proved from Maxwell's equations, and is treated in texts on electromagnetic theory. We give here a simplified version.

For current density $\mathbf{J}$, electric field $\mathbf{E}$ and magnetic induction $\mathbf{B}$, two of Maxwell's equations, neglecting displacement current, give in MKS

$$\nabla \times \mathbf{E} = -\partial \mathbf{B}/\partial t \qquad \nabla \times \mathbf{B} = \mu_0 \mathbf{J}$$

Ohm's law states

$$\mathbf{J} = \sigma(\mathbf{E} + \mathbf{v} \times \mathbf{B})$$

Putting $\mathbf{v} = 0$, and eliminating $\mathbf{J}$ and $\mathbf{E}$, we obtain

$$\partial \mathbf{B}/\partial t = \nabla^2 \mathbf{B}/\mu_0 \sigma = \epsilon_0 c^2 \cdot \nabla^2 \mathbf{B}/\sigma \qquad (2.23)$$

c being the speed of light.

This is a diffusion equation (Section 2.2.4) with diffusion coefficient $1/\mu_0 \sigma$ having dimensions of $(\text{length})^2/(\text{time})$. If the conductor occupies a space characterized by length L, the time taken for a magnetic field to enter or leave it is approximately $\tau = (\mu_0 \sigma)L^2$. For times smaller than τ the field and the plasma can be considered to move together. For times greater than τ they can be considered to move independently. Representative values of τ are given in Table 2.3.

Table 2.3 Some Magnetic Decay Times

Object	τ
Copper sphere of 1 cm radius	1 s
Core of the Earth	10^4 years
Sun	10^{10} years

2.3.7 Fermi acceleration

In Section 2.3.3 we saw that charged particles moving at right angles to a magnetic field can be accelerated by the betatron process. For particles travelling along a magnetic field the main energization process is Fermi acceleration. It is assumed that a particle is reflected back and forth between two reflection regions that are moving closer together. If the particle speed is v and the path has length l, one reflection occurs every l/v s. If the path is shortening at $\mathrm{d}l/\mathrm{d}t$ the particle speed increases by $\mathrm{d}l/\mathrm{d}t$ at each reflection, and therefore at a rate

$$\frac{\mathrm{d}v}{\mathrm{d}t} = \frac{\mathrm{d}l}{\mathrm{d}t} \cdot \frac{v}{l}$$

Therefore

$$\mathrm{d}v/v = \mathrm{d}l/l$$

and the rate of increase of energy of a particle of mass m is

$$\frac{\mathrm{d}E}{\mathrm{d}t} = \frac{\mathrm{d}}{\mathrm{d}t}\left(\frac{mv^2}{2}\right) = mv \cdot \frac{\mathrm{d}v}{\mathrm{d}t} = \frac{mv^2}{1} \cdot \frac{\mathrm{d}l}{\mathrm{d}t}$$

Also, for an increment of l,

$$\frac{\mathrm{d}E}{\mathrm{d}l} = \frac{\mathrm{d}}{\mathrm{d}l}\left(\frac{mv^2}{2}\right) = mv \cdot \frac{\mathrm{d}v}{\mathrm{d}l} = \frac{mv^2}{1} = \frac{2E}{1}$$

17

Hence the relative increment of energy (dE/E) due to a relative increment of path shortening (dl/l) is simply

$$dE/E = 2 \, dl/l \qquad (2.24)$$

If the path shortens by 5% the energy increases by 10%. This affects geomagnetically trapped particles if the field lines become shorter.

2.4 Waves in a plasma

We shall be concerned with the propagation of certain kinds of waves in plasmas, and give here some fundamentals, beginning with some general wave properties.

2.4.1 Refractive index and phase velocity

If a source generates a displacement $A = A_0 \cos \omega t$, which then propagates in the x direction at velocity v, the displacement at distance x from the source is

$$A = A_0 \cos \omega (t - x/v) = A_0 \cos (\omega t - 2\pi x/\lambda) \qquad (2.25a)$$

It is sometimes convenient to use instead the j notation, writing

$$A = A_0 \exp j (\omega t - 2\pi x/\lambda) \qquad (2.25b)$$

where $j \equiv \sqrt{-1}$; it being understood that the real part is taken (since $e^{j\theta} \equiv \cos \theta + j \sin \theta$).

The angular frequency is $\omega = 2\pi f$, where f is the frequency in Hertz (cycles per second) as measured by a stationary observer. The wavelength, λ, is the distance between wave crests as measured from an instantaneous snapshot of the wave. In the three-dimensional case, the wavelength depends on the direction of propagation in relation to the direction in which the wavelength is measured. For example, if a plane wave travels in the y direction, the wavelength in the x direction, at right angles to y, will be infinite (see Section 6.5.1 and Fig. 6.17). λ is often expressed as a wavenumber, $\mathbf{k} = 2\pi/\lambda$, which can be regarded as a vector along the direction of propagation, with components k_x, k_y, k_z, giving $\lambda_x, \lambda_y, \lambda_z$ respectively.

In Equation 2.25 v is the phase velocity, $v = \lambda f$. (In the three-dimensional case, the phase speeds in the x, y and z directions are $v_x = \lambda_x f, v_y = \lambda_y f, v_z = \lambda_z f$. In an isotropic medium v_x, v_y and v_z are all greater than v measured along the propagation direction.) The velocity depends in general on the nature of the propagation medium. Thus for an electromagnetic wave in vacuum $v = c$, the speed of light. For any other medium we write

$$v = c/n \qquad (2.26)$$

where n is the *refractive index*, considered to be a property of the medium rather than a property of the wave. If the refractive index depends on the wave frequency, the medium is said to be *dispersive*. Then waves of different frequency will travel at different speeds. Section 2.4.3 discusses an important consequence of dispersion.

In some media the refractive index will be complex, written

$$n = \mu - j\chi \qquad (2.27)$$

Then, using the j notation, the displacement is

$$A = A_0 \exp j\,(\omega t - n\omega x/c) = A_0 \exp j\,(\omega t - \mu\omega x/c)\,e^{-\chi\omega x/c} \qquad (2.28)$$

This wave travels at speed c/μ, but the last term represents an exponential attenuation, i.e. absorption in the medium, the amplitude reducing by a factor e in a distance $c/\chi\omega$. Writing the attenuation as $e^{-\kappa x}$, the absorption coefficient is $\kappa = \omega\chi/c = 2\pi\chi/\lambda_0$ where λ_0 is the free-space wavelength (c/f).

In some cases the refractive index will be purely imaginary, so that

$$n = -j\chi$$

and $\qquad\qquad\qquad\qquad\qquad\qquad\qquad\qquad\qquad\qquad\qquad\qquad$ (2.29)

$$A = A_0\, e^{j\omega t}\, e^{-\chi\omega x/c}$$

This is an *evanescent* wave, which extends a distance of about $c/\chi\omega$ into the medium but does not propagate because its phase does not vary with distance. When a propagating wave is totally reflected at the interface between two media an evanescent wave exists just inside the second medium.

2.4.2 Polarization

The wave displacement discussed above may be the distance of an air molecule from its rest position, the perturbation of a water surface, the air pressure in a longitudinal sound wave, the electric or magnetic fields in an electromagnetic wave. In general, since space is three-dimensional, it is necessary to specify the direction as well as the magnitude of the displacement. In transverse waves — where by definition the displacement is normal to the direction of phase propagation — the simplest case is plane polarization in which the displacement is confined to a plane, e.g. vertical or horizontal. Circular polarization is a little more difficult to envisage. The displacement rotates by $360°$ for every wavelength travelled. Fig. 2.6 illustrates how a circularly polarized wave can be considered as the sum of two plane-polarized waves at right angles with a $90°$ phase difference. If the plane waves are not of equal amplitude or the phase difference is not $90°$, the resultant is elliptically polarized. Conversely, a plane-polarized wave can be considered to be composed of two circularly polarized waves rotating in opposite directions (e.g. Fig. 3.9). The polarization of a transverse wave is quantified by comparing the displacements in two directions at right angles, $R = A_x/A_y$. The wave is linearly polarized if R is real, elliptical if R is complex, and circular if R is purely imaginary. In a right-handed coordinate system, with the wave propagating in the z direction, $R = +j$ means a circular wave in which the displacement at a given value of z rotates clockwise.

Linear, circular and elliptical transverse waves are met in the theory of radio-wave propagation in the ionosphere. In other kinds of wave, e.g. acoustic–gravity waves, which in general include both transverse and longitudinal air movements, the displacement includes the direction of propagation. Such a case requires a second polarization parameter, $S = A_z/A_x$ — though we shall generally avoid the deeper complexities of such waves in the present text.

2.4.3 Group velocity

Fourier analysis shows that a wave can be monochromatic only if it is of infinite duration. If the duration is finite (ΔT) the waves must occupy a finite bandwidth

19

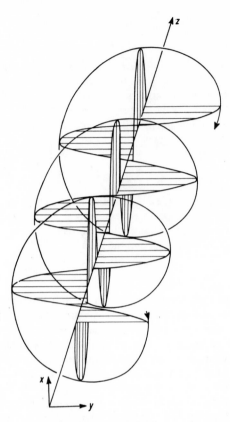

Figure 2.6 Wave with left-handed circular polarization as the sum of two plane-polarized waves.

$(\Delta f \sim 1/\Delta T)$ so that there can be cancellation before and after the interval ΔT. Similarly, any change of amplitude of the wave implies that the bandwidth must be finite. The simplest form of modulation to treat is sinusoidal amplitude modulation

$$A = A_0 (1 + m \cos \omega_m t) \cos \omega_c t \qquad (2.30)$$

where ω_c and ω_m are respectively the angular frequencies of the carrier and the modulation, and m is the depth of modulation (m is between 0 and 1). Equation 2.30 can be written

$$A = A_0 \cos \omega_c t + \tfrac{1}{2} m A_0 \cos (\omega_c - \omega_m) t + \tfrac{1}{2} m A_0 \cos (\omega_c + \omega_m) t \qquad (2.31)$$

showing that a modulated sinusoid is identical to a monochromatic carrier (ω_c) plus two monochromatic sidebands at frequencies ($\omega_c \pm \omega_m$). If the phase or the frequency is modulated instead of the amplitude, the modulated wave can still be represented as a carrier plus a spectrum of sidebands. In the case when the wave is of finite duration, the equivalent spectrum is a continuum rather than a number of discrete lines.

In Section 2.4.1, the phase velocity and the refractive index refer to an un-

20

modulated, and therefore monochromatic, wave of infinite duration. If a propagating wave is modulated we also need to known the velocity of the modulation. Known as the *group velocity*, this is not necessarily equal to the phase velocity. Each carrier and sideband will propagate at the phase velocity appropriate to its frequency. If all these velocities are the same, the wave as a whole travels at the phase velocity, but this is a special case. In general the medium will be dispersive, all three waves represented by Equation 2.31 will travel at different speeds, and the modulation will travel at some other velocity – the group velocity.

To find the relation between group and phase velocities in a dispersive medium, consider three sinusoidal waves travelling in the x direction:

$$\left.\begin{array}{l} a_0 = \cos(\omega t - kx) \\ a_1 = \cos[(\omega - \delta\omega)t - (k - \delta k)x] \\ a_2 = \cos[(\omega + \delta\omega)t - (k + \delta k)x] \end{array}\right\} \tag{2.32}$$

where k is the wavenumber ($= 2\pi/\lambda$). We have allowed k to vary with ω as $\delta k/\delta\omega$. The sum of the three monochromatic waves in Equation 2.32 is

$$A = \cos(\omega t - kx)[1 - 2\cos(\delta\omega \cdot t - \delta k \cdot x)]$$

Thus, just as the carrier with frequency $f (= \omega/2\pi)$ travels at speed $f\lambda (= \omega/k)$, so the modulation with frequency $\delta f (= \delta\omega/2\pi)$ travels at speed $\delta\omega/\delta k$. Taking a small frequency increment, it follows that the group velocity is

$$u = \partial\omega/\partial k \tag{2.33}$$

By definition the group refractive index is

$$n_g = \frac{c}{u} = c\left(\frac{\partial k}{\partial\omega}\right) = \frac{\partial}{\partial\omega}(n\omega) = \frac{\partial}{\partial f}(nf) \tag{2.34}$$

where n is the phase refractive index ($n = c/f\lambda$).

Although derived here for a simple amplitude modulation, Equations 2.33 and 2.34 apply to any plane waves. It is important to appreciate the distinction between group and phase velocities. For example, a pulse of waves travels at the group speed, while waves within the pulse travel at the phase speed. Some kinds of measurement will give the group speed and others the phase speed. Energy, and also information, are carried at the group velocity. (A little thought will show that a strictly monochromatic wave is of no use for signalling, since all the waves are identical.) This is perhaps fortunate, since the phase velocity of an electromagnetic wave can exceed the speed of light in the ionosphere; the group velocity, however, is always less than c.

2.4.4 Wave energy

In a sinusoidal wave the restoring force (F) is proportional to the displacement (A), $F = CA$, C being a constant. The work done by increasing the displacement by dA is $dW = F \cdot dA = CA \cdot dA$. Thus, at the maximum displacement (A_0) the potential energy is

$$W = \int_0^{A_0} dW \cdot dA = C\int_0^{A_0} A \cdot dA = CA_0^2/2 \tag{2.35}$$

The energy density of the wave is potential to the square of the amplitude. At other phases of the cycle at least some of the energy is kinetic. (When the displacement is zero the medium has maximum velocity and the energy is all in kinetic form). Equation 2.35 gives the energy density of the medium due to the wave propagating through it, if the constant C is defined in terms of unit volume of the medium. The energy flux of the wave is $CA_0^2 u/2$, where u is the group velocity.

According to Equations 2.15 and 2.16, the energy density due to electric and magnetic fields is $(\epsilon E^2 + \mu H^2)/2$. This applies equally well to the varying fields of an electromagnetic wave, in which the energy oscillates between electric and magnetic forms. It can be shown that, if the electric and magnetic amplitudes are E_0 and H_0, then $E_0/H_0 = (\mu/\epsilon)^{1/2}$. Thus the energy density can be written

$$\epsilon E_0^2/2 = \mu H_0^2/2 \qquad (2.36)$$

the peak electric and magnetic energy densities being equal. Note the dependence on the square of the amplitude.

2.4.5 Electromagnetic waves in an ionized medium

The propagation of electromagnetic waves in an ionized medium is treated by magneto-ionic theory. This was developed during the first part of the present century, following Marconi's experiments in long-distance radio propagation and Kennelly and Heaviside's suggestions in 1902 that the waves were reflected from an electrified layer in the upper atmosphere. Contributions were made by several theorists, including Eccles who treated the layer as a metallic reflector in 1912, and Larmor who treated it as a dielectric in 1924. The dielectric approach is in fact the appropriate one for most radio frequencies; the absorption is relatively small inside the medium, and the wave is returned to the ground by gradual refraction. In the form now used the theory is mainly due to the work of E. V. Appleton between 1927 and 1932.

The formula for the refractive index of an ionized medium is generally known as the Appleton equation or the Appleton—Hartree equation:

$$n^2 = 1 - \frac{X}{1 - jZ - [Y_T^2/2(1 - X - jZ)] \pm \{[Y_T^4/4(1 - X - jZ)^2] + Y_L^2\}^{1/2}} \qquad (2.37)$$

n is complex in the general case, $n = \mu - j\chi$ (Equation 2.27). X, Y, Z are dimensionless quantities as follows:

$$X = \omega_N^2/\omega^2 \quad Y = \omega_B/\omega \quad Y_L = \omega_L/\omega \quad Y_T = \omega_T/\omega \quad Z = \nu/\omega \qquad (2.38)$$

in which ω_N is the angular plasma frequency (Equation 2.20), ω_B is the electron gyrofrequency (Equation 2.18), and ω_L and ω_T are respectively the longitudinal and transverse components of ω_B with respect to the direction of propagation. That is, if θ is the angle between the propagation direction and the geomagnetic field,

$$\omega_L = \omega_B \cos \theta$$

and

$$\omega_T = \omega_B \sin \theta$$

22

ν is the electron collision frequency and ω is the angular wave frequency. The derivation of the Appleton—Hartree equation is beyond our present scope. The reader is referred to the book by Ratcliffe (1959) for derivations and a detailed exposition of applications.

Several special cases of the Appleton—Hartree equation will be needed, particularly in Section 3.7. In the simplest case, when there are no collisions $(Z = 0)$ and the magnetic field is neglected $(Y_L = Y_T = 0)$, the equation becomes

$$n^2 = 1 - X = 1 - \omega_N^2/\omega^2 = 1 - Ne^2/\epsilon_0 m\omega^2 \qquad (2.39)$$

Since the ionosphere is a nonmagnetic medium, Equation 2.39 also expresses the dielectric constant (κ)

$$n^2 = c^2/v^2 = \mu_0 \epsilon/\mu_0 \epsilon_0 = \epsilon/\epsilon_0 = \kappa$$

If the magnetic field is included the refractive index becomes double valued. The two waves are called the *characteristic waves*, the upper sign giving the so-called *ordinary* wave, and the lower sign the *extraordinary* wave. When collisions are significant $(Z \neq 0)$ the refractive index is complex.

Another magneto-ionic equation gives the polarization, defined as the ratio $-H_y/H_x$. (For our purposes we may consider that $-H_y/H_x = E_x/E_y$, though this is not rigorously true in all cases.) The ratio is

$$R = \frac{j}{Y_L} \left\{ \frac{Y_T^2}{2(1 - X - jZ)} \mp \left[\frac{Y_T^4}{4(1 - X - jZ)^2} + Y_L^2 \right]^{1/2} \right\} \qquad (2.40)$$

(This formula may be expressed differently if a different definition is taken for Y_L. In our case Y_L is positive if $\theta = 0$.) The two values again refer to the characteristic waves, for which the polarization remains constant as the wave propagates. A wave of arbitrary polarization entering the ionosphere is resolved into the two characteristic waves, which are considered to propagate independently within the ionosphere, to be recombined on leaving the medium. It is left as an exercise for the reader to investigate some special cases of Equation 2.40.

2.4.6 Hydromagnetic and magnetosonic waves

In a highly conducting magnetoplasma the electron—ion gas cannot cross the magnetic field lines (Section 2.3.6), but the field and the plasma can move together. Sound waves can also propagate, and for sound propagation directly along the magnetic field the field has no effect since the gas displacement is longitudinal in a sound wave. In general, if an element of the magnetoplasma is displaced, both the gas pressure and the magnetic pressure contribute to the restoring force. The existence of magnetic pressure makes possible a range of *hydromagnetic waves* that are of some importance in the magnetosphere.

The basic hydromagnetic wave is the *Alfvén wave*, in which the displacement is transverse to the magnetic field and propagation is directly along the field. This wave can be likened to that on a taut string, for which the velocity of propagation is

$$v = (T/\rho)^{1/2}$$

where T is the tension in the string and ρ is its mass per unit length. In a magnetoplasma ρ is just the mass density of the plasma $(= N_i m_i + N_e m_e$, where N_i, N_e are the number densities and m_i, m_e the masses of the ions and electrons respectively). T is the tension in the magnetic field, given by $B^2/4\pi$ in cgs and B^2/μ_0 in MKS.

Thus

$$v_A = B/(4\pi\rho)^{1/2} \qquad \text{in cgs (e.m.u.)}$$
$$v_A = B/(\mu_0\rho)^{1/2} \qquad \text{in MKS} \qquad\qquad\qquad (2.41)$$

v_A is called the *Alfvén speed.*

Along the magnetic field the Alfvén wave and the sound wave propagate independently, the first being a transverse wave and the second longitudinal. The sound speed is given by the usual formula:

$$s = (\gamma P/\rho)^{1/2} \qquad\qquad\qquad (2.42)$$

P being the gas pressure and γ the ratio of specific heats. At other angles to the magnetic field the situation is more complicated. The waves are mixed, in that both magnetic and gas pressures are involved in each of the possible waves. These mixed waves are often called *magnetosonic waves* (see Fig. 2.7). Perpendicular to the magnetic field only one wave propagates, a longitudinal wave with velocity $(v_A^2 + s^2)^{1/2}$.

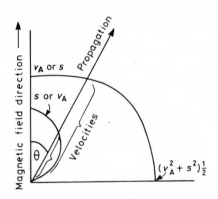

Figure 2.7   Velocities of magnetosonic waves at various angles to the magnetic field. At angle θ to the field the speed is represented by the distance from the origin. The nature of the waves depends on whether v_A or s is the greater. (After S.J. Bauer, *Physics and Chemistry in Space, Vol. 6, Physics of Planetary Ionospheres*, Springer-Verlag, 1973).

2.5 Further reading

Bauer, S. J. (1973) *Physics of Planetary Ionospheres* (Springer-Verlag) Chapter 7.

Ratcliffe, J. A. (1972) *An Introduction to the Ionosphere and Magnetosphere* (Cambridge) Chapters 6, 7, 8.

Ratcliffe, J. A. (1959) *The Magneto-ionic Theory and its Applications to the Ionosphere* (Cambridge).

3
Observational Techniques

3.1 Upper-atmosphere sensing — direct, indirect and remote

Science proceeds by its techniques. A new method of experiment, a fresh approach to observations, a new computational technique, nearly always brings a major advance of knowledge. Upper-atmosphere geophysics is no exception. Indeed, being concerned with one of the more inaccessible regions of the terrestrial environment, it provides some striking examples of the interrelation between scientific knowledge and experimental technique, and of the importance of the collaboration — long traditional in this particular field — between the physicist and the engineer.

In upper-atmosphere studies, the techniques available at any one time determine what properties, parameters and behaviour can be observed. Availability here includes the cost factor, since a known technique can remain unavailable to experimenters simply because it is too expensive. When techniques are lacking, for whatever reason, a whole region may remain in obscurity for lack of information. The first observations to become feasible will, in all probability, serve to verify some behaviour that had already been anticipated, but at the same time it is likely that some puzzles will be revealed. The progressive solution of the puzzles through observation, interpretation and further experimentation, leads eventually to a reasonably full understanding of the region and the phenomena found there — though some of the problems may well remain outstanding for many years.

The purpose of this chapter is not to go into the hardware details of equipment used in upper-atmosphere investigations, but rather to indicate the physical principles underlying the main techniques and the way they are put to use. Generally speaking, there are three approaches to upper-atmosphere sensing, which we will call direct sensing, indirect sensing and remote sensing.

(a) A direct sensor is an instrument that is placed physically in the atmospheric medium to make measurements of its immediate surroundings. Most direct sensors of the upper atmosphere are variations on standard physical laboratory instruments, carried on a rocket, a satellite or a space probe.

(b) An indirect sensor is also placed in the medium. The object does not itself make measurements, but properties of the medium may be deduced by observing the behaviour from afar — usually from a tracking station on the ground. When atmospheric winds are determined by tracking a meterological balloon by radar, the balloon is serving as an indirect sensor.

(c) In remote sensing, no instrument is placed in the medium, but its properties are deduced from the effects on a wave propagating through it. The wave can be electromagnetic or sonic. It might be transmitted into the medium from a sender (active remote sensing), or it might be a wave arising in the medium by natural processes (passive remote sensing). The study of the dust in the atmosphere by lidar is an example of active remote sensing. Studies of the luminous aurora may be considered as passive remote sensing, since the light is generated by natural processes in the upper atmosphere.

Each of the approaches has advantages and disadvantages. In general, direct methods are more specific, but they are relatively expensive because of the vehicle required. The geographical coverage is usually limited. Indirect methods also tend to be expensive but they are very good for certain measurements, e.g. air density. Remote methods are usually cheaper, often much cheaper, but they tend to suffer from difficulties of interpretation, being less specific in that a given result may have more than one possible cause. They usually have excellent continuity in time, but are limited to one geographical location at a time. However, because of the lower cost, it is often practicable to obtain geographical coverage by deploying an array of identical ground-based instruments at a number of different sites.

3.2 Direct methods for the neutral atmosphere

At high altitudes the pressure and density of the air are similar to those in a laboratory vacuum, and direct measurements of the neutral upper atmosphere are based on laboratory vacuum techniques. The transposition of these standard instruments to the new environment would be straightforward enough but for the need for a vehicle, and a moving vehicle at that. A satellite moves through the medium at a speed ($8 \, \text{km s}^{-1}$) well in excess of the r.m.s. velocity of the gas molecules ($400 \, \text{m s}^{-1}$ for oxygen). Rocket speeds are also relatively high over most of the trajectory. Vehicles are also subject to rolling and spinning. Thus if a gauge points along the trajectory it sweeps up molecules, and if it points the other way it receives only the high energy tail of the velocity distribution. Gauges therefore make use of the ram effect. One version is a ribbon microphone preceded by a rotating shutter. The periodic fluctuation of the ribbon, amounting to a few angstroms, is converted to an alternating voltage which is amplified and telemetred to ground. It will be shown in Chapter 4 (Equation 4.3) that the height variation of a gas of constant composition depends on its temperature; this provides a way of determining the temperature of the neutral atmosphere. Above 100 km the composition departs from that of the lower region, which is well mixed by turbulence and essentially like the air at ground level, at least where the major constituents are concerned. The composition can be determined with a mass spectrometer, and some types suitable for rocket and satellite use are discussed in Section 3.3.3. The problems of rocket-based mass spectrometry include contamination from the rocket exhaust and the possibility of reactions occurring on the walls of the instrument.

3.3 Direct methods for the ionized component

In studies of the ionized components of the upper atmosphere one is interested in the concentrations of electrons and ions (positive and negative), their temperatures, and the nature of the ion species. The determination of composition requires the use of a mass spectrometer, but a good deal of information about concentrations and temperatures can be obtained from simpler devices usually described as probes, traps or analysers. Many such devices have been developed for rocket and satellite applications. A probe projecting into the medium from the vehicle draws a current of ions or electrons when a potential is applied to it. A trap collects ions from the medium because of the high speed of the vehicle relative to the gas velocities. Sometimes additional electrodes are added to enable a more detailed analysis of the plasma to be made in real time, the information being transmitted to a ground station by telemetry. Another approach regards the medium around the vehicle as a dielectric, and sets out to measure the bulk electrical properties, which can be related theoretically to the properties of the individual particles. Such devices are generally known as resonance probes or impedance probes. Some typical devices are discussed in the following sections.

3.3.1 The Langmuir probe and related devices

Consider an open-wire grid in a plasma. It would be possible, in principle, to adjust the potential of the grid (relative to some large conductor at a distance, such as the Earth) so that the flow of ions and electrons through the holes of the grid was not affected. In this condition the potential of the grid can be regarded as the potential of the plasma. In ionospheric measurements it is called the *space potential,* V_s. Now replace the grid by a plate which is maintained at the same potential. Due to the electron motions in the plasma, an electron current proportional to $-N|e|v_e$ would flow to the plate, N being the electron density in the medium, $|e|$ the magnitude of the electronic charge, and v_e the electron velocity. Similarly there will be an ion current proportional to $+N|e|v_i$, it being assumed that the ion and electron densities are equal. Thus the total current to the plate at potential V_s depends on $N|e|(v_i - v_e)$. The current is not zero, but is actually negative since $v_e > v_i$ (the electron mass being much smaller than the ion mass — see Section 2.2.2).

The negative current represents a net accumulation of electrons on the plate, and if the plate is now disconnected from the external source of potential it will become more negative. Eventually the total current will fall to zero, and at this point the plate is at the *floating potential,* V_f. When the total current is zero the electron (N_e) and ion (N_i) densities around the plate cannot be equal, but are related by $N_e v_e = N_i v_i$. In this condition (since $v_e > v_i$) the plate is negative relative to the space potential and is surrounded by a positive sheath in which $N_i > N_e$. The sheath controls the current to the plate, like the space charge in a simple electronic diode. Spacecraft in general are subject to the sheath effect, which means that the presence of the spacecraft alters the properties of the plasma in its immediate vicinity. The thickness of the sheath is the Debye length (Section 2.3.5).

The simple Langmuir probe is a flat plate which is extended into the plasma. The current to or from the plate is measured as a function of the potential of the probe.

The curve of current against potential has the general form shown in Fig. 3.1, where the space potential is V_s and the floating potential (where $I = 0$) is V_f. The operation is as follows.

(a) If $V > V_s$ ions are retarded while all the electrons are accepted. The electron density can be determined from this saturation current (provided the electron temperature is known).

27

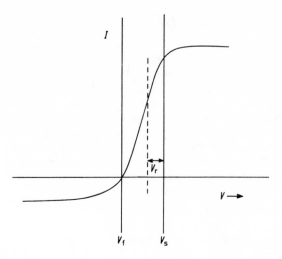

Figure 3.1   Current-potential characteristic of a Langmuir probe.

(b) If V lies between V_s and V_f there is some retardation of the electrons but most of the current is still due to them. If the retarding potential with respect to V_s is V_r, i.e. $V_r = V_s - V$, only the faster electrons, those with velocity $v > (2 V_r e/m_e)^{1/2}$ can reach the probe. (This relationship can be derived by comparing potential energy, $V_r e$, and kinetic energy, $mv^2/2$.) In a Maxwellian distribution (Equation 2.7) the number of electrons with velocity between v and $v + dv$ is proportional to $\exp(-mv^2/2kT) \, dv$. Thus the number arriving at the probe in unit time, proportional to the current, is

$$\int_{(2 V_r e/m)^{1/2}}^{\infty} v \exp(-mv^2/2kT) \, dv$$

Thus, by integration,

$$I = I_0 \exp(-eV_r/kT) \tag{3.1}$$

(c) If $V < V_f$ the electrons are strongly retarded but all the ions are collected. In this condition it must be recognized that the velocity of the ions with respect to the probe owes more to the orbital velocity of the spacecraft than to the thermal speed of the ions. The condition for ion current is then $|e| V_r \leqslant m_i v_{sat}^2/2$. The ionic masses and abundances can be determined in this way. Information about the ion temperatures can also be obtained from a detailed analysis of the curve.

The Langmuir probe, which has been in laboratory use since about 1923, is the basis of many devices used on spacecraft in the 1960s and 1970s. In form the probe may be planar, cylindrical or spherical. Sometimes one or more grids are placed before the collecting electrode and biased at a different potential. The field between grid and collector modifies the operation of the probe by excluding electrons in certain energy ranges from the collector current. Such a device may be called an electrostatic analyser or a retarding potential analyser, and again various geometrical forms are possible. In the ion mode, the ions are swept up by the moving spacecraft and the collector becomes an ion trap. Additional grids may again be used. Fig. 3.2 shows grid arrangements used in retarding potential analysers, vintage 1965, capable of measuring ion and electron density and temperature, and ion composition.

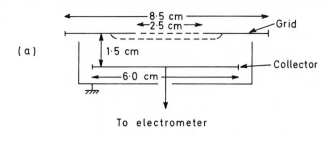

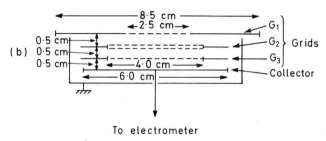

Figure 3.2 Typical grid and collector arrangements in traps: (a) for ions, (b) for electrons. (After J.L. Donley, *Proc. Inst. elect. electronics Engr.* **57**, 1061, 1969.)

3.3.2 Impedance and resonance probes

At a radio frequency ω (rad s^{-1}), a plasma containing N electrons m^{-3} has dielectric constant (Section 2.4.5)

$$\kappa = (1 - Ne^2/\epsilon_0 m\omega^2)$$

A standard method of determining the dielectric constant of a material is to measure the capacitance of a parallel-plate capacitor with a slab of the material between the plates:

$$C = \epsilon_0 A\kappa/d$$

The same method can be used in the ionosphere, the capacitor being part of a tuned circuit, whose resonant frequency therefore varies with the dielectric constant.

One technique is to use the tuned circuit (which includes the capacitance of the resonance probe) to control the frequency of an oscillator. The frequency would typically be set initially within a band 100 kHz to 5 MHz. The signal would be compared with a standard oscillator, and the beat frequency, equal to the difference between the two frequencies, is telemetered to a ground station. If such a device is carried on a rocket, the electron density profile may be determined. Fig. 3.3 shows such a flight into the auroral ionosphere. The electron density was determined simultaneously by a radio propagation experiment (Section 3.7.3). It will be noted that the probe has resolved fine structure missed by the propagation method, though in general it would be less reliable for determining the absolute value of the electron density.

If a radio transmitter and receiver are carried into the ionosphere, it is found that various characteristic resonances of the medium may be excited. The resonances

29

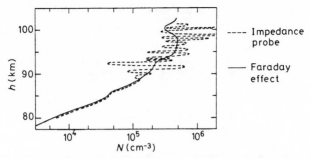

Figure 3.3 Electron density profiles determined by two methods during a rocket flight into auroral ionosphere. (After O. Holt and G.M. Lerfald, *Rad. Sci.* **2**, 1283, 1967, copyrighted by American Geophysical Union.)

appear as spikes on topside ionograms (see Section 3.7.1), and they indicate a local storage of energy at radio frequencies where the ionosphere has a natural resonance. The fundamental frequencies are the plasma frequency (Section 2.3.4), $f_N = (Ne^2/4\pi^2\epsilon_0 m)^{1/2}$, and the gyrofrequency (Section 2.3.2.), $f_B = eB/2\pi m$, but resonances also occur at combination frequencies such as $(f_N^2 + f_B^2)^{1/2}$. Harmonics are also observed. Systems of this kind are *resonance probes*. Resonances can be seen in Fig. 3.7.

3.3.3 Mass spectrometers

A mass spectrometer for ions in the upper atmosphere starts with some advantages. There is no need for an ion source because the sample is already ionized. Then, if used above 100 km the spectrometer does not need to be evacuated; with the mean free path greater than 1 m, the necessary vacuum is already present. The standard laboratory mass spectrometer, using magnetic deflection, is not very suitable for use in space vehicles because of the weight of the permanent magnet. Some have been carried on rockets, but many satellites support several experiments and there is the additional problem that the magnet might interfere with the operation of the other experiments. Most ionospheric measurements have therefore been made with other forms of mass spectrometer.

The Bennett, or time-of-flight, spectrometer has been in laboratory use since 1950. The ions have to pass through a number of grids (Fig. 3.4) between source and detector. Some grids have steady potentials applied to them, and the centre one carries an oscillating potential. Between grids 1 and 2 an ion mass m_i and charge e is accelerated to a velocity $(2eV_1/m_i)^{1/2}$ which depends on the mass/charge ratio, m_i/e, of the ion. The oscillating potential on grid 3 ensures that further acceleration is given to ions that pass grid 2 with velocities close to a value d/t, where d is the spacing between electrodes and t is the time between field reversals. The final electrode (5) is reverse biased to reject all ions having less than the maximum speed. In this way the ion beam reaching grid 2 is velocity filtered and only ions with a selected mass/charge ratio may reach the detector.

In the lower part of the ionosphere a 'quadrupole' mass spectrometer, first introduced by Narcisi and Bailey in 1965, has been used on rockets. The theory of this instrument is too complicated to go into here, but physically it consists of

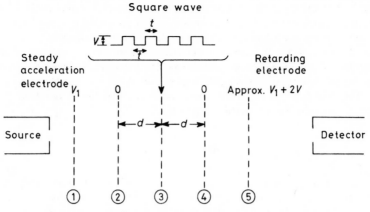

Figure 3.4   Principle of the time-of-flight mass spectrometer.

four parallel cylinders surrounding the ion drift path. A combination of steady and alternating potentials is applied between the cylinders, and it can be shown that only ions with a certain mass/charge ratio can survive the journey to the detector at the end of the instrument, the others being lost to the cylinders on the way.

3.4 Detectors for energetic particles and the radiation environment

3.4.1 Energetic particles

Several different kinds of detector are used for particles with energies above the thermal range. We are concerned here with energies in the range from about 1 keV to 1000 MeV or more. The most commonly used detectors, again based on laboratory techniques, are the Geiger counter, the plastic scintillator and the channel multiplier. In each case the detector registers the arrival of individual particles, rather than the current collected from a large number. The Geiger counter relies on the electrical breakdown of a gas under a high potential gradient when the ionized track from an energetic particle is formed in it. In a scintillator the particle produces a flash of light when it enters a transparent solid, and the flash is then detected by a photomultiplier tube. The channel multiplier is a modern device, which uses the secondary electrons produced when an energetic primary particle strikes a solid surface. The multiplier has the form of a narrow tube, along which a large potential gradient is applied. Each secondary electron is accelerated sufficiently to produce further secondaries when it, in turn, strikes the surface of the tube. Thus an avalanche is produced, which is readily detected at the end of the tube. The size of the pulse is proportional to the energy of the original particle. Some kind of energy selector is commonly placed before a particle detector, and collimators may be added so that the direction of arrival can be measured.

3.4.2 Electromagnetic radiation

The upper atmosphere is irradiated by solar emissions as well as by signals of more local origin. The solar spectrum, particularly in the X-ray and ultraviolet regions,

31

is important in studies of the formation of the ionosphere. Various kinds of spectro-meter, and photometers consisting essentially of a photocell covered by a narrow-band optical filter, are used on rockets and satellites for this purpose.

The electromagnetic environment extends down to radio frequencies. Signals at frequencies of several kilohertz and below are particularly important in studies of the magnetosphere, since they give information about plasma processes occurring there. The detectors are essentially radio receivers. The use of radio for studying the upper atmosphere is discussed in Section 3.7.

3.4.3 Steady magnetic and electric fields

At the low-frequency limit of electromagnetic signals we come to steady magnetic and electric fields. The geomagnetic field has been observed from satellites for many years in order to map the field of internal origin, to study the more change-able and distorted fields in the magnetosphere, and to investigate the fields in space. In the more distant regions these fields are very weak (a few gamma) and it is essential that the spacecraft itself should be magnetically clean. If the space-craft carries several experiments, as is usually the case, there must be no stray fields from the other equipment. (The general requirement that space measurements should not interfere with each other is obvious in principle but not always easy to meet in practice). Satellite magnetometers are based on the same principles as ground-based instruments, the most common being the fluxgate and the alkali-vapour (cesium or rubidium) magnetometers. A detailed description would be beyond our present scope.

Steady electric fields are generated in the upper atmosphere by various processes (see Chapters 4, 6 and 8) and in recent years there has been much interest in the form and magnitude of these fields. It might be thought that if two probes were extended from a spacecraft a simple measurement of the potential difference between them would immediately lead to the component of the electric field in the line of the probes. In fact there are two major obstacles to the direct measure-ment of electric fields, the first being the plasma sheath that forms around the probes (Section 3.3.1), and the second the induced electric field that appears if the spacecraft is moving through the geomagnetic field (Equation 8.2). Consequently, direct measurements are rather difficult and most of the available information on electric fields has come from an indirect method, the ion cloud technique which also measures the neutral-air wind. This will be described in Section 6.3.1.

3.5 Satellite drag and related methods

The most widely used method for finding the density of the air at heights of several hundred kilometres is an indirect one, which relies on the drag of the air on a moving object. Satellite tracking by optical and radio techniques is quite accurate enough to detect the effects of the atmosphere on the orbit of a satellite, and tracking data have been applied to air density determinations ever since the first satellites were launched in the late 1950s. The height range that may be covered is 200–1100 km, and as well as the average density, seasonal and short-term changes, and diurnal and latitudinal variations can be determined. We give here a simple version of the theory, applicable to satellites in circular orbits, the atmosphere being assumed constant and spatially uniform.

Equating the centripetal and gravitational accelerations of the satellite,

$$\frac{m_s v^2}{r} = \frac{G M_E m_s}{r^2}$$

where m_s is the mass of the satellite, M_E the mass of the Earth, r the distance of the satellite from the centre of the Earth, and G the gravitational constant.

Therefore

$$GM_E = v^2 r \qquad (3.2)$$

The kinetic energy of the satellite is $KE = mv^2/2 = GM_E m_s/2r$, and the potential energy is $PE = -GM_E m_s/r$, the potential energy being taken as zero when $r = \infty$. Thus $|PE| = 2|KE|$, and the total energy given by $PE + KE = -KE$. Note that if r is decreased the kinetic energy, and therefore the orbital velocity, increases. However the potential energy decreases twice as much, and so the total energy does in fact decrease. Therefore the observation that a satellite moves faster as it descends into the atmosphere is not inconsistent with a loss of energy.

If the drag force on the satellite is D (defined as the rate of loss of momentum, $-m_s \cdot dv/dt$), the rate of loss of kinetic energy is

$$\frac{-d}{dt}(mv^2/2) = -mv \cdot \frac{dv}{dt} = Dv$$

Also

$$\frac{d}{dt}(KE) = \frac{d}{dr}(KE) \cdot \frac{dr}{dt}$$

From Equation 2.11,

$$D = \frac{\rho v^2 A C_D}{2}$$

where ρ is the air density, A is the cross-sectional area of the satellite, and C_D is a drag co-efficient depending on the shape of the satellite. Putting values into the relation

$$\frac{d}{dr}(KE) \cdot \frac{dr}{dt} = -Dv$$

gives

$$\frac{-dr}{dt} = \frac{\rho v A C_D r}{m_s} \qquad (3.3)$$

for the rate of decrease of altitude. In practice the period of a satellite can be measured more accurately than the altitude. The orbital period is given by

$$P = 2\pi r/v$$

Therefore

$$P^2 = 4\pi^2 r^3/GM_E \qquad \text{(since } v^2 = GM_E/r \text{ from Equation 3.2)}$$

$$2P \frac{dP}{dt} = \left(\frac{4\pi^2}{GM_E}\right) 3r^2 \frac{dr}{dt}$$

$$\frac{dP}{dt} = -\frac{3\pi A C_D r}{m_s} \rho \qquad (3.4)$$

The rate of decrease in the orbital period of a satellite in circular orbit is proportional to the air density at the height of the satellite.

A more elaborate analysis is needed for the more general case of an elliptical orbit. As such an orbit covers a range of altitudes it is possible to find the scale height in this case. Moreover, since most of the drag in an elliptical orbit is encountered near perigee, where the air density is greatest, it is possible to determine the variation of density with time of day and with latitude. Much of the evidence about the heating and cooling of the thermosphere diurnally (Section 6.3.3) and with the sunspot cycle (Section 5.3) has come from studies of satellite drag. Some examples are given in Fig. 3.5.

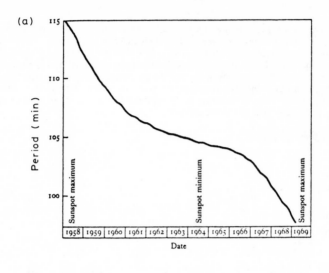

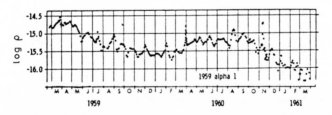

Figure 3.5 Results of satellite drag measurements. (a) Period of satellite in its orbit, 1958–1969. The decay rate was greater at sunspot maximum, indicating a greater air density. (b) Deduced variations of air density, indicating a 27-day periodicity. (After J.A. Ratcliffe, *An Introduction to the Ionosphere and Magnetosphere*, Cambridge University Press, 1972; and I. Harris and N.W. Spencer, *Introduction to Space Science* (Ed. Hess and Mead) Gordon and Breach, 1968.)

Closely related to the satellite drag technique is the falling sphere method used from rockets. This works well in the altitude range 90–130 km, below the height where a satellite orbit can be maintained. The principle of the falling sphere technique is simply to eject a sphere from a rocket and observe the rate at which it falls. The sphere is retarded by the air drag force

$$D = \frac{\rho v^2 A C_D}{2}$$

opposing the gravitational force $m_s g$, and hence it accelerates towards the ground more slowly than it would in a vacuum. The drag is measured with an accelerometer carried within the sphere, and the readings from the accelerometer are telemetered to a ground station.

3.6 Remote sensing of the neutral atmosphere

Remote sensing depends on the interaction between the medium and a propagating wave. Two methods that give useful information about the neutral upper atmosphere use sound waves and light.

3.6.1 Sound propagation

The speed of sound relative to the air of density ρ at pressure P is

$$s = (\gamma P/\rho)^{1/2} \tag{3.5}$$

where γ is the ratio of specific heats. If the composition is constant, $P/\rho \propto T$, the absolute temperature, and thus the temperature can be determined by measuring the speed of sound through the air. The method is to eject a grenade every few kilometres from an ascending rocket. The visible flash and the arrival of the acoustic signal are timed at a network of ground stations, and thus the speed of sound may be determined as a function of height. The technique is useful up to about 100 km. Since the velocity with respect to observers at the ground depends also on the neutral-air wind, the wind profile can be determined at the same time.

3.6.2 Light propagation

Visible light is subject to Rayleigh scattering by particles much smaller than a wavelength. In clean air the scattering is by individual molecules. Thus, by observing the intensity of Rayleigh scattering with altitude the atmospheric density can be found. By application of the barometric equation (derived in Chapter 4) it is possible to obtain the temperature and pressure as well (see Equations 4.1 and 4.2) provided the composition of the air is known at the altitudes concerned. The first applications of light propagation used searchlights, and the observations reached nearly 70 km. With lasers the range has been extended to above 100 km, and the height may be resolved by lidar technique in which the laser is pulsed as in a radar system. Pulses of a few microseconds duration give a height resolution less than 1 km. A ruby laser (emitting red light at 6943 Å) might transmit 10 J of energy during the pulse, and the receiver will be a telescopic system with a photomultiplier as detector.

In addition to measuring the air density from the backscattered intensity, in which the accuracy may be 20% at 70 km and 10% at 100 km, the temperature may be deduced from the Doppler broadening of the spectrum and the wind component in the line of sight found from the mean spectral shift. The vertical profile of a specific constituent can be determined by Raman scattering, in which the returned signal is shifted in frequency from the transmitted signal by an amount characteristic of the scattering molecule. In this way water vapour can be measured to heights of several kilometres. An enhanced return is obtained by transmitting at a resonance line, where a species both absorbs and emits strongly. In this situation the energy received at the centre of the line comes from a volume relatively close to the lidar, whereas the signal from the edges of the line, where the absorption coefficient is less, comes from further away. Thus an analysis of the line shape can be interpreted as a profile of concentration against height.

3.6.3 Remote temperature sensing

Another group of techniques depends on the emission of light by atmospheric species. Some aspects will be met in later chapters, e.g. Sections 4.3.5 and 8.2.1, but here we shall outline an important new method of measuring the temperature of the neutral air remotely from a satellite.

The intensity of emission from a gas depends on the temperature and on the wavelength. Within an emission band some wavelengths are emitted more strongly than others, but by Kirchhoff's law these same wavelengths are most strongly absorbed from radiation passing through the gas. Now if a radiation detector looks down into the atmosphere the radiation it receives at a given wavelength comes preferentially from some altitude. The existence of a peak can be shown by considering that there will be relatively less emission higher up because of the reduced gas density, whereas radiation originating at a low altitude passes through a great depth of the atmosphere on its way to the detector and so suffers heavy absorption. The altitude of the peak depends on the absorption coefficient, and so it is possible to study the temperature at different heights by selecting suitable wavelengths.

To prove these points we borrow some results from Chapter 4. Consider an exponential atmosphere of scale height H, so that the molecular concentration at height h is given by $n = n_0 \exp(-h/H)$ where n_0 is the concentration at the ground (from Equation 4.5). Suppose that radiation of intensity I_h is emitted from the gas at height h. The radiation is attenuated by absorption as $dI/dh = -I\sigma n$ (Equation 4.17), where σ is the absorption cross-section per molecule, and therefore the intensity at the top of the atmosphere is $I' = I_h \exp(-\sigma nH)$ by Equation 4.22.

By Kirchhoff's law the emission from a slab of gas of thickness Δh containing n emitting molecules per unit volume is $I_h = B\sigma n \cdot \Delta h$, and hence

$$I' = B\sigma n \exp(-\sigma nH) \Delta h$$

Substituting for n and differentiating with respect to h shows that I' is a maximum where

$$h = H \ln(\sigma n_0 H)$$

The maximum is at greater height if σ is larger. B is the Planck radiation formula which depends on the temperature. Thus it is possible to derive the temperature from the observed intensity, knowing the concentration and scale height of the emitting gas, and the absorption cross-section as a function of wavelength.

These principles have been successfully applied in satellite experiments using the carbon dioxide band at 15 μ in the infrared. Originally applied to temperature measurements in the stratosphere, the technique has recently been extended as high as 90 km.

36

3.7 Remote sensing by radio propagation

The most important technique for observing the ionized regions of the upper atmosphere makes use of the propagation of radio waves. Whereas most of the devices flown on a spacecraft for direct sensing have been developed from standard physical laboratory techniques, the radio methods were developed specifically for upper atmosphere studies. We shall therefore consider them in some detail.

The refractive index (n) of an ionized gas is given by the Appleton–Hartree equation (Section 2.4.5). A full study of the equation and its many implications would be lengthy and involved, but for our purpose it will be sufficient to examine some of the simpler special cases. The approach that has to be taken depends on the change of the refractive index over one wavelength of the radio wave. If the change is not large, the Appleton–Hartree equation can be applied to each part of the medium as though it was of infinite extent. If the refractive index changes within a wavelength, the situation is analogous to mirror-like reflection (which may be total or partial) at a discontinuity. The classical techniques of ionospheric sounding are based on the first assumption.

3.7.1 Ionospheric sounding

The principle of ionospheric sounding is to transmit a pulse of radio waves vertically, and to measure the time which elapses before the echo is received. Neglecting collisions $(Z = 0)$ and the geomagnetic field $(Y = 0)$ in the Appleton–Hartree equation, the refractive index (n) is given by

$$n^2 = 1 - X = 1 - \omega_N^2 / \omega^2 \tag{3.6}$$

ω_N^2 is proportional to the electron density (Equation 2.20). As the wave penetrates into the ionosphere the refractive index becomes smaller. If $\omega_N > \omega$ the wave cannot propagate because the refractive index is imaginary. Therefore the energy carried by the radio wave is reflected back to the ground from the level where $\omega_N = \omega$, i.e. where the plasma frequency equals the wave frequency.

The standard piece of equipment using these principles is the ionospheric sounder, or *ionosonde,* in which a transmitter and a receiver are slowly swept in frequency, the echo time delay being recorded as a function of the radio frequency. The recording is usually made photographically from a cathode ray tube. The resulting ionogram (Fig. 3.6) may be interpreted as a profile of electron density against altitude, the electron density being obtained from the formula for plasma frequency, which, putting in numerical values, can be written

$$[f_N \,(\text{kHz})]^2 = 80.5 \, N \,(\text{cm}^{-3}) \tag{3.7}$$

and the height is given by the time delay of the echo, it being assumed that the radio pulse travels at the speed of light, c.

This simple interpretation is not quite correct for two reasons. First, a radio pulse travels at the group velocity (Section 2.4.3), which is related to the speed of light by the group refractive index, n_g, $(u = c/n_g)$. If the phase refractive index (n) is given by Equation 3.6, it can be shown that the group refractive index is $n_g = 1/n$. Therefore the radio pulse travels more slowly than light, and this *group retardation* in the ionosphere below the reflection level means that in general the

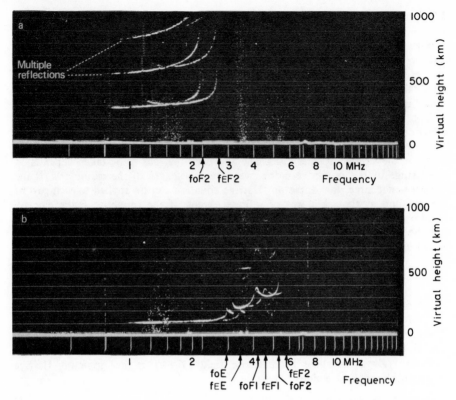

Figure 3.6 Ionograms from Port Stanley, Falkland Islands: (a) 8 July 1976 at 0200, typical of the winter night; (b) 20 January 1977 at 1545, typical of the summer day. f_OE, f_EE, f_OF1, f_EF1, f_OF2 and f_EF2 denote the ordinary and extraordinary critical frequencies of the E, F1 and F2 layers. (Many people use X instead of E, but your present author still spells 'extraordinary' with an E!) (From the Science Research Council, Appleton Laboratory, Slough. Crown copyright.)

virtual height, $h'(f)$, worked out on the assumption of a constant velocity throughout, is always too large. The cusps on Fig. 3.6 are due to strong group retardation in the structure of the lower ionosphere. It should be noted, also, that at the reflection level, where $\omega = \omega_N$, n is zero but the group refractive index goes to infinity, meaning that the forward velocity of the radio pulse falls to zero — as would be expected at the reflection level. It is possible (if somewhat tedious) to correct for the group retardation and thereby produce a *real-height profile*. The ionosphere often contains several layers, and the maximum plasma frequency of a given layer is called the *critical frequency* of that layer. The maximum critical frequency of the whole ionosphere is known as the *penetration frequency*, since waves of higher frequency may propagate right through the ionosphere to be lost to space. Conversely, only waves whose frequency exceeds the penetration frequency can be received from space by a ground-based receiver.

The second weakness of our simple interpretation of an ionogram is the neglect of the geomagnetic field. If $Y \neq 0$, Equation 2.37 gives two values for the refractive index. These waves are designated 'ordinary' and 'extraordinary', and their refractive indices are obtained by taking the positive and negative signs respectively. Fig. 3.6 shows the traces due to the O- and E-waves, and further consideration of Equation 2.37 shows that the reflection conditions are

$$
\begin{aligned}
\omega_N^2 &= \omega(\omega - \omega_B) && \text{for the E-wave} \\
\omega_N^2 &= \omega^2 && \text{for the O-wave}
\end{aligned}
\qquad (3.8)
$$

If $\omega_B \ll \omega_N$, the difference between the E- and O-wave critical frequencies is $\omega_E - \omega_O \approx \omega_B/2$. Sometimes a second reflection condition, $\omega_N^2 = \omega(\omega + \omega_B)$, is observed for the O-wave, and the ionogram then shows three traces.

The pulse method of ionospheric sounding was first used by Breit and Tuve in 1924, and has remained the bedrock technique for ionospheric observations ever since. Although complemented by many newer methods, the ionosonde has not been supplanted as the basic tool for ionospheric monitoring, and does not seem likely to be.

Since the electron density passes through a maximum (usually between 200 and 400 km) the upper part of the ionosphere cannot be observed by an ionosonde on the ground. To study the 'topside', an ionosonde may be carried on a satellite. The first of the *topside sounders* was the Canadian satellite, Alouette 1, launched in 1962. A topside sounder typically orbits at 1000 or 1500 km and sounds downwards towards the maximum of the layer. A topside ionogram is shown in Fig. 3.7. Interpretation is much as for a bottomside ionogram, but a valuable bonus is the resonance phenomenon already referred to in Section 3.3.2. The sounder is immersed in the ionospheric medium, and the resonances give the local plasma and gyromagnetic frequencies.

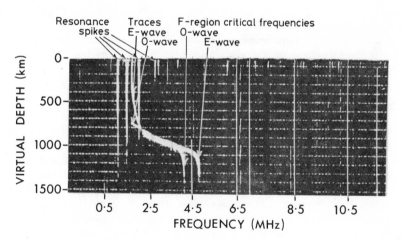

Figure 3.7 A topside ionogram typical of middle latitudes near noon. (From the Science Research Council, Appleton Laboratory, Slough. Crown copyright.)

If collisions are not negligible ($Z \neq 0$) the refractive index formula becomes

$$
n^2 = 1 - \frac{X}{1 - jZ} = 1 - \frac{\omega_N^2}{\omega(\omega - j\nu)}
\qquad (3.9)
$$

39

Then the refractive index is complex ($n = \mu - j\chi$), and a wave propagating through the medium varies as $\exp(-x\chi\omega/c) \cdot \cos \omega (t - x\mu/c)$. The first term of this expression represents a decay of the wave amplitude with distance (x). The absorption coefficient is $\kappa = \omega\chi/c$ (in units of nepers per metre), and the amplitude falls to $1/e$ of its original value in a distance $1/\kappa = c/\omega\chi$. By substituting the expressions for ω_N^2 in Equation 3.9, it can be shown that (using MKS units)

$$\kappa = \frac{\omega}{c} \cdot \frac{1}{2\mu} \cdot \frac{XZ}{1 + Z^2} = \frac{e^2}{2\epsilon_0 mc} \cdot \frac{1}{\mu} \cdot \frac{N\nu}{\omega^2 + \nu^2} \qquad (3.10)$$

Useful special cases are $\mu = 1$ and $\mu \ll 1$. The first is called *non-deviative* absorption, since the velocity of the wave is not affected. Then

$$\kappa = \frac{e^2}{2\epsilon_0 mc} \cdot \frac{N\nu}{\omega^2 + \nu^2} \qquad (3.11)$$

If, further, the collision frequency (ν) is small compared with the angular wave frequency (ω), it is seen that the absorption coefficient is proportional to the product of electron density and collision frequency. The practical unit of absorption is the decibel, which is defined in terms of the ratio of initial (P_1) and final (P_2) power densities, the power density in a radio wave being proportional to the square of the amplitude (Equation 2.36). Thus, absorption A (decibels) = $10 \log_{10}(P_1/P_2)$. The decibel and the neper are related by the formula, 1 neper = 8.69 decibels. Putting values into Equation 3.10

$$A = 4.5 \times 10^{-5} \int \frac{N\nu}{\nu^2 + \omega^2} \, dx \qquad dB \qquad (3.12)$$

all quantities being in MKS units. Measurement of the non-deviative absorption can provide an estimate of the electron density in the lower ionosphere, where the collision frequency is relatively large. In practice the magnetic field should also be taken into account because the amount of absorption is not quite the same for the two magneto-ionic modes.

In the second special case, when μ is small (strictly, $1 - \mu^2 \gg \chi^2$)

$$\kappa = \frac{\nu}{2c} \left(\frac{1}{\mu} - \mu \right) \qquad (3.13)$$

Known as *deviative absorption*, this condition applies near the reflection level where $\mu \to 0$ and the group velocity is small.

There are several ways of measuring the absorption of radio waves. The basic method is to transmit radio pulses and receive the echoes, as with an ionosonde, but now the frequency is held constant (usually within the range 2–6 MHz) and the equipment is designed for accurate measurement of the echo amplitude. A variation is to transmit continuously and to monitor the intensity of the signal received at a point at least 100 km away from the transmitter. A considerable separation of transmitter and receiver is needed to avoid interference from the ground wave propagating directly across the intervening distance. Such methods have been in regular use since about 1935, and have produced much valuable information, particularly about the lower ionosphere (see Sections 5.2.3 and 11.4.1). One difficulty that tends to arise in ground-based absorption studies is deciding between the deviative and non-deviative contributions, which may be imposed in quite different parts of the ionosphere.

An important variation on ionospheric sounding is the *HF Doppler* technique, which has proved valuable for studying small perturbations of the reflecting layer. Consider a monostatic arrangement, with transmitter and receiver co-located so that the radio signal is reflected at normal incidence. If the reflection level is moving upward at speed W, the frequency of the reflected signal is Doppler shifted by $\Delta f = -2fW/c$, where f is the transmitted frequency and c is the velocity of the radio wave. Thus, $W = -2\lambda \cdot \Delta f$, where λ is the radio wavelength. If $\lambda = 100$ m, and $\Delta f = 0.5$ Hz, then $W = 100$ m s^{-1}. A typical HF Doppler system uses continuous, not pulsed, signals. The received signal is mixed with a reference signal at the transmitter frequency, and the resulting beat frequency is measured and recorded. The output gives the layer vertical velocity in the first instance, but this may be integrated to give the relative height changes. The absolute reflection level would normally be determined from an ionosonde. Some examples of HF Doppler results will be met in Section 6.5.

3.7.2 Partial-reflection technique and VLF propagation

If the refractive index of the ionospheric medium changes abruptly in relation to the wavelength of the radio wave, an entirely different approach has to be taken. As is well known from optics, if there is a sharp interface between two media with respective refractive indices n_1 and n_2, a wave normally incident on the interface is partially reflected, a fraction $(n_1 - n_2)^2/(n_1 + n_2)^2$ of the incident power being reflected. Up to about 100 km the atmosphere is turbulent and contains spatial irregularities from which partial reflections may be obtained at medium frequencies (for instance 2–3 MHz, wavelengths about 1 km). The echoes are weak, a mere 10^{-3} to 10^{-5} of the amplitude of total reflections, and thus the transmitter is typically of high power (up to 100 kW) and large antenna arrays are needed. Such an installation becomes a substantial engineering project, and relatively few of them are in use at the present time.

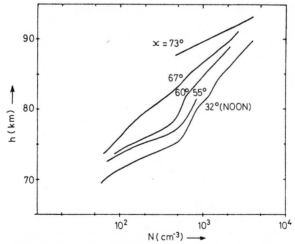

Figure 3.8 D-region profiles of electron density, measured by the partial-reflection technique for various solar zenith angles. Crete, 11 September 1965, afternoon. (From E.V. Thrane *et al., J. atmos. terr. Phys.* **30**, 135, 1968, by permission of Pergamon Press.)

The great value of the partial reflection technique is that the ionosphere can be studied in the 60—90 km height region, where the electron density is too small to be detected by an ionosonde. An example is shown in Fig. 3.8, which includes electron density values down to 100 cm^{-3}. The usual method is to transmit pulses of the circularly polarized O- and E-waves alternately and to compare the amplitudes of the received echoes as a function of the reflection height. Two factors contribute to the ratio of the O- and E-mode echoes: the relative intensity of the partial reflections, and the relative absorption of the waves in the ionosphere below the reflection level. Both factors may be shown to depend on the profiles of electron density and collision frequency. If the second is assumed, the first may be deduced through application of magneto-ionic theory.

At very low radio frequencies (3—30 kHz, corresponding to wavelengths 10— 100 km) the base of the ionosphere itself appears as a sharp boundary. The collision frequency (2×10^6 s^{-1} at 70 km) is large relative to the wave frequency, and thus

$$n^2 = 1 - jX/Z = 1 - \frac{\omega_N^2}{j\omega\nu} \qquad (3.14)$$

At such frequencies the ionosphere appears as a metal rather than a dielectric, the conductivity being $\sigma_i = Ne^2/m\nu = \epsilon_0 \omega_N^2/\nu$ (see Section 4.4.1, Equation 4.68). The lower ionosphere can be studied by observing the amplitude and phase of VLF signals at various distances from VLF transmitters. An effective reflection height and conductivity can be readily deduced, and variations occurring diurnally and during disturbances are observed. Reflection coefficients are typically 0.2 to 0.5.

Although it makes for a convenient first approximation, the assumption of a sharp ionospheric boundary is not strictly true, even at very low radio frequencies. Fortunately, with the ready availability of fast and large computers it has become practicable to solve the boundary problem exactly, and this can be done for any assumed vertical profile of electron density and collision frequency, and for any frequency, polarization, and angle of incidence of the radio wave. The approach is to solve as a function of altitude the equations governing radio propagation in the medium. The electric and magnetic fields are derived, many points being evaluated within each wavelength, subject to the condition that at great height there may only be an upward propagating wave. The wavefields below the ionosphere are resolved into two waves, propagating respectively upward and downward. The relationship between these two waves then gives the required set of complex reflection coefficients (of which there are in general four, to include polarization changes on reflection). Solutions derived by these methods are called *full-wave solutions*. The method is different from that of the Appleton—Hartree magneto-ionic theory, but the solutions become identical to the Appleton—Hartree characteristic waves in the limiting case of a slowly varying medium.

3.7.3 Propagation through the ionosphere

If the frequency of a radio wave exceeds the penetration frequency of the ionosphere the wave will propagate right through the medium, but will still be measurably affected by it if the frequency is not too high. Consequently there is a group of techniques for ionospheric study that are based on trans-ionospheric radio propagation. The radio source may be the natural radio emission from the galaxy, or a transmitter carried on an Earth satellite. Some studies have used a radar system, with the Moon or other extraterrestrial body serving as reflector.

If $X \ll 1$ and the magnetic field and collisions are both neglected, Equation 3.6 is simplified to

$$n \approx 1 - \omega_N^2/2\omega^2 = 1 - 1.6 \times 10^3 N/\omega^2 \tag{3.15}$$

in MKS units. The departure of the refractive index from unity now varies linearly with the electron density and inversely with the square of the radio frequency. If the wave travels a distance x in the medium, the phase becomes

$$-\frac{\omega}{c} \int n \, dx = -\frac{x\omega}{c} + \frac{1.6 \times 10^3}{c\omega} \int N \, dx \tag{3.16}$$

which depends on the quantity $\int N \, dx$, generally known as the *electron content, I*. The electron content is the number of electrons in a column of unit cross-section taken along the line of propagation. In the practical application a wave is also transmitted at a higher frequency to provide a phase reference. The higher frequency is less affected by the ionosphere, and the relative change of phase between the two signals can be measured with a suitable receiver to give the electron content. There are several variations of the method.

If the geomagnetic field is taken into account with the assumption that $Y \ll 1$, and it is assumed that propagation is directly along the field ($Y_T = 0$), Equations 2.37 and 2.38 give

$$n^2 = 1 + \frac{\omega_N^2}{\omega(\omega \pm \omega_B)} \tag{3.17}$$

The refractive index for the O-wave (n_O) is obtained by taking the positive sign in the denominator, and that for the E-wave by taking the negative sign. In the conditions being considered, both waves are circularly polarized. As seen by a stationary observer looking along the direction of the geomagnetic field, the O-wave rotates anticlockwise and the E-wave rotates clockwise. The sum of circular components of equal amplitude is a wave with linear polarization, and if at some instant the electric vectors of the O- and E-waves make angles θ_O and θ_E respectively

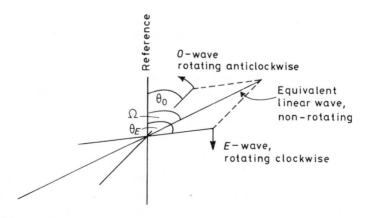

Figure 3.9   Addition of two circularly polarized waves to give a linear wave, as seen by a stationary observer looking along the geomagnetic field.

43

with some reference direction, the electric vector of the equivalent linear wave is at angle

$$\Omega = (\theta_O + \theta_E)/2 \tag{3.18}$$

to the same reference direction (see Fig. 3.9).

According to Equation 3.17 the two circular waves travel at slightly different speeds, depending on the ratio ω_B/ω. If $\theta_O = \theta_E = 0$ at the source, then at a distance x from the source

$$\theta_O = \omega t + \frac{\omega n_O}{c} \cdot x \qquad \theta_E = \omega t - \frac{\omega n_E}{c} \cdot x$$

Thus

$$\Omega = (\theta_O + \theta_E)/2 = \frac{\omega x}{2c} (n_O - n_E) \tag{3.19}$$

In the case where $\omega_B \ll \omega$, which is true at radio frequencies exceeding 50 MHz, Equation 3.17 gives

$$(n_O - n_E) = \frac{\omega_N^2 \omega_B}{\omega^3}$$

Substituting in Equation 3.19

$$\Omega = \frac{x}{2c} \cdot \frac{\omega_N^2 \omega_B}{\omega^2} \tag{3.20}$$

The polarization angle therefore changes as the wave propagates through the ionospheric medium. This is the *ionospheric Faraday effect*. Substituting expressions for ω_N and ω_B, inserting numerical values, and allowing the electron density to vary along the propagation path

$$\Omega = \frac{2.97 \times 10^{-2}}{f^2} \int H_L N \, dx \qquad \text{rad} \tag{3.21}$$

where N is the electron density in m^{-3}, dx is the element of path in m, f is the radio frequency in Hz, and H_L is the longitudinal component of the geomagnetic field in $A\,m^{-1}$. The use of H_L means that ω_B has been replaced by ω_L (the quasi-longitudinal approximation), a slight extension of our argument that allows for propagation somewhat across the geomagnetic field. In the Faraday effect, the QL approximation is valid to within a few degrees of the normal to the geomagnetic field.

Equation 3.21 includes a version of the electron content weighted by the geomagnetic field. The form of the geomagnetic field at ionospheric heights is well known, and thus measurements of the received polarization angle of a satellite transmission may be used to derive values of the electron content, I. The method is both convenient and inexpensive because of the large number of transmitting satellites now in orbit, and it has been widely applied to ionospheric monitoring. The method is not exact, however. Variations in the height distribution of the ionization alter the magnetic-field weighting, and the derived values of electron content may be in error by 10 or 20%. If the phase and polarization methods are

44

operated together, something may be learned about the height distribution of electron density.

Electron content data are often combined with ionosonde data. The ratio of electron content to peak electron density is the *slab thickness*

$$\tau = I/N_{\mathrm{m}} \tag{3.22}$$

This is the thickness of a hypothetical ionosphere with uniform electron density N_{m} and electron content I. The slab thickness depends physically on the temperature and ionic composition of the ionosphere.

The third basic technique of trans-ionospheric propagation measures the absorption of radio waves from an extraterrestrial source. Here the integral in Equation 3.12 is taken along the propagation path through the ionosphere. It would be possible to use a satellite transmission as the source, but nature has provided a convenient source in the form of radio stars and emitting hot gas in the galaxy, which together make up the cosmic radio noise. The stable receiver used to monitor the received intensity of the emissions is a *riometer* (relative ionospheric opacity meter). A riometer record is shown in Fig. 3.10. Since the cosmic noise is incoherent, with the statistical properties of wide-band electrical noise, it is not a suitable source for electron content studies by the techniques described above. However, some information may be obtained about the electron content by observing the changes in the apparent position of a single radio star, a result of the refraction of the radio signal as it passes through the ionosphere.

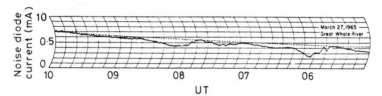

Figure 3.10 Absorption events on a riometer chart. The noise diode current is proportional to the received power, and the events appear as depressions of the trace below the trend marked by the dotted line. (From J.K. Hargreaves, *Proc. Inst. elect. electronics Engr.* 57, 1348, 1969.)

The methods of trans-ionospheric propagation can be adapted to studies of the lower ionosphere by carrying the source on a rocket. As the signal need not now penetrate the densest part of the ionosphere its frequency can be reduced to make observations more sensitive to the smaller electron densities of the lower ionosphere. Moreover, by recording signals as the rocket ascends and descends, the electron density and the collision frequency can be determined as a function of height — thus overcoming a weakness of simple trans-ionospheric propagation methods. There are many versions of the rocket approach, including the use of VLF transmissions to study the details of the wavefield near the reflection level (Section 3.7.2).

3.7.4 Incoherent scatter

Ionospheric sounding, as carried out by the ionosonde or the topside sounder, depends on the refractive index of the ionized medium, and the echo is returned from the altitude at which the refractive index is zero (collisions being neglected) for the frequency of the exploring radio wave. In a plane-stratified atmosphere the critical condition is reached simultaneously over a wide area, and thus the return is coherent. The energy is reflected as from a smooth mirror. On the other hand, if the medium is irregular and contains spatial variations smaller than the wavelength of the radio wave, there will be scattering from each irregularity; and since each irregularity will scatter independently of the others the returned signal will be incoherent. The ionosphere contains irregularities of all sizes (see Chapter 6), but the limit is reached with the individual electron.

The scattering cross-section of a single electron, i.e. the fraction of the incident energy flux re-radiated, is $8\pi r_e^2/3$, where r_e is the classical radius of the electron, $r_e = e^2/\epsilon_0 m_e c^2 = 2.82 \times 10^{-15}$ m. The electrons are in random thermal motion, so that random phase shifts are imparted to the scattered signals. Therefore the scattered signal is incoherent and the total power collected by a receiver is the sum of the powers received from individual electrons. It is therefore proportional to the electron density in the scattering region. Since r_e does not depend on the wavelength of the radiation being scattered, it is possible to use a radio frequency above the ionospheric penetration frequency and so to explore the whole depth of the ionosphere from the ground. This is the basis of the technique known as *incoherent scatter* and sometimes as *Thomson scatter.*

It can readily be appreciated that the incoherent scatter signal will be very weak. According to the classical treatment, the energy scattered by one electron at polarization angle ψ is $(r_e \sin \psi)^2$ per unit solid angle per unit incident flux (in W m^{-2}). The polarization angle is the angle between the electric vector of the incident flux and the line from the electron to the observer. For backscattering, $\psi = 90°$, and the scattered energy is just r_e^2 for unit incident flux. In radar, targets are quantified by the cross-sectional area of a sphere which would give the same return as the actual target, and thus the *radar cross-section* of an electron is $\sigma_e = 4\pi r_e^2 \approx 10^{-28}$ m^2. At the peak of the ionosphere the electron density may be as large as 10^{12} m^{-3} and the total volume sampled by a radar at any one time may be about 10^{12} m^3. Thus the total cross-section is about 10^{-4} m^2 or 1 cm^2. The radar therefore has to detect an object no bigger than a British new penny or a United States cent at a range of 400 km or more.

The possibility of detecting the incoherent scatter signal was first raised by W. E. Gordon in 1958 and tested by K. L. Bowles in the same year. The early experiments, however, revealed a discrepancy between theory and observation — a discrepancy that for once came out to the benefit of the experimenter! The point concerns the spectrum of the scattered signal. Since the electrons are moving with thermal velocities, not only are the phases of the scattered signals independent but also their frequencies are shifted by Doppler effect. It can be shown that the returned spectrum should have a Gaussian form with a half-power width of 0.71 Δf_e, where $\Delta f_e (= 2v/\lambda)$ is the Doppler shift of an electron approaching the radar at the mean thermal speed $v = (2kT_e/m_e)^{1/2}$. That is

$$\Delta f_e = (8kT_e/m_e)^{1/2}/\lambda \tag{3.23}$$

where λ is the radio wavelength, T_e the temperature of the electron gas, m_e the mass of the electron, and k Boltzmann's constant. Putting in values

$$\Delta f_e = 11 \ (T_e)^{1/2}/\lambda \qquad \text{kHz}$$

where λ is in metres. For typical ionospheric temperatures and a wavelength of 1.5 m, the bandwidth would be about 200 kHz. The observations, on the other hand, find that the power is returned in a band nearly 200 times narrower than that. The ratio of expected to observed bandwidth is equal to $(m_i/m_e)^{1/2}$ where m_i is the mass of the positive ions in the ionosphere, and this points the way to the explanation: that the return is from the electrons, but the motion of the electrons is controlled by the thermal speed of the ions. If the ions are O$^+$, the reduction in bandwidth is a factor of 170. The narrowing of the bandwidth makes the signal much easier to detect.

To see the connection between electronic and ionic motion we must consider the Debye length (Section 2.3.5)

$$\lambda_D = (\epsilon_0 kT_e/4\pi Ne^2)^{1/2}$$

numerically equal to $69\,(T_e/N)^{1/2}$ m, with values of one to several centimetres in the ionosphere. The Debye length indicates the distance to which the charge on a single electron is effective in the ionospheric plasma, and hence it indicates the distance over which electron motions are cooperative. To observe the electronic component of the scatter signal it would be necessary to use a radio wavelength small in relation to the Debye length. Present radar technique is not sufficiently sensitive to detect the return at these frequencies. In practice $\lambda > \lambda_D$, and most of the signal is the ionic component from irregularities of the size of the Debye length moving according to the thermal motions of the ions. It can be shown that for the ionic component

$$\Delta f_i = (8kT_i/m_i)^{1/2}/\lambda \qquad\qquad (3.24)$$

which reduces to $65\,(T_i)^{1/2}/\lambda$ Hz when the positive ions are atomic oxygen. Fig. 3.11 shows how the spectral shape depends on a parameter $\alpha = 4\pi\lambda_0/\lambda$.

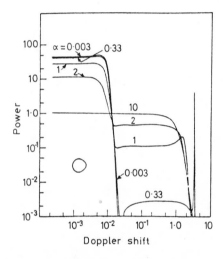

Figure 3.11 Spectral shape of the incoherent-scatter return, as a function of a parameter relating Debye length to radio wavelength. (After J.V. Evans, *Proc. Inst. elect. electronics Engr.* 57, 496, 1969).

The strength of the incoherent scatter method is that a wide range of parameters can be measured. The electron density is obtained from the intensity of the return and from the Faraday effect. Fig. 3.12 shows a profile out to 7000 km determined from the echo intensity. Most other measurements of ionospheric variables depend on the spectrum. The width of the spectrum depends on both the ion temperature and the ion mass. If the nature of the dominant ion is known, its temperature, T_i, can be determined. The ions and electrons in the ionosphere are not usually at the same temperature, and if $T_e \neq T_i$ the scatter spectrum has two maxima. From the detailed shape of this feature and particularly the depth of the central minimum, the ratio T_e/T_i may be determined. It is also possible to find the nature of the ions in some circumstances. It will also be observed in Fig. 3.11 that when α is small a sharp line appears as a sideband. This line is displaced by the plasma frequency, f_N, and provides in principle another method of measuring the electron density (Equation 2.20). Unfortunately the plasma lines, which arise from the collisions of

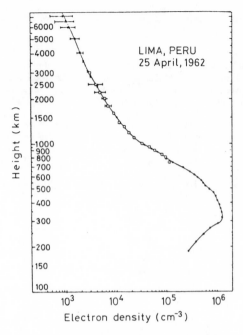

Figure 3.12   An electron-density profile to 7000 km determined from incoherent scatter. Jicamarca, Lima, Peru. 23 April 1962. (From K.L. Bowles, *NBS Boulder Tech. Note* 169, 1963.)

electrons with other electrons, are relatively weak. They are intensified if hot 'photoelectrons' are present, when it is possible to measure both the plasma frequency and the influx of photoelectrons. Bulk motions of the plasma in the line of sight shift the whole spectrum by Doppler effect with respect to the frequency of the transmitted signal, and this provides a basis for measuring the ionospheric electric fields and the effects on the ionization of large-scale atmospheric waves and winds.

The first major incoherent scatter facility was constructed at Jicamarca in Peru in 1960–61. The scale of the project is demonstrated by the size of the antenna – a broadside array of 96 × 96 dipoles, covering an area of 84 000 m² – and by the high power of the radio transmitter – 6 MW. The magnitude of the construction and running costs of such a large system has limited the number constructed. Table 3.1 shows a list of current and recent facilities. One other major installation is being constructed, and should come into operation in 1979. A second, in North America, is being considered. These two installations are at locations that reflect the growing interest in using scatter technique to study the complexities of the ionosphere at high latitudes (see Sections 5.4.2 and 8.2).

Two modes of operation are used in incoherent scatter. In the monostatic mode the transmitter and the receiver are installed at the same site. The transmitter is pulsed and the height resolution depends on the pulse length. Measurements can be made at several heights simultaneously by gating out the echoes at appropriate time delays. In a bistatic or multistatic mode the transmitter and one or more receivers are separated. The transmitter can now send a continuous signal, and the

Table 3.1   Incoherent Scatter Facilities

Location	Frequency (MHz)	Power (MW)	Antenna	Dip latitude ($^{\circ}$N)
Jicamarca, Peru	50	6 Pulsed	290 × 290 m array	1
Aricibo, Puerto Rico	430	2 Pulsed	300 m spherical reflector	30
St. Santin France	935	0.15 Continuous	20 × 100 m reflector	47
Malvern, England	400	0.04 Continuous	43 m parabola	50
Millstone Hill, USA	440 1300	3 4 Pulsed	68 m 25 m parabola	57
Chatanika, Alaska	1300	5 Pulsed	27 m parabola	65
EISCAT* Norway Sweden Finland	224 933	5 Coded 2 Coded	70 m parabola or equivalent Several parabolas, 25 m, 32 m	67

*European Incoherent SCATter facility.

altitude to be studied is selected by the elevation of the receiving antennas. With two receivers at different sites it is possible to obtain two components of horizontal motion simultaneously, to gain information about the electric fields in the ionosphere (Section 4.4.3).

3.7.5 The whistler mode

If we take the quasi-longitudinal approximation to the Appleton–Hartree equation (2.37), put $X/Y_L \gg 1$ to represent wave frequencies much smaller than the plasma frequency but not too small in relation to the gyrofrequency, and neglect collisions ($Z = 0$), we obtain

$$n^2 = -\frac{X}{1 \pm |Y_L|} \tag{3.25}$$

By taking the negative sign in the denominator we obtain a real refractive index and therefore a propagating wave provided $Y_L > 1$

$$\mu^2 = n^2 = f_N^2/f(f_L - f) \tag{3.26}$$

The velocity of a pulse is determined by the group refractive index

$$\mu_g = \frac{d}{df}(\mu f) = \frac{f_N f_L}{2f^{1/2}(f_L - f)^{3/2}} \tag{3.27}$$

Because μ varies with f the medium is dispersive, the velocity being higher at the higher frequencies.

In this mode radio signals of very low frequency can propagate through the ionosphere. It can be shown that they tend to be guided along the geomagnetic field lines and thereby travel from one hemisphere to the other. A natural source of suitable radio signals in the lightning flash, which radiates a short pulse of energy covering a wide band of the radio spectrum. In the opposite hemisphere, the higher frequencies are received first and the lower frequencies later. Thus the signal is received as a falling tone described by the name *whistler*. Some of the energy can be reflected back to the hemisphere where the signal originated, and the process may be repeated several times. Time—frequency curves for whistler propagation are shown in Fig. 3.13. The whistler mode is not restricted to waves of natural origin, but can also be launched from VLF transmitters used in communications. Propagation between hemispheres in the whistler mode can be applied to the study of the electron density of the magnetosphere, as will be seen in Section 7.5.1.

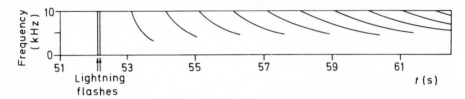

Figure 3.13   A long train of whistler echoes, with the dispersion increasing at each reflection. (After *Whistlers and Related Ionospheric Phenomena* by Robert A. Heliwell with permission of the publishers, Stanford University Press. © 1965 by the Board of Trustees of the Leland Stanford Junior University.)

3.7.6 Summary

Table 3.2 summarizes the major radio techniques that are routinely used to observe the ionized upper atmosphere.

Table 3.2   The Principal Radio Techniques for Observing Ionosphere and Magnetosphere

Technique	Reflected in medium or transmitted	Propagation treatment	Measured
Ground-based ionosonde	R	A–H*	$N(h)$ to max.
Topside sounder	R	A–H	$N(h)$ to max.
Signal strength	R	A–H	Absorption
HF Doppler	R	A–H	Vertical movements
Partial reflections	R	Sharp boundary and A–H	$N(h)$, D region
VLF propagation	R	Sharp† boundary and A–H	D region
Faraday effect Phase effect	T	A–H	Electron content
Riometer	T	A–H	Absorption
Whistler reception	T	A–H	N in magnetosphere
Incoherent scatter	(Scattered)	Scatter	$N(h)$, T_e, T_i, ion composition, plasma motions

    * A–H, Appleton–Hartree equation.
    † With full-wave solutions in more exact treatments.

3.8 Further reading

Davies, K. (1966) *Ionospheric Radio Propagation* (Dover).

Ratcliffe, J. A. (1972) *An Introduction to the Ionosphere and Magnetosphere* (Cambridge University Press) Chapters 9, 10.

Ratcliffe, J. A. (1959) *The Magneto-ionic Theory and its Applications to the Ionosphere* (Cambridge University Press).

Pfotzer, G. (1972) History of the use of balloons in scientific experiments, *Space Sci. Rev.* **13**, 199.

Willmore, A. P. (1974) Exploration of the ionosphere from satellites, *J. atmos. terr. Phys.* **36**, 2255.

Ness, N. F. (1970) Magnetometers for space research, *Space Sci. Rev.* **11**, 459.

Mozer, F. S. (1973) Analysis of techniques for measuring DC and AC electric fields in the magnetosphere, *Space Sci. Rev.* **14**, 272.

Anderson, K. A. (1972) Instrumentation for balloon and rocket experiments, *Space Sci. Rev.* **13**, 337.

Evans, J. V. (1974) Some post-war developments in ground-based radiowave sounding of the ionosphere, *J. atmos. terr. Phys.* **36**, 2183.

Special Issue: Topside sounding (1969) *Proc. Inst. elect. electronics Engr.* **57**, 859.

Evans, J. V. (1969) Theory and practice of ionosphere study by Thomson scatter radar, *Proc. Inst. elect. electronics Engr.* **57**, 496.

Beynon, W. J. G. (1974) Incoherent scatter sounding of the ionosphere, *Contemp. Phys.* **15**, 329.

Walker, A. D. M. (1976) The theory of whistler propagation, *Rev. Geophys. Space Phys.* **14**, 629.

Utlaut, W. F. (1975) Ionospheric modification induced by high-power HF transmitters — a potential for extended range VHF–UHF communications and plasma physics research, *Proc. Inst. elect. electronics Engr.* **63**, 1022.

4
Vertical Structure of the Undisturbed Upper Atmosphere

4.1 The neutral air

4.1.1 Nomenclature

The four basic properties of the upper atmosphere are pressure (P), density (ρ), temperature (T) and composition. These properties are not independent, but are related by the gas law $PV = RT$ for one mole of gas (R being the gas constant) or $PV = NRT$ for N moles. Also, since $R = N_0 k$, where N is Avogadro's number $(6.02 \times 10^{23}\,\mathrm{g^{-1}\,mol^{-1}})$ and k is Boltzmann's constant $(1.38 \times 10^{-16}\,\mathrm{erg}\ \mathrm{^\circ K^{-1}})$, and $\rho = NM/V$, where M is the molecular weight, the gas law can also be written in the useful form

$$P = nkT \tag{4.1}$$

where n is the number of molecules per unit volume. 'n' is properly called the concentration, but it may also be termed the *number density* or, particularly if referring to free electrons or ions, simply *density*.

The regions of the atmosphere are named according to various schemes based, in particular, on temperature, composition, and state of mixing. Fig. 4.1 illustrates the nomenclature that is in most common use. The primary classification is by temperature. In this system individual temperature regions are named *sphere* and each upper boundary is called a *pause*. Thus, the *troposphere*, in which the temperature falls off at a rate of 10 $^\circ$K km^{-1} or less, is bounded by the *tropopause* at a height of 10–12 km. The *stratosphere* above that was originally thought to be isothermal but in fact is a region of increasing temperature. A maximum due to ozone absorption appears at about 50 km and this is the *stratopause*. The temperature decreases again in the *mesosphere* (or *middle atmosphere*) to the temperature minimum at the *mesopause* at 80–85 km. With the temperature around 180 $^\circ$K this is the coldest part of the atmosphere. Above the mesopause dT/dh remains positive, and this is the *thermosphere*. Eventually the thermospheric temperature becomes constant with altitude at a value which changes with time but is usually over 1000 $^\circ$K, the hottest part of the atmosphere.

The temperature classification is generally the most useful one, but schemes based on the state of mixing, the composition and the state of ionization are also indispensable in their respective contexts. The well mixed part of the atmosphere, in which the composition does not change with height, can be called the *turbosphere* or the *homosphere*. The upper region where mixing is absent and the composition changes with height is the *heterosphere*. The boundary between the regions, at about 100 km, is the *turbopause*. Within the heterosphere there are regions where the dominant gas is helium or hydrogen, and these regions are the

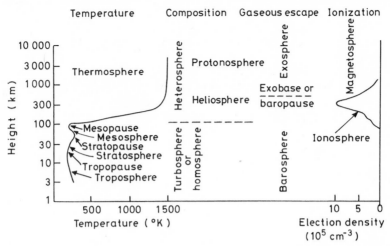

Figure 4.1 Nomenclature of the upper atmosphere based on temperature, composition, gaseous escape and ionization.

heliosphere and the *protonosphere* respectively. At the highest levels, above about 600 km, individual molecules can escape the Earth's gravitational attraction and this region is the *exosphere*. The base of the exosphere is the *exobase* or the *baropause*. For reasons that will become clear in the next section, the lower region, below the baropause, is the *barosphere*.

Finally, we must recall the *ionosphere* and the *magnetosphere*, the first applying to the ionized regions (above about 70 km) and the second to the outermost part where motions are controlled by the geomagnetic field. The outer termination of the geomagnetic field, at distances of the order of 10 earth radii, is the *magnetopause*.

4.1.2 Hydrostatic (barometric) equation

The most basic physical feature of the atmosphere is the decrease of pressure and density with increasing height. The height variation is described by the hydrostatic equation, derived as follows.

If a gas contains n molecules, each of mass m, per unit volume, a cylinder of unit cross-section contains total mass $nm \cdot dh$, where dh is the height of the cylinder (see Fig. 4.2). Under gravity the cylinder experiences a downward force $nmg \cdot dh$ which, in static equilibrium, is balanced by the net upward force due to the difference of pressure between the top and bottom faces:

$$(P + dP) - P = -nmg \cdot dh$$

therefore

$$dP/dh = -nmg$$

But, by the gas law, $P = nkT$. Hence

$$\frac{1}{P} \cdot \frac{dP}{dh} = -\frac{mg}{kT} = -\frac{1}{H} \tag{4.2}$$

where $H = kT/mg$ is defined as the *scale height*.

53

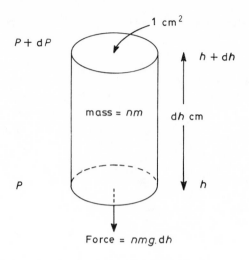

Figure 4.2 Forces on a cylinder of gas.

If H is constant, Equation 4.2 is integrated to give

$$P = P_0 \exp(-h/H) \tag{4.3}$$

where P_0 is the pressure at a height $h = 0$.

The scale height is the vertical distance in which P changes by a factor e. Even if g, T and m are not independent of height, H can be defined by Equation 4.2 in terms of the relative rate of change of concentration with height. In the terrestrial atmosphere H increases with height from about 5 km at height 80 km, to 70–80 km at 500 km. A useful approximate formula, accurate to about 10% over the height range 200–900 km is

$$H(\text{km}) \sim T(^\circ\text{K})/M(\text{mass units}) \tag{4.4}$$

Instead of Equation 4.3, it is sometimes convenient to put

$$\frac{P}{P_0} = \exp\left[-\frac{h - h_0}{H}\right] = e^{-z} \tag{4.5}$$

where $P = P_0$ at $h = h_0$, and z is the *reduced height* defined by $z = (h - h_0)/H$. The hydrostatic equation can also be written in terms of density (ρ) or concentration (n), since $n/n_0 = \rho/\rho_0 = P/P_0$ if T, g and m are constant. The ratio k/m can be replaced by R/M, where R is the gas constant and M the molecular weight.

Whatever the height distribution of a gas, its pressure P_0 at height h_0 is just the weight of a gas in a column of unit cross-section above h_0. Hence

$$P_0 = N_T mg = n_0 k T_0 \tag{4.6}$$

where N_T is the total number of molecules in the column above h_0, and n_0 and T_0 are the concentration and temperature at h_0. Hence

$$N_T = n_0 k T_0 / mg = n_0 H_0 \tag{4.7}$$

H_0 being the scale height at h_0. Equation 4.7 means that if all the atmosphere

54

above h_0 were compressed until it had the density of the real atmosphere at h_0, it would occupy a column of one scale height. Note also that Equation 4.6 implies that the total mass of the atmosphere above unit area of the Earth's surface is equal to the surface pressure divided by g.

The acceleration due to gravity varies with altitude as $g(h) \propto 1/(R_E + h)^2$. The effect of changing gravity is sometimes taken into account by defining a *geopotential height*

$$h* = R_E h/(R_E + h) \tag{4.8}$$

A molecule at height h over the spherical earth, and one at height $h*$ over a flat earth with gravitational acceleration $g(0)$, have the same potential energy.

Within the homosphere, where the atmospheric gas is well mixed, the mean molecular mass, $\bar{m}$, is used to determine the scale height. In the heterosphere the partial pressure of each component is determined by the molecular mass of that species. The different species take up individual distributions, each with its own scale height, the total pressure being the sum of the partial pressures in accordance with Dalton's law.

4.1.3 The exosphere

The hydrostatic equation depends on the gas law, which is valid when there are sufficient collisions between molecules for thermodynamic equilibrium, and a Maxwellian velocity distribution, to be established. At about 600 km the mean free path between collisions becomes comparable with the scale height of the atmosphere. The hydrostatic equation therefore holds strictly only in the barosphere. Higher up, in the exosphere, the question of the vertical distribution needs to be re-examined in terms of the trajectories of individual molecules and atoms. This is not an easy problem. It has been shown that a hydrostatic form will still hold if the velocity distribution is Maxwellian, and it is common practice to use such an equation up to 1500–2000 km — even if only for lack of a better one!

However, one certainly cannot expect to take this liberty if there is significant escape of gas from the atmosphere, because the loss depletes the velocity distribution of its faster components and the distribution cannot then be Maxwellian. Neglecting collisions, a molecule moving vertically at velocity v will escape if $mv^2/2 > mgr$, the right-hand side being the potential energy of a particle of mass m at distance r from the centre of the Earth. Hence, $v^2 > 2gr$ for escape. At the surface the escape velocity is 11.2 km/s, whether the particle be a molecule or a rocket. The mean thermal velocity of gas molecules depends on their mass and temperature, $mv^2/2 = 3kT/2$. Thus, corresponding to an escape velocity, an *escape temperature* T_e can be defined, having the following values for common atmospheric species:

Atom	O	He	H
$T_e (^\circ K)$	84 000	21 000	5200

Exospheric temperatures are 1000–2000 °K, much less than the escape temperatures. Loss of gas occurs at the high speed end of the gas velocity distribution, and is significant for H, slight for He, and insignificant for O. Detailed calculations show that the resulting vertical distribution of H departs seriously from hydrostatic at distances more than one earth radius above the surface, but for He the departure is small.

4.1.4 Heat balance and temperature profile of the upper atmosphere

The main source of heat in the upper atmosphere is solar ultraviolet radiation. When an energetic photon is absorbed, to dissociate or ionize a molecule or atom, the energy of the photon ($h\nu$) will normally exceed that needed for the reaction and the excess appears as kinetic energy of the reaction products. For example, a newly created photoelectron may have a kinetic energy between 1 and 100 eV and this energy becomes distributed in the medium as the electron subsequently makes collisions and in turn excites other molecules. The secondary excitation can be optical, electronic, vibrational, rotational, or by elastic collisions, depending on the energy. The last process, acting mainly on other electrons, distributes energy less than 2 eV and keeps the electrons hotter than the ions. In the ionosphere it can be assumed that the rate of heating in a given region is proportional to the rate of ionization. Other sources of heat are the dissipation of tidal and other atmospheric motions, and electric currents in the ionosphere. These, however, are secondary to the absorption of ultraviolet radiation.

The temperature maximum at the stratopause results from the radiation in the 2000–3000 Å part of the solar spectrum, which is absorbed by ozone at heights between 20 and 50 km. Molecular oxygen, which is relatively abundant up to 95 km, absorbs radiation between 1027 and 1750 Å and in consequence is dissociated into atomic oxygen. All the major species, O_2, O and N_2, absorb ultraviolet (912–1027 Å) and X-rays (8–140 Å) at ionospheric heights. Of the three regions, by far the greatest amount of energy, 1.8×10^4 erg cm^{-2} s^{-1} or 18 W m^{-2}, is absorbed in the ozone layer. Absorption by oxygen at wavelengths greater than 1027 Å accounts for 30 erg cm^{-2} s^{-1}, and the higher levels receive only 3 erg cm^{-2} s^{-1}. Of course the air density is much smaller at the greater heights and thus the temperature can be raised considerably by a small input of energy.

Energy is lost from the heated upper atmosphere by radiation at wavelengths where the atmosphere is transparent. The infrared region is the most effective for heat loss, and this emission forms part of the airglow, to be discussed in Section 4.3.5.

Very important, however, are the transport processes by which heat is moved from one level to another. Conduction, convection and radiation all come into play. At the lowest levels radiation is the most efficient process, and between 30 and 90 km the atmosphere is in radiative equilibrium. At the highest levels the thermal conductivity is large because of the reduced pressure and the increased proportion of free electrons. In the thermosphere above 100 km heat is transported mainly by conduction and, as shown in Fig. 4.1, the temperature becomes independent of height above 300 or 400 km — though, as we shall see, it varies greatly with time. Between these extreme cases is an intermediate region, within the turbosphere, where convection is important. The process of turbulent mixing is generally known as *eddy diffusion*, and by definition it is more efficient than molecular diffusion below the turbopause. Eddy diffusion carries heat from the thermosphere down into the mesosphere, providing a major heat loss mechanism for the thermosphere but a minor heat source, relative to others, for the mesosphere.

4.1.5 Composition

The turbulent mixing of the atmosphere to about 100 km tends to maintain a constant composition through the turbosphere. Basically, the composition is like that at ground level shown in Table 4.1.

Table 4.1   Composition of the Atmosphere at the Ground

Molecule	Mass	Percent of volume	Concentration per cm^3
Nitrogen	28.02	78.1	2.1×10^{19}
Oxygen	32.00	20.9	5.6×10^{18}
Argon	39.96	0.9	2.5×10^{17}
Carbon dioxide	44.02	0.03	8.9×10^{15}
Neon	20.17	0.002	4.9×10^{14}
Helium	4.00	0.0005	1.4×10^{14}
Water	18.02	Variable	

However, complete uniformity cannot be maintained if there are sources or sinks for particular species, and such is the case. As indicated in the previous section, molecular oxygen is dissociated into atomic oxygen by ultraviolet radiation between 1027 and 1750 Å

$$O_2 + h\nu \rightarrow O + O \tag{4.9}$$

$h\nu$ representing a quantum of radiation. In consequence an increasing amount of atomic oxygen appears above 90 km. O and O_2 are present in equal concentrations at 125 km, and above this the atomic form increasingly dominates over the molecular form. Nitrogen, on the other hand, is not directly associated into the atomic form in the atmosphere, though atomic nitrogen does appear as a product of other chemical reactions.

Relatively low in the atmosphere, between 15 and 35 km, ozone is produced by the reaction

$$O + O_2 + M \rightarrow O_3 + M \tag{4.10}$$

This is a 'three-body' reaction, in which the third body 'M' serves to carry away excess energy. Such a reaction requires a relatively high pressure because its rate depends crucially on the probability that three atoms or molecules come together at the same time. The ozone-producing reaction, moreover, requires both the molecular and the atomic forms of oxygen, and the latter is again produced by the reaction of Equation 4.9, except that a more penetrating solar radiation (2000–2400 Å) is the effective band. The ozone, once formed, may be dissociated again by radiation over a wide wavelength range, but particularly by 2100–3100 Å

$$O_3 + h\nu \rightarrow O_2 + O \tag{4.11}$$

or destroyed by the reaction

$$O_3 + O \rightarrow O_2 + O_2 \tag{4.12}$$

On balance a small concentration of ozone remains, which, though it amounts to less than 10 parts per million of the total concentration, is important for its absorption in the ultraviolet and as an atmospheric heat source. The region maximizes near 20 km, and may be called the *ozonosphere*.

The turbosphere contains many minor constituents in addition to ozone, and in spite of their low concentrations they may play important roles in the photo-chemistry of the region. Nitric oxide, which is significant in the production of the lower ionosphere, is formed in the mesosphere by the reaction

$$N + O_2 \rightarrow NO + O \tag{4.13}$$

where the atomic nitrogen has been formed in other reactions. NO is removed by a further reaction involving N:

$$NO + N \rightarrow N_2 + O \tag{4.14}$$

Using square brackets for the concentrations of reactions, and k_1 and k_2 for reaction-rate coefficients, we can say that the rate of production of NO is

$$k_1 [N] [O_2]$$

and the rate of loss is

$$k_2 [N] [NO]$$

At equilibrium these rates must be equal. Thus

$$[NO] / [O_2] = k_1/k_2 = K \tag{4.15}$$

provided there is sufficient N present for reaction 4.13 to occur at a reasonable rate. This is satisfied above 90 km where, putting in values for k_1 and k_2, it can be shown that $[NO]/[O_2] \approx 10^{-8}$. Thus a minute concentration of nitric oxide is predicted. Nevertheless the amount is sufficient to make it an important source of free electrons in the lower ionosphere, as will be discussed in Section 4.3.4.

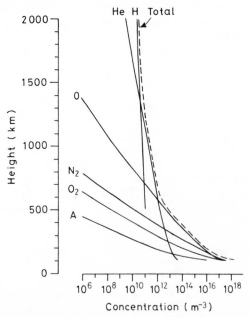

Figure 4.3   Distribution of major gases in the thermosphere for an exospheric temperature of 800°K. (Data from COSPAR International Reference Atmosphere, 1972).

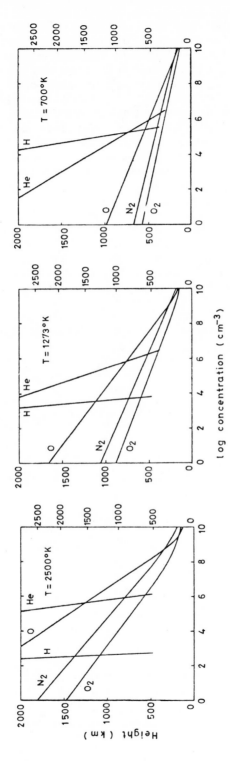

Figure 4.4 Effect of temperature on exospheric composition. Geopotential heights are given on the left of the diagrams, geometric heights on the right. (From L.G. Jacchia, *COSPAR International Reference Atmosphere*, North-Holland, 1965.)

59

Above the turbosphere mixing is unimportant, each component takes an independent height distribution governed by its own scale height ($H = RT/Mg$), and the composition changes with altitude. The scale heights of the common gases differ greatly because of the range of molecular weights concerned, H = 1, He = 4, O = 16, N_2 = 28, O_2 = 32. Hence the lighter gases become relatively more abundant at greater heights, as shown in Fig. 4.3. Atomic oxygen is the most abundant species at several hundred kilometres. Above that is the heliosphere where helium is the most abundant, and eventually hydrogen is dominant in the protonosphere. The exact altitudes of these various regions depends on the temperature of the exosphere. As Fig. 4.4 shows, the protonosphere begins much higher in a hot exosphere than in a cool one.

4.2 The ionized air – physical aeronomy

The ionized part of the upper atmosphere, comprising free electrons and ions, is the *ionosphere,* the name having been coined in 1926 by R. Watson-Watt. The vertical structure of the ionosphere has been observed and studied for many years, and is understood fairly well in terms of the physical and chemical processes at work in the upper atmosphere. Typical ionospheric vertical structures are illustrated in Fig. 4.5. The various layers were originally identified from ionograms, which tend to emphasize the inflections of the profile. In fact the layers are not necessarily separated by minima. The main regions can be summarized thus:

D region: 60–90 km, $N \sim 10^2 - 10^4$ cm^{-3} by day
E region: 105–160 km, $N \sim$ several $\times 10^5$ cm^{-3} by day

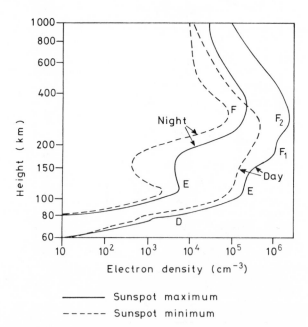

Figure 4.5 Typical vertical profiles of the mid-latitude ionosphere. (From wallchart *Aerospace Environment*, W. Swider, Air Force Geophysics Laboratory, Hanscom Air Force Base, Massachusetts.)

F region: above 180 km, N up to several $\times 10^6$ cm^{-3} at peak of day, factor of 10 smaller by night, height of maximum ~ 300 km but variable.

For a detailed explanation of the structure one has to take account of the atmospheric composition, of the solar spectrum, and of the detailed photochemistry. However, the first step is to examine the formation of an ionosphere in the more general terms of the physical processes. The principles so demonstrated should apply to ionospheric formation on any planet having a gaseous atmosphere.

The rate of change of electron density (N) at a given place is given by the *continuity equation*,

Rate of increase = rate of production — recombination rate — loss by movements

$$\frac{\partial N}{\partial t} \quad = \quad q \quad - \quad L \quad -\mathrm{div}(N\mathbf{v}) \quad (4.16)$$

where v is the mean drift velocity of the electrons. Equilibrium is reached when the production and loss processes balance, and when $\partial N/\partial t = 0$. Sometimes a further simplification can be affected when one of the terms on the right-hand of Equation 4.16 can be shown to be negligible by comparison with the others. We consider first the production rate.

4.2.1 The Chapman production function

The Chapman theory, dating from 1931, starts with a few simplifying assumptions and works out the form of an ionospheric layer and how it should change with time. In essence the assumptions are as follows:

(a) an exponential atmosphere, with only one gaseous component and of constant scale height
(b) plane stratification
(c) absorption of solar radiation in proportion to the number density of gas particles
(d) constant absorption coefficient, i.e. monochromatic radiation and a single absorbing component.

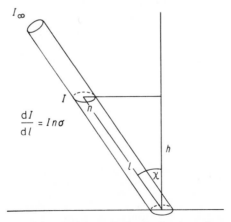

Figure 4.6 Ionizing radiation in the atmosphere.

Ionizing radiation enters the atmosphere at a zenith angle χ. Its intensity (energy flux per unit area) is I_∞ above the atmosphere and I at altitude h, corresponding to slant distance l from the ground (Fig. 4.6). If there are n absorbing atoms per unit volume, and each has an absorption cross-section σ, the intensity of the radiation changes with distance as

$$dI/dl = -I\sigma n \qquad (4.17)$$

If the probability that the absorbed radiation will ionize the atom is η (ionization efficiency), the rate of ionization at height h is

$$q = \eta\sigma nI \qquad (4.18)$$

since σnI is the energy absorbed by unit volume of the gas. q maximizes where

$$\frac{dq}{dl} = 0 = \eta\sigma\left(n\frac{dI}{dl} + I\frac{dn}{dl}\right)$$

Thus

$$\left(\frac{1}{I}\cdot\frac{dI}{dl}\right)_m + \left(\frac{1}{n}\cdot\frac{dn}{dl}\right)_m = 0 \qquad (4.19)$$

where the subscript m denotes values at the maximum of q. Now

$$\frac{1}{n}\cdot\frac{dn}{dl} = \frac{\cos\chi}{n}\cdot\frac{dn}{dh} = \frac{\cos\chi}{H}$$

by definition of the scale height (H). Thus, from Equations 4.17 and 4.19

$$\left(\frac{dI}{dl}\right)_m = I_m\sigma n_m = \frac{I_m\cos\chi}{H_m}$$

Therefore

$$\sigma n_m H_m\sec\chi = 1 \qquad (4.20)$$

Note that when the Sun is overhead $\sigma n_m H_m = 1$, and if H is independent of height

$$n_m \propto \cos\chi \qquad (4.21)$$

By integration of Equation 4.17

$$\int_l^\infty \frac{dI}{I} = -\sigma n\,dl$$

Therefore

$$ln\,(I/I_\infty) = -\sigma\int_l^\infty n\,dl = -\sigma\sec\chi\int_l^\infty n\,dh = -\sigma nH\sec\chi$$

(since

$$\int_l^\infty n\,dh = nH).$$

Thus

$$I/I_\infty = e^{-\sigma nH\sec\chi} = e^{-\tau} \qquad (4.22)$$

by definition of the optical depth (τ).

Thus, $\sigma nH\sec\chi$ is the optical depth and is unity at the level of maximum production. I is reduced to $1/e$ of its original value at this level. Except as noted, the foregoing results hold for any single component atmosphere, irrespective of its height variation.

In an exponential atmosphere, $n = n_0\exp(-h/H)$, where H is now constant. The maximum rate of production now occurs (from Equation 4.20) where

$$\sigma n_0\exp(-h_m/H)\cdot H\sec\chi = 1,$$

whence

$$h_m = H \ln (\sigma n_0 H \sec \chi) = \text{constant} + \ln (\sec \chi) \tag{4.23}$$

From Equations 4.22 and 4.20

$$I/I_\infty = \exp (-n/n_m) \tag{4.24}$$

Therefore

$$\frac{q}{q_m} = \frac{\eta \sigma n I}{\eta \sigma n_m I_m} = \frac{n}{n_m} \cdot \frac{\exp (-n/n_m)}{\exp (-1)}$$

But

$$\frac{n}{n_m} = \exp [-(h - h_m)/H]$$

Hence

$$\frac{q}{q_m} = \exp [-(h - h_m)/H] \cdot \exp [-e^{-(h - h_m)/H}] \cdot e^{+1}$$

$$= \exp \{1 - (h - h_m)/H - \exp [-(h - h_m)/H]\}$$

$$= \exp (1 - y - e^{-y}) \tag{4.25}$$

where $y = (h - h_m)/H$ and is the reduced height referred to the height of maximum production. Equation 4.25 is the *Chapman production function*.

The expression can also be derived in terms of the maximum rate of production when the Sun is overhead (q_{m0}), the height of maximum then being h_{m0}:

$$q/q_{m0} = \exp (1 - z - \sec e^{-z}) \tag{4.26}$$

where $z = (h - h_{m0})/H$

The Chapman production function serves as a basis of reference for ionospheric behaviour, and several points about it are worthy of special note.

(a) The result, $q = q_{m0} \exp (1 - z - \sec e^{-z})$ (Equation 4.26), where z is the reduced height $[z = (h - h_{m0})/H]$ and χ is the zenith angle of the Sun. q_{m0} is the production rate at the maximum if h_{m0} is the height of the maximum when the Sun is overhead.

(b) The shape of the curve (Fig. 4.7). At great altitudes, z is large and positive and

$$q \to q_0 e^{-z} \tag{4.27}$$

independent of the zenith angle. At altitudes well below the maximum, z is large and negative

$$q \to q_0 \exp (-\sec e^{-z}) \tag{4.28}$$

with a rapid cut-off. Physically, the production rate is limited by a lack of ionizable molecules at high altitude and a lack of ionizing radiation at low altitude. When plotted as $\log (q/q_0)$ against z, all curves have the same shape and move upwards and to the left as χ increases (Fig. 4.7).

(c) Maximum production rate occurs where the optical depth is unity (Equation 4.22).

(d) From Equation 4.23, it is readily shown that

$$z_m = \ln \sec \chi \tag{4.29}$$

where z_m is the reduced height of the maximum, reckoned from the maximum with overhead Sun.

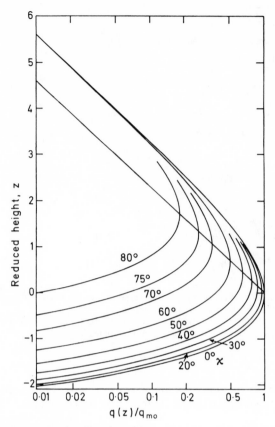

Figure 4.7 Chapman production function. (From T.E. Van Zandt and R.W. Knecht, *Space Physics* (Ed. LeGalley and Rosen) Wiley, 1964.)

(e) From this and Equation 4.26,

$$q_m = q_{m0} \cos \chi \qquad (4.30)$$

where q_m is the production rate at the maximum, and q_{m0} is the production rate at the maximum when the Sun is overhead. Equations 4.29 and 4.30 provide useful tests of whether or not an observed ionospheric layer obeys the Chapman law.

4.2.2 Recombination laws

The Chapman theory neglects movements but considers two kinds of loss term, L. First we assume that the electrons recombine directly with positive ions

$$X^+ + e \rightarrow X \qquad (4.31)$$

The rate of recombination can be written as

$$L = \alpha [X^+] N = \alpha N^2 \qquad (4.32)$$

where N is the electron density, and we put $[X^+] = N$ so that the bulk neutrality of the medium is maintained. α is the *recombination coefficient*. Thus at equilibrium

$$q = \alpha N^2 \tag{4.33}$$

and if q is given by the Chapman production function

$$N = N_{m0} \exp \tfrac{1}{2}(1 - z - \sec \chi \, e^{-z}) \tag{4.34}$$

where $z = (h - h_{m0})/H$. From Equation 4.30,

$$N_m = N_{m0} \cos^{1/2} \chi \tag{4.35}$$

showing that the maximum electron density of the layer varies as $\cos^{1/2} \chi$.

The alternative assumption is that electrons are removed by becoming attached to neutral molecules:

$$M + e \to M^- \tag{4.36}$$

The rate of attachment is

$$L = \beta N \tag{4.37}$$

where β is an *attachment coefficient*. The rate is now linear with N because it is assumed that the species M is numerous compared to the number of free electrons. However, β is a function of altitude. At equilibrium

$$q = \beta N \tag{4.38}$$

and, from Equation 4.26,

$$N = N_{m0} \exp(1 - z - \sec \chi \, e^{-z}) \tag{4.39}$$

Also

$$N_m = N_{m0} \cos \chi \tag{4.40}$$

Such a layer is a *β-Chapman layer*.

The height variation of the attachment coefficient, β, has important consequences in the terrestrial ionosphere because of the nature of the photochemical processes. It is known that electron loss in most of the ionosphere occurs not directly but by a two-stage process

$$\left.\begin{array}{ll} X^+ + A_2 \to AX^+ + A & \text{(i)} \\[2ex] AX^+ + e \to A + X & \text{(ii)} \end{array}\right\} \tag{4.41}$$

Here A_2 is a common molecular species such as O_2 or N_2. In the first step the positive charge is exchanged between the atomic and molecular species, and in the second one the charged molecule is dissociated by recombining with an electron — a *dissociative recombination* reaction. The rate of (i) is $\beta[X^+]$ and of (ii) is $\alpha[AX^+] N$. At low altitude β is large and so all X^+ is rapidly converted to AX^+. The overall rate is then controlled by (ii) and one expects to see an α-type process overall because $[AX^+] = N$ for neutrality. At a high level β is small and then (i) is relatively slow and controls the overall rate. Now $[X^+] = N$, and we have a β-type process. The behaviour therefore changes from α-type to β-type as the height increases. It can be shown that in general the reaction scheme of Equation 4.41 in equilibrium can be expressed by

$$\frac{1}{q} = \frac{1}{\beta(h)N} + \frac{1}{\alpha N^2}$$ (4.42)

where q is the production rate. The transition from α- to β-behaviour is considered to occur at a height h_t where

$$\beta(h_t) = \alpha N$$ (4.43)

4.2.3 Vertical transport

We now consider what happens if changes of electron concentration are caused by the movement of ionization from one place to another rather than by photo-chemical production and loss. Since our present purpose is to study the distribution of ionization with height in the atmosphere we will confine the discussion to vertical movements. The treatment is readily generalized to three dimensions where necessary.

If we think of a region of the atmosphere, it can be shown (Section 2.2.1) that the rate of increase of electron density within it due to vertical motion is

$$\frac{dN}{dt} = -\frac{\partial}{\partial h}(WN)$$ (4.44)

where N is the electron density, W is the drift velocity in the vertical (upward) direction, and h is the height. (In the general case where the velocity is a vector $\mathbf{v}$, $dN/dt = -\text{div}(\mathbf{v}N)$.) In the present example we suppose the vertical drift to be due to diffusion (Section 2.2.4), in which case the vertical drift is given by

$$W = -\frac{D}{N} \cdot \frac{\partial N}{\partial h}$$ (4.45)

where D is the diffusion coefficient. In the simplest case the diffusion coefficient is given by

$$D = kT/mv$$ (4.46)

where k is Boltzmann's constant, T the temperature, m the particle mass and v the collision frequency.

For vertical motion, the gravitational force must be included. Following the treatment of Section 2.2.3, the drag force due to collisions between the minority gas (ions and electrons) and the majority gas (neutral air) is

$$NmvW = -\frac{dP}{dh} - Nmg$$ (4.47)

from which, since $P = NkT$, $D = kT/mv$ and $H_N = kT/mg$,

$$NW = -D\left(\frac{dN}{dh} + \frac{N}{H_N}\right)$$

H_N is here the scale height of the minority gas. By differentiation, the continuity

66

equation becomes

$$\frac{dN}{dt} = -\frac{\partial(WN)}{\partial h} = \frac{\partial}{\partial h}\left\{D\left(\frac{dN}{dh} + \frac{N}{H_N}\right)\right\} \tag{4.48}$$

In a plasma containing equal numbers of electrons and ions, an additional consideration is introduced by the electric charge. Initially the ions, because of their greater mass, will tend to settle away from the electrons. For a separation δh, an electric field $E = Ne\delta h/\epsilon_0$ is produced, and a restoring force $Ee = Ne^2\delta h/\epsilon_0$ acts on each charged particle in such a direction as to lift the ions and depress the electrons. This force is very large, being comparable to the gravitational force on an atomic oxygen ion when the separation is less than 10^{-4} mm (regarding the ionosphere as a slab with sharp boundaries). At these separations the electrostatic force balancing the gravitational force is larger than 1 $\mu V/m$; we see that even this minute field is sufficient to prevent the significant separation of electrons from positive ions, but each particle behaves as though it had charge $\pm e$ and mass $(m_e + m_i)/2 \sim m_i/2$. A plasma therefore has scale height

$$H_p = 2kT/m_i g = 2H \tag{4.49}$$

and diffusion coefficient

$$D_p = 2kT/m_i \nu_i = 2D \tag{4.50}$$

provided the electrons and ions are at the same temperature. If this is not so, T is replaced by $(T_i + T_e)/2$ in these equations. Equation 4.48 for a plasma should therefore be written

$$\frac{dN}{dt} = \frac{\partial}{\partial h}\left\{D_p\left(\frac{dN}{dh} + \frac{N}{H_p}\right)\right\} \tag{4.51}$$

(This result can be obtained formally by writing Equation 4.47 as a pair of equations, one for electrons and one for ions, a term $+NeE$ being added to the right-hand side of the ion equation, and a term $-NeE$ being added similarly to the electron equation.)

We see from Equation 5.51 that at equilibrium $dN/dh = -N/H_p$, so that under thermal diffusion the plasma is distributed exponentially with height as

$$N/N_0 = \exp(-h/H_0)$$

with scale height H_p. This gives the important result that in diffusive equilibrium the scale height of the ionization is twice that of the neutral gas.

In the upper atmosphere the significance of vertical diffusion by comparison with other processes is governed by its relative speed, and, as may be seen from Equation 4.51, this depends on the magnitude of the diffusion coefficient D_p. This comparison will be made in Section 4.3.3, but we note here that diffusion should become more important at greater heights because of the decrease of collision frequency (Equation 4.46). If ν is proportional to the pressure

$$D = D_0 \exp(h/H) \tag{4.52}$$

H being the scale height of the neutral gas.

4.3 The ionized air — chemical aeronomy

4.3.1 Photochemical principles

The preceding section covered physical principles that may be expected to govern the intensity and height distribution of an ionosphere. To account for the ionosphere observed on the Earth or any other planet it is necessary to look in detail at the photochemistry. The problem is to determine what gas can be ionized, to measure the energetic radiation available for ionization, and to compute production rates; then to assess the loss rates, which requires a knowledge of all possible reactions and their rate coefficients, so as to arrive at the equilibrium electron density. This is a formidable problem in the general form. Fortunately it is usually possible to identify certain major sources for a given altitude range, though it should be pointed out that the photochemistry of even the terrestrial ionosphere still contains some unsolved problems. Before considering any details related to particular ionospheric layers we outline some general principles.

The constitution of the terrestrial atmosphere was considered in Section 4.1.5. The atmosphere consists mainly of molecular nitrogen and oxygen up to 100 km, but above that level atomic species, particularly oxygen, helium and hydrogen, come into dominance in turn as the composition changes with height due to diffusive separation in a non-turbulent atmosphere. Various minor constituents are also present.

These species may be ionized by capturing a quantum of radiation whose energy exceeds the ionization potential of the species. Ionization potentials of various molecular and atomic gases are shown in Table 4.2.

Table 4.2 Ionization Potentials and Equivalent Wavelengths for Gaseous Species

Constituent	I (eV)	λ_{max} (Å)
NO	9.25	1340
O_2	10.08	1027
H_2O	12.60	985
O_3	12.80	970
H	13.59	912
O	13.61	911
CO_2	13.79	899
N	14.54	853
H_2	15.41	804
N_2	15.58	796
A	15.75	787
Ne	21.56	575
He	24.58	504

Thus, only radiation of wavelength less than λ_{max} can produce ionization. This immediately defines the part of the solar spectrum with which we are concerned as being the X-ray (1–170 Å) and extreme ultraviolet or EUV (170–1750 Å) regions. These emissions originate in the solar chromosphere and corona, and some of them can be greatly enhanced during solar flares. From rockets and satellites fairly good measurements of the spectrum are now available, and the energy flux in various bands and lines is shown in Fig. 4.8.

However, not every photon with sufficient energy will actually produce an ionization. The chance that the photon will be absorbed is given by the absorption

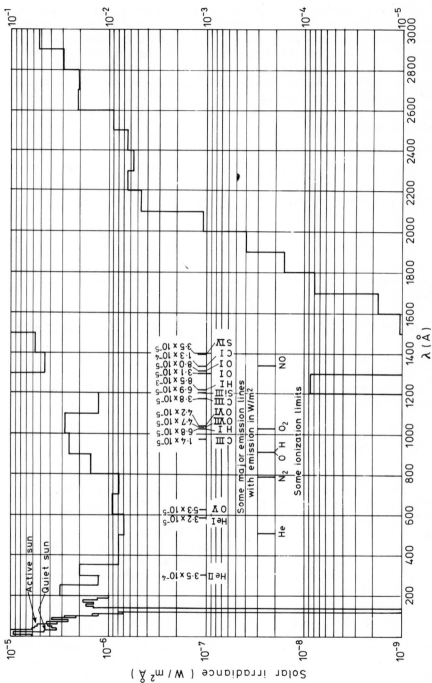

Figure 4.8 Solar irradiance in the X-ray and ultraviolet regions. Note the change of scale over 1200–1500 Å. (Data from Smith and Gottlieb, *Space Sci. Rev.* **16**, 77, 1974.)

cross-section σ, and the chance that an absorbed photon produces an ionization is given by the ionization efficiency η. These parameters were encountered in the discussion of the Chapman theory in Section 4.2.1. The product $\sigma\eta$ is the ionization cross-section σ_i. Absorption cross-sections depend on the wavelength, and usually fall off at wavelengths considerably smaller than λ_{max}. In the EUV range maximum values of 10^{-17} to 10^{-18} cm^2 are attained; in the X-ray range the values are much lower. For atomic species the ionization efficiency is unity, so that all the absorbed energy goes into producing ion–electron pairs. For molecules, $\eta < 1$. A convenient though approximate rule is that one ion–electron pair is produced for every 34 eV of energy absorbed, no matter what the wavelength of the radiation. This is equivalent to the simple formula

$$\eta = 360/\lambda \quad (\text{Å}) \tag{4.53}$$

The height at which the ionosphere is formed depends on the absorption cross-section and the density of the gas being ionized, since by Chapman theory the maximum production rate occurs where the optical depth, $\sigma n H \sec \chi$, is unity. If several gases contribute to the absorption at wavelength λ, the condition is

$$\sum \sigma_i(\lambda) n_i H_i \sec \chi = 1$$

Fig. 4.9a shows how this height varies with wavelength in the terrestrial atmosphere when $\chi = 0°$. The production rate at this altitude is

$$q_m = \frac{\eta \cos \chi}{eH} \cdot I_\infty \tag{4.54}$$

where I_∞ is the intensity of ionizing radiation above the atmosphere (from Equations 4.18, 4.20 and 4.22).

Fig. 4.9b summarizes in a simplified form the roles of various solar radiations in the upper atmosphere.

4.3.2 E and F1 regions

The E region, peaking near 105 km, and the F1 region, whose maximum usually occurs between 160 and 180 km, form the central parts of the ionosphere and are both fairly well understood. The F1 region is created by that part of the EUV spectrum which is most heavily absorbed. Radiation of 500–600 Å reaches unit optical depth at about 160 km, and the range 200–910 Å probably contributes to the F1 ionization (see Fig. 4.9). The band includes an intense solar emission line at 304 Å in addition to the general continuum. The radiations form O_2^+, N_2^+, O^+, He^+ and N^+ as primary ionization products. Due to subsequent reactions, NO^+ and O_2^+ remain as the most abundant.

The E region comes from the more penetrating part of the spectrum: EUV between 800 and 1026 Å, absorbed by molecular oxygen to form O_2^+, and X-rays from 10 to 100 Å. The X-rays would ionize all constituents present to form N_2^+, O_2^+ and O^+ in particular, but the intensity of the incident X-rays varies considerably during the solar cycle and it is likely that the X-ray contribution to the E region is relatively small at solar minimum. The most abundant ions are again observed to be NO^+ and O_2^+.

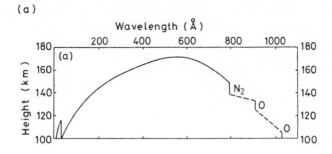

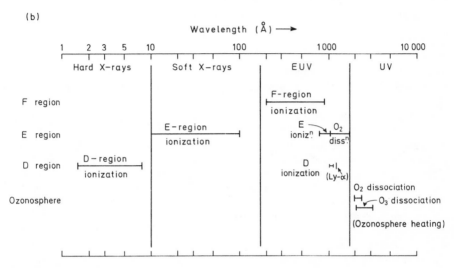

Figure 4.9 (a) Height of unity optical depth. (From H. Rishbeth and O.K. Garriott, *Introduction to Ionospheric Physics,* Academic Press, 1969.) (b) The role of various solar emissions in the upper atmosphere.

Recombination processes which remove an electron are of three types:

radiative recombination: $e + X^+ \rightarrow X + h\nu$
dissociative recombination: $e + XY^+ \rightarrow X + Y$
attachment: $e + Z \rightarrow Z^-$

Attachment is not important in the E and F regions because the pressure is too low. Of the other reactions, the second is much quicker than the first because energy and momentum are more easily conserved if two bodies are produced. The respective recombination coefficients are 10^{-13} and 10^{-18} m^3 s^{-1}. If molecular ions are present, as in the E region, dissociative recombination occurs as follows:

$$e + O_2^+ \rightarrow O + O$$
$$e + N_2^+ \rightarrow N + N$$
$$e + NO^+ \rightarrow N + O$$

71

If the primary ions are atomic, as in the F1 region, they are first converted to molecular ions by charge exchange or rearrangement reactions:

$$O^+ + O_2 \rightarrow O_2^+ + O$$

$$O^+ + N_2 \rightarrow NO^+ + N$$

and the molecular ion then undergoes dissociative recombination with an electron as before. This is the basis of the two-stage process referred to in Section 4.2.2 (Equation 4.41). Another important charge exchange reaction is

$$O + N_2^+ \rightarrow NO^+ + N$$

which also contributes to the abundant NO^+ observed in both E and F1 regions.

4.3.3 F2 region

The F2 region comes to a maximum anywhere between 200 and 600 km, but there is no peak of electron production to account for it. To account for the F2 peak we have to consider how the loss rate varies with height. Section 4.2.2 showed how the characteristic recombination law changed from α-type to β-type at a transition level h_t if electron loss occurs by a two-stage process of ion–atom rearrangement followed by dissociative recombination. This is just the situation in the F region, and to explain the F2 layer we consider heights above h_t where recombination is essentially β-type. That is, the loss rate can be expressed as $\beta(h) N_e$ (where we now use N_e for electron density to avoid confusion with the chemical symbol for atomic nitrogen). Note that β is a function of height because it depends on the concentration of neutral molecules. In the F region the relevant reaction is

$$O^+ + N_2 \rightarrow NO^+ + N$$

Thus, $\beta \propto [N_2]$. Most of the F-region ionization comes from atomic oxygen, however, and thus the production rate $q \propto [O]$. Therefore the equilibrium electron density

$$N_e = q/\beta \propto [O]/[N_2] \propto \exp\left[-\frac{h}{H(O)} + \frac{h}{H(N_2)}\right]$$

where $H(O)$ and $H(N_2)$ are the scale heights of O and N_2 respectively. Thus, the electron density varies with height as

$$N_e \propto \exp\left[-\frac{h}{H(O)}\left(1 - \frac{H(O)}{H(N_2)}\right)\right] = \exp\left[+0.75\frac{h}{H(O)}\right] \qquad (4.55)$$

because the masses of N_2 and O are in the ratio $1.75:1$. Equation 4.55 represents a layer whose electron density increases with height because the loss rate falls off more quickly than the production rate. This is often known as a *Bradbury layer*.

We know that the F2 layer has a maximum and that at a sufficiently great altitude the electron density again decreases with height. Above the F2-layer peak the atmospheric density is so low that chemical recombination (depending on $[N_2]$) becomes less important as a loss process than vertical transport, whose efficacy increases with height because of the change of diffusion coefficient with height (see Equation 4.52). In the topside ionosphere, above the peak of the F2 layer, the electron density is therefore governed by diffusion and is distributed with a scale height H_p numerically equal to twice the scale height of the neutral gas.

The maximum will occur where the two loss processes — β-type chemical recombination and transport — are equally important. This amounts to a comparison of characteristic times, on the principle that of two competing processes the more rapid one will effectively control the total loss rate. For recombination this is relatively straightforward. If the source of ionization is removed, the electron density will decay at a rate

$$dN_e/dt = -\beta N_e = -N_e/\tau_\beta$$

Therefore the characteristic time (in this case the 'time constant') is

$$\tau_\beta = 1/\beta \tag{4.56}$$

For a diffusion process let the instantaneous distribution of electron density be expressed by $N_e = N_0 \exp(-h/\delta)$, where δ is a characteristic height that will tend to H_p, the plasma scale height, at equilibrium. Then a little manipulation of Equation 4.51 shows that

$$\frac{dN_e}{dt} = DN_e \left(\frac{1}{\delta} - \frac{1}{H} \right) \left(\frac{1}{\delta} - \frac{1}{H_p} \right)$$

where D is the diffusion coefficient, whose height variation is determined by the scale height (H) of the neutral gas. If we start with a plasma distribution such that $\delta \gg H, H_p$, the time constant is simply $\tau_D = HH_p/D$. If $\delta \ll H, H_p$, $\tau_D = \delta^2/D$. We can therefore say that in most cases

$$\tau_D \sim H_1^2/D \tag{4.57}$$

where H_1 is a typical scale height for the region being considered.

Comparison of Equations 4.56 and 4.57 then places the F2 maximum where

$$\beta \sim D/H^2 \tag{4.58}$$

The peak electron density is given by

$$N_m \sim q_m/\beta_m \tag{4.59}$$

q_m and β_m being the values of production rate and attachment coefficient at the height of the maximum. This approximate treatment gives a satisfactory result in comparisons with observations. For exact solutions it is necessary to solve the relevant equations exactly by computer methods.

We note in passing that the method of comparing time constants is also useful for finding the level of the turbopause (Section 4.1.4), for which purpose one would compare the time constant for eddy diffusion $\tau_K = H^2/D_1$, where D_1 is the eddy diffusion coefficient, with that for molecular diffusion, $\tau_m = H^2/D_2$ where D_2 is the molecular diffusion coefficient. Since D_2 is inversely proportional to the concentration of neutral gas (Section 2.2.4) molecular diffusion becomes less important at lower heights. The level where $\tau_K = \tau_m$ is the turbopause.

At some level above the F2 peak the region dominated by O^+ gives way to the protonosphere dominated by H^+, this transition being within the O^+ diffusion regime. However the boundary is strongly influenced by a charge exchange reaction

$$H^+ + O \rightleftharpoons H + O^+$$

which is rapid because the ionization potentials of H and O are almost the same (see Table 4.2). Thus it may be assumed that there is an equilibrium

$$[H^+] [O] = 9/8 [H] [O^+]$$

(The factor 9/8 arises for statistical reasons that we need not go into here.)

Just as there are photochemical and diffusion regimes within the F region, separated by the level where $\beta \sim D/H^2$, so similar regimes can be defined for the protons. In the lower region the distribution of H^+ is determined by the above reaction, which means that the H^+ distribution is tied to the O^+ distribution. The height variation of $[H^+]$

$$\propto \frac{[H] [O^+]}{[O]} = \frac{\exp[-h/H(H)] \cdot \exp[-h/H(O^+)]}{\exp[-h/H(O)]} = \exp[+ 7h/H(H)]$$

where $H(H)$, $H(O)$ and $H(O^+)$ are respectively the scale heights of hydrogen atoms, oxygen atoms and oxygen ions. Note that the proton density increases upwards with scale height $H(H)/7$ in this region.

If the rate coefficient of the above reaction is k, the lifetime of a proton is $(k [O])^{-1}$. Since the time constant for diffusion is H^2/D, the boundary between the photochemical regimes is where $k [O] \sim D/H^2$. This occurs at 700 km or above — well above the maximum of the F region.

4.3.4 D region

The D region of the ionosphere does not include a maximum, but is that ionization below about 90 km which is not accounted for by E-region processes. Pressures are a million times greater than in the F2 region. Because of the relatively high pressure the photochemistry is complex, and is not yet fully understood. The D region is particularly interesting from the photochemical point of view, and we will dwell there at somewhat greater length even though its small electron density might make it seem a minor part of the ionosphere as a whole. Fig. 4.10 shows some D region profiles of electron density.

Four sources produce ionization in the D region:

(a) The Lyman-α line of the solar spectrum (1215 Å) penetrates into the D region and ionizes nitric oxide (NO), a minor constituent with ionization limit at 1340 Å. The concentration of NO is 10^8 cm^{-3} or less at 70 km.
(b) The EUV spectrum between 1027 and 1118 Å ionizes excited oxygen molecules in the state described as $O_2(^1\Delta_g)$.
(c) Hard X-rays of 2–8 Å ionize all constituents, thereby acting mainly on O_2 and N_2.
(d) Cosmic rays similarly ionize all constituents.

Theoretical profiles of the ionization rates due to these sources and also to the E-region sources (Lyman-β, EUV and soft X-rays) are shown in Fig. 4.11. Lyman-α appears to be the major source below the E region down to about 70 km, below which cosmic rays are important. Source (b) is less important. The hard X-rays (c) are not important at solar minimum but make a larger contribution at solar maximum when their flux is between a hundred and a thousand times greater.

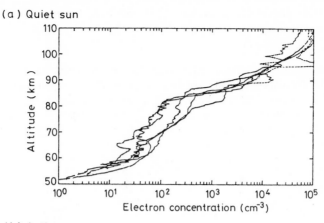

(a) Quiet sun

(b) Active sun

Figure 4.10 Profiles of electron concentration in the D region, from rocket measurements: (a) quiet Sun; (b) active Sun. (After E.A. Mechtly *et al., J. atmos. terr. Phys.* **34**, 1899, 1972, by permission of Pergamon Press.)

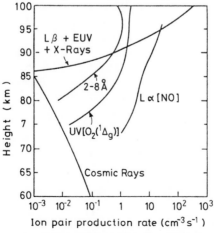

Figure 4.11 Ionization rates in the quiet day-time D region at solar minimum. (After L. Thomas, *Rad. Sci.* **9**, 121, 1974, copyrighted by American Geophysical Union.)

The production of electrons leaves primary ions, particularly NO^+, O_2^+ and N_2^+. The N_2^+ is rapidly converted by a charge-exchange reaction to O_2^+:

$$N_2^+ + O_2 \rightarrow O_2^+ + N_2$$

removing N_2^+ and leaving O_2^+ and NO^+ as the major positive ions. Mass spectrometer measurements from rockets find mass numbers 19 and 37 below 80 or 85 km, which correspond to the hydrated species $H^+ \cdot H_2O$ and $H_3O^+ \cdot H_2O$. Higher hydrates, $H_3O^+(H_2O)_n$, are also present. These hydrates occur up to the level of the mesopause, and are generally found in the region where the water vapour concentration exceeds about 10^9 cm^{-3}.

The hydrates are thought to be responsible for a marked ledge at about 85 km in the electron density profile of the D region in summer (Fig. 4.10). A comparison between the calculated production rate and the observed electron density above and below the ledge leads to values for the effective recombination coefficient ($\alpha_e = q/N_e^2$) of 5×10^{-7} and 4×10^{-5} cm^3 s^{-1} respectively. The hydrated ions react more readily with free electrons and so reduce the electron density to a much smaller value than would be the case if hydration did not occur.

The full explanation is probably more complicated than this, because the transition level between hydrated and non-hydrated ions falls to about 82 km in winter (from 87 km in summer), whereas the ledge in the electron density profile drops 10 km, to about 75 km in winter. It is thought that there must be a large seasonal change in the concentration of nitric oxide at these altitudes.

The hydrated ions $H_3O^+(H_2O)_n$ arise from a series of reactions involving O_2 and H_2O. Thus

$$O_2^+ + 2O_2 \rightarrow O_4^+ + O_2$$
$$O_4^+ + H_2O \rightarrow O_2^+ \cdot H_2O + O_2$$
$$O_2^+ \cdot H_2O + H_2O \rightarrow H_3O^+ \cdot OH + O_2$$
$$\rightarrow H_3O^+ + OH + O_2$$
$$H_3O^+ \cdot OH + H_2O \rightarrow H_3O^+ \cdot (H_2O) + OH$$

and so on. However, there are problems concerning the fate of the NO^+ ions both below and above the 80 km level.

An important feature of the D region is the formation of negative ions such as O_2^-. These are produced by a three-body reaction

$$e + O_2 + M \rightarrow O_2^- + M \tag{4.60}$$

in which M is any other molecule, serving to increase the chance of reaction by removing excess kinetic energy from the reactants. The electron affinity of O_2 is small, only 0.45 eV, and so the electron may be detached again by a quantum of low energy such as infrared or visible light:

$$O_2^- + h\nu \rightarrow O_2 + e \tag{4.61}$$

Thus, even in the presence of a constant level of primary ionization, there should be a balance between the concentrations of O_2^- and free electrons:

$$[O_2^-]/N_e = \lambda \tag{4.62}$$

λ being the ion/electron ratio (not to be confused with wavelength!).

Negative ions may also combine with positive ions, and the overall balance between production and loss is therefore

$$q = \alpha_e N_e N_+ + \alpha_i N_- N_+$$

where N_e, N_+, N_- are the concentrations of electrons, positive ions and negative ions, and α_e and α_i are recombination coefficients for positive ions with electrons and with negative ions respectively. Putting $N_- = \lambda N_e$ and $N_+ = (1 + \lambda)N_e$ (for overall neutrality)

$$q = (1 + \lambda)(\alpha_e + \lambda \alpha_i) N_e^2 \tag{4.63}$$

which simplifies to

$$q = (1 + \lambda)\alpha_e N_e^2 \tag{4.64}$$

if $\lambda \alpha_i \ll \alpha_e$.

This description is correct as far as it goes, but observations of the behaviour of the D region, coupled with laboratory measurements of reaction rates, show that reaction 4.61 cannot be the controlling one. The crucial evidence concerns the way in which the D region builds up at sunrise. If visible sunlight were the agent detaching the electron from O_2^-, the detachment seems to occur while the Earth's shadow sweeps through the range 0–50 km. However, primary ionizing radiation does not penetrate so deep into the atmosphere. To make the two regions overlap it is necessary to assume a screening layer for 30–40 km above the Earth's surface, and it seems clear that this must be the ozone layer. However, ozone does not absorb visible light strongly enough. Therefore the process must be one requiring ultraviolet, and thus simple photodetachment will not do.

The explanation of this anomaly, and an appreciation of the full complexity of the D-region chemistry, first emerged with laboratory measurements in the mid-1960s. It was found that the reaction with atomic oxygen

$$O_2^- + O \rightarrow O_3 + e \tag{4.65}$$

is more efficient than photons in detaching electrons. The atomic oxygen is produced by photodissociation of molecular oxygen, a reaction requiring ultraviolet, as already discussed. At night the O combines with O_2 to form ozone (O_3), and is no longer available as a detaching agent. Hence we obtain a diurnal variation of ion/electron ratio (λ), as required by the observations. Another detaching reaction is

$$O_2^- + O_2(^1\Delta_g) \rightarrow 2O_2 + e$$

The addition of two efficient detachment reactions introduces a numerical difficulty, since the predicted electron densities then come out too large. However, further reactions have been discovered which can remove O_2^- and convert them in stages to very stable ions such as CO_3^-, NO_2^- and NO_3^-. These are called *terminal ions*, from which O_2^- cannot be recovered. One possible reaction scheme is shown in Fig. 4.12. It is observed that the most abundant negative ion in the D region is NO_3^-. Cluster ions such as $O_2^- \cdot O_2$, $O_2^- \cdot CO_2$ and $O_2^- \cdot H_2O$ are also present.

Calculated height profiles of the total negative ion/electron ratio are shown in Fig. 4.13.

4.3.5 Airglow

During the photochemical reactions of the upper atmosphere, quanta of radiation may be emitted in the ultraviolet, the visible or the infrared parts of the spectrum.

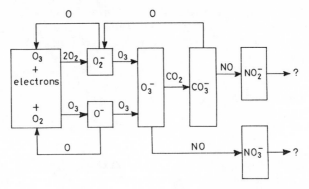

Figure 4.12   Scheme of reactions in the D region. (From F.C. Fehsenfeld *et al.*, *Planet Space Sci.* **15**, 373, 1967.)

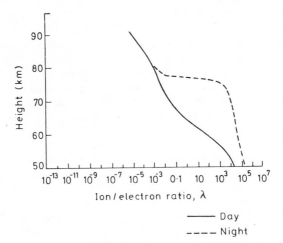

Figure 4.13   Ion/electron ratios. (Data from R.P. Turco and C.F. Sechrist, *Rad. Sci.* **7**, 725, 1972, copyrighted by American Geophysical Union.)

These emissions comprise the *airglow*. The subject is now a well developed speciality within upper-atmosphere studies, and the relevant literature is very extensive. We cannot delve deeply into airglow in the present text, but a brief introduction is appropriate to indicate the relevance of airglow to aeronomy.

Airglow should be distinguished from aurora. Both phenomena arise in photochemical processes, but there the connection ends. The aurora is a high-latitude phenomenon, energized by the entry of energetic charged particles into the upper atmosphere during solar and geophysical disturbances. It is highly structured in space and time. Airglow, by contrast, covers all latitudes and is virtually unstructured. For the most part it is energized by the steady solar ultraviolet and X-ray emissions. Though always present, airglow is difficult to observe by day. By night it provides the major contribution to the total skylight, exceeding the integrated starlight in intensity.

Scientific studies of airglow are relatively young, having been pioneered by R. J. Strutt, the fourth Baron Rayleigh, in the 1920s. In his honour the unit of airglow emission is called the Rayleigh:

$$1\ R = 10^6\ \text{photons cm}^{-2}\ \text{s}^{-1}$$

The Rayleigh measures the height-integrated emission rate, as detected by a photometer at the ground. Chapman also made notable contributions to airglow studies, and in 1931 was responsible for the suggestion that airglow is produced by the recombination of ionized and dissociated species. The name 'airglow' dates from 1950.

Basically the airglow emissions depend on the creation of an excited species and its subsequent return to the ground state. The immediate causes are of several kinds, including the following:

(a) radiative recombination reactions emit a quantum of radiation;
(b) reaction products are often left in an excited state, and they then emit in returning to the ground state;
(c) excited species can be produced by hot electrons (from ionization processes) and by electric fields;
(d) the excitation can be by solar radiation, giving rise to resonance emissions.

One important group of lines, the atomic oxygen emissions, will serve to illustrate the principles involved. The transition scheme, shown in Fig. 4.14, gives rise in particular to the prominent *green line* at 5577 Å and doublet *red line* at 6300/6364 Å (actually a triplet, but the third line at 6391 Å is very weak). Atomic oxygen in the ^1S state passes to the ground state in two steps, the first in a time of 0.74 s and the second, producing the red line, in a longer time of 110 s. (The line at 2972 Å, produced by direct transition to the ground state, is too weak to be of interest in airglow.) If the ^1D state is produced in the first instance, only the red line can be emitted. In the ionospheric F region the excited atoms result from the dissociative recombination reaction

$$O_2^+ + e \rightarrow O^* + O^*$$

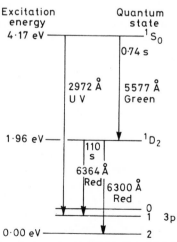

Figure 4.14   Scheme of atomic oxygen transitions. (After S.J. Bauer, *Physics and Chemistry in Space, Vol. 6, Physics of Planetary Ionospheres,* Springer-Verlag, 1973.)

79

Energy of 7 eV is also released, and the excited atoms can be in the ^1S or ^1D states. At E-region altitudes, the main source is the reaction first suggested by Chapman:

$$O + O + O \rightarrow O_2 + O^*$$

where O^* is in the ^1S state. At this lower altitude the green line is seen but the red one is missing. The reason is that the intermediate level ^1D has such a long lifetime that the energy is removed by collisions with other atoms and molecules before the second quantum (6300 Å) can be emitted. The red line is said to be *quenched*. The F-region reaction is part of the recombination scheme for O^+, and, as would be expected, the 6300 Å intensity is found to be closely related to measurable F-region parameters. The Chapman reaction, operating at lower altitudes, appears to have no direct connection with the ionosphere.

Another interesting feature of the 6300 Å airglow is the *stable auroral red (SAR) arc*, that encircles the Earth at latitudes near 40° — also known sometimes as the M (for 'mid-latitude') arc. The SAR arc was discovered over France by Barbier in 1956, and it appears at times when auroral activity appears at high latitude and, notably, during geomagnetic storms when the planetary magnetic index K_p reaches 5 or more. The emissions extend several hundred kilometres in the meridional direction, and between heights of 300 and 700 km, peaking near 400 km in the upper F region. The emission is normally the doublet 6300/6364 Å, and the absence of substantial 5577 Å enhancement indicates that the excitation source is of low energy (< 4 eV). Various theories have been put forward to explain the SAR arc. It is currently thought that the energy comes from the magnetospheric ring current that forms during magnetic storms (see Section 8.3.4). Considering the auroral connection and the relative uniformity of the emission, we see that the distinction between airglow and aurora is less clear with this feature.

Many series of airglow emissions are observed in addition to those of atomic oxygen, e.g. molecular oxygen, molecular and atomic nitrogen, hydrogen, hydroxyl radicals, nitric oxide and alkali metals. The hydroxyl emissions in the infrared arise at heights near 90 km and, while of no ionospheric relevance, are important as a mechanism of heat loss. The hydroxyl emissions are so intense that if they were in the visible most of the stars would be obliterated from view.

The emissions from alkali metals provide an example of resonance scattering, visible most strongly at twilight when the upper atmosphere is illuminated by the Sun but an observer at the ground is in darkness. The emissions reveal the presence of sodium, potassium, lithium and calcium at the 90 km level.

4.4 Electrical conductivity of the upper atmosphere

As it contains free electrons and ions, the upper atmosphere has an electrical conductivity much greater than that of the lower atmosphere. The geomagnetic field inhibits the motion of charged particles in directions normal to the field lines, and thus the conductivity is anisotropic. The conductivity varies with altitude, both in magnitude and in direction. The motions of both ions and electrons must be taken into account, and in combination they can produce an electric current, where ions and electrons move in opposite directions, or a plasma drift, where they move together. The particle motions can be produced by an electric field or by a wind in the neutral air.

4.4.1 Motion of a charged particle in a magnetic field

A particle of mass m, when acted on by a force F, accelerates at F/m. If the collision frequency is ν we can say that on average the particle travels a distance $F/2m\nu^2$ between collisions. We imagine that the particle stops momentarily at each collision, after which it again accelerates at the same rate as before. The overall speed is then

$$v = F/m\nu \qquad (4.66)$$

In the absence of a magnetic field the velocity is in the same direction as the applied force.

For a particle with charge e in an electric field E, the force is $F = Ee$. If the plasma contains N such particles per unit volume, the electric current density (current per unit cross-section)

$$J = Nev \qquad (4.67)$$

and therefore the conductivity

$$\sigma = J/E = Ne^2/m\nu \qquad (4.68)$$

The current flow is parallel to the applied field in this simple case. Adding the currents due to electrons and ions

$$J = J_e + J_i = N|e|(V_i - V_e) \qquad \sigma = \sigma_e + \sigma_i = N|e|^2 \left(\frac{1}{m_i \nu_i} + \frac{1}{m_e \nu_e} \right)$$

We saw in Section 2.3.2 that in a magnetic field of induction B, a particle with charge e and mass m gyrates with angular frequency

$$\omega_B = \frac{B|e|}{m}$$

where ω_B is the gyrofrequency. We disregard the sign of the charge (to avoid the embarrassment of a negative frequency), but note that to an observer looking along the direction of the magnetic field a positive particle gyrates anticlockwise and a negative one goes clockwise.

Consider a positive particle at the origin of the coordinate system of Fig. 4.15. In vector notation, if the particle moves at velocity $\mathbf{v}$, it experiences a force $\mathbf{F} = e\mathbf{v} \times \mathbf{B}$ because of the magnetic field, which is directed at right angles both to the field and to the velocity vector. To this force must be added the applied force F_x. If the particle happens to be moving with velocity $v_x = 0$, $v_y = -F_x/eB$ as it passes the origin, the total force in the x-direction will be

$$F_x + ev_y B = F_x - eF_x B/eB = 0$$

Since the particle would not be aware of any net force, it would continue along the same path at the same speed.

If the particle was initially at rest at the origin, it would at first be accelerated in the x-direction by the applied force F_x; but then the magnetic field would deviate the path in the $-y$-direction, and the particle would move in a loop, coming to rest again further along the $-y$-axis. The average velocity during such a loop is approximately

$$\frac{F}{m} \times \frac{P}{4}$$

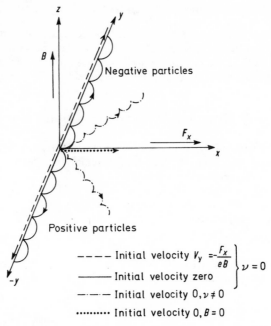

Initial velocity $V_y = -\dfrac{F_x}{eB}$
Initial velocity zero
$\left.\vphantom{\begin{array}{c}a\\b\end{array}}\right\} \nu = 0$
Initial velocity $0, \nu \neq 0$
Initial velocity $0, B = 0$

Figure 4.15 Charged particle motions in a magnetic field

where P is the gyroperiod $(2\pi m/B|e|)$, because the particle is accelerated for only half the loop and is retarded during the second half. Therefore

$$v_{av} \sim \frac{\pi}{2} \cdot \frac{F}{B|e|}$$

A more exact treatment gives

$$v_y = -F_x/B|e| \tag{4.69}$$

as before. It can be shown that this is the drift velocity due to F_x whatever the initial velocity might be. If the particle has negative charge it moves in the opposite direction, along the $+y$-axis, but the drift speed is the same as for a positive particle.

The foregoing treatment neglects collisions. If the collision frequency is ν, the particle stops every $1/\nu$ s instead of every $P/2$ (provided $1/\nu < P/2$.) This happens before the particle has returned to the y-axis, and so a component of drift in the x-direction is introduced at the expense of motion in the y-direction. A full treatment of the motion gives

$$v_y = -\frac{F_x}{|e|B} \cdot \frac{\omega_i^2}{\nu_i^2 + \omega_i^2} \qquad v_x = \frac{F_x}{|e|B} \cdot \frac{\omega_i \nu_i}{\nu_i^2 + \omega_i^2} \qquad \text{for a positive ion}$$

and

$$v_y = \frac{F_x}{|e|B} \cdot \frac{\omega_e^2}{\nu_e^2 + \omega_e^2} \qquad v_x = \frac{F_x}{|e|B} \cdot \frac{\omega_e \nu_e}{\nu_e^2 + \omega_e^2} \qquad \text{for an electron} \tag{4.70}$$

82

We have used ω_e and ω_i for the electron and ion gyrofrequencies, and the absolute value $|e|$ for the charge. ν_e and ν_i are the electron and ion collision frequencies (with neutral particles) respectively. We see that these equations revert to the simpler forms, Equations 4.69 and 4.66, if $\omega_B \gg \nu$ or $\nu \gg \omega_B$ respectively, using $\omega_B = |e|B/m$. Motion in the z-direction depends only on the presence of a force in the z-direction, when, by Equation 4.66,

$$v_z = \frac{F_z}{m\nu} = \frac{F_z}{|e|B} \cdot \frac{\omega_i}{\nu_i} \qquad \text{for a positive ion}$$

or

$$v_z = \frac{F_z}{|e|B} \cdot \frac{\omega_e}{\nu_e} \qquad \text{for an electron}$$

(4.71)

Values of electron collision frequency and gyrofrequency are given as a function of altitude in Fig. 4.16. The electron drifts at angle $45°$ to F_x when $\omega_e = \nu_e$. The corresponding point for ions occurs higher up because of the lower gyrofrequency.

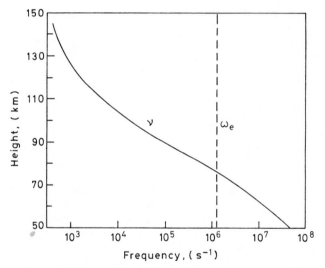

Figure 4.16 Electron collision frequency (ν_e) and gyrofrequency (ω_e), 50–150 km. The ion collision frequency is 10–100 times smaller than ν_e, and the ion gyrofrequency is 30–50×10^3 times smaller than ω_e depending on the nature of the ion. (After E.V. Thrane and W.R. Piggott, *J. atmos. terr. Phys.* **28**, 721, 1966, by permission of Pergamon Press.)

4.4.2 Effects of a neutral wind

One kind of force affecting upper-atmosphere plasma is the wind in the neutral air. A neutral-air wind will exert a force

$$F = m\nu U$$

(4.72)

on a charged particle, where U is the velocity of the neutral air (relative to the plasma), and m and ν are the mass and collision frequency of the charged particle

83

as before. Equation 4.72 gives the force simply as the rate of change of momentum due to collisions with particles moving at velocity U. This value can now be substituted for F_x in Equations 4.70. The following special cases are worthy of note.

At high levels ω_i, $\omega_e \gg \nu_i$, ν_e

$$v_y = -F_x/|e|B = -\nu_i U/\omega_i \qquad v_x = 0 \qquad \text{for positive ions}$$

$$v_y = F_x/|e|B = \nu_e U/\omega_e \qquad v_x = 0 \qquad \text{for electrons}$$

At low levels ω_i, $\omega_e \ll \nu_i$, ν_e

$$v_y = 0 \qquad v_x = F_x\omega_i/|e|B\nu_i = F_x/m_i\nu_i = U \qquad \text{for positive ions}$$

$$v_y = 0 \qquad v_x = F_x\omega_e/|e|B\nu_e = F_x/m_e\nu_e = U \qquad \text{for electrons}$$

Thus at high altitude the ions and electrons move in opposite directions and at right angles to the neutral wind, producing an electric current. At low altitudes they move together in the direction of the wind.

4.3.3 Effects of an electric field

The force F_x due to the action of an electric field E_x on a particle with charge e is $F_x = eE_x$. Substituting in Equation 4.70 gives, for high levels,

$$v_y = -E_x/B \qquad v_x = 0 \qquad \text{for positive ions}$$

$$v_y = -E_x/B \qquad v_x = 0 \qquad \text{for electrons}$$

At low levels

$$v_y = 0 \qquad v_x = E_x\omega_i/B\nu_i = E_x|e|/\nu_i m_i \qquad \text{for positive ions}$$

$$v_y = 0 \qquad v_x = -E_x\omega_e/B\nu_e = -E_x|e|/\nu_e m_e \qquad \text{for electrons}$$

It can be seen that an electric field produces a plasma drift normal to the field at high levels, and an electric current in the direction of the field at low levels. Table 4.3 summarizes the effects of wind and electric field at high and low levels, and includes also the intermediate region where $\nu_e < \omega_e$ but $\nu_i > \omega_i$. (This can occur because

$$\omega_i/\omega_e = m_e/m_i = \frac{1}{1836M_i}$$

where M_i is the atomic weight of the ion.)

Table 4.3 Effects of Wind and Electric Field

Altitude	Wind (U)	Electric field (E)
High ($\nu \ll \omega_B$)	Current $\perp$ U	Drift $\perp$ E
Low ($\omega_B \ll \nu$)	Drift $\parallel$ U	Current $\parallel$ E
Intermediate		
($\nu_e < \omega_e$)	Electron current $\perp$ U	Electron current $\perp$ E
($\omega_i < \nu_i$)	Small ion current $\parallel$ U	Small ion current $\parallel$ E

4.4.4 Total conductivity

To get the total conductivity we add the contributions from the ions and the electrons

$$\sigma = \sigma_i + \sigma_e = \frac{N|e|}{E}(v_i - v_e)$$

If the electric field is applied in the x-direction, there are in general components of velocity in the x- and y-directions, and thus x- and y-components of the conductivity. The conductivity in the direction of the applied electric field is the *Pederson conductivity:*

$$\sigma_1 = \left[\frac{N_e}{m_e v_e}\left(\frac{v_e^2}{v_e^2 + \omega_e^2}\right) + \frac{N_i}{m_i v_i}\left(\frac{v_i^2}{v_i^2 + \omega_i^2}\right) \right] |e|^2 \tag{4.73}$$

and that perpendicular to the applied field is the *Hall conductivity:*

$$\sigma_2 = \left[\frac{N_e}{m_e v_e}\left(\frac{\omega_e v_e}{v_e^2 + \omega_e^2}\right) - \frac{N_i}{m_i v_i}\left(\frac{\omega_i v_i}{v_i^2 + \omega_i^2}\right) \right] |e|^2 \tag{4.74}$$

σ_1 and σ_2 are standard symbols for these two conductivities (except that the sign of σ_2 is reversed by some authors). In our terms we could have called them σ_x and σ_y. The important point is that the conductivities and the applied electric field are all in the plane perpendicular to the magnetic field. The conductivity along the magnetic field due to a parallel electric field is the *direct* or *longitudinal* conductivity,

$$\sigma_0 = \left[\frac{N_e}{m_e v_e} + \frac{N_i}{m_i v_i} \right] |e|^2 \tag{4.75}$$

These expressions are readily derived from the foregoing discussion, but note that $\omega_B = |e|B/m$.

If the electric field is written as a vector

$$\mathbf{E} = \mathbf{i}E_x + \mathbf{j}E_y + \mathbf{k}E_z$$

the current vector is given by

$$\mathbf{J} = \mathbf{i}J_x + \mathbf{i}J_y + \mathbf{k}J_z = \begin{pmatrix} \sigma_1 & \sigma_2 & 0 \\ -\sigma_2 & \sigma_1 & 0 \\ 0 & 0 & \sigma_0 \end{pmatrix}\begin{pmatrix} E_x \\ E_y \\ E_z \end{pmatrix} \tag{4.76}$$

where the conductivity is a tensor. This form enables any component to be written down if all components of the electric field are known.

Fig. 4.17 shows the height distribution of conductivities in the mid-latitude ionosphere at local noon. Note that σ_1 and σ_2 maximize in the E region.

Since the ionosphere has a layered structure, flow of current in the vertical direction is inhibited. The extreme case of inhibition occurs over the magnetic equator, where the geomagnetic field is horizontal. Then, from our definitions of σ_1 and σ_2 (and directly from Equation 4.76)

$$\left. \begin{aligned} J_x &= \sigma_1 E_x + \sigma_2 E_y \\ J_y &= -\sigma_2 E_x + \sigma_1 E_y = 0 \end{aligned} \right\} \tag{4.77}$$

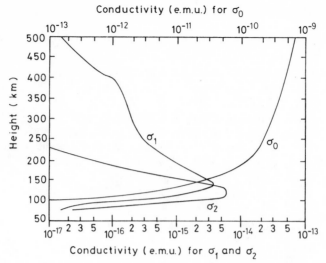

Figure 4.17   Conductivity profiles for middle latitude at noon. (From S.-I. Akasofu and S. Chapman, *Solar-Terrestrial Physics,* Oxford University Press, 1972, after Maeda and Matsumoto *Rep. Ionos. Space Res. Japan,* **16**, 1, 1962.)

because the y-direction is now vertical and no current can flow. Hence,

$$E_y = E_x \sigma_2/\sigma_1$$

and by substitution

$$J_x = (\sigma_1 + \sigma_2^2/\sigma_1)E_x = \sigma_3 E_x \tag{4.78}$$

σ_3, defined thus, is called the *Cowling conductivity.* A more general treatment along these lines derives conductivities σ_{xx}, σ_{xy}, σ_{yy}, as a function of latitude. This takes a different coordinate system, in which the x- and y-axes are taken to the magnetic south and east respectively and the z-axis is vertical. The subscripts indicate the directions of the applied field and the current component, and the formulae for σ_{xx}, etc., include the dip angle (I) of the geomagnetic field. Then

$$J_x = \sigma_{xx}E_x - \sigma_{xy}E_y$$
$$J_y = -\sigma_{xy}E_x + \sigma_{yy}E_y \tag{4.79}$$

which are a generalization of Equations 4.77. It can be shown that

$$\sigma_{xx} \approx \sigma_1/\sin^2 I \qquad \sigma_{xy} \approx \sigma_2/\sin I \qquad \sigma_{yy} \approx \sigma_1$$

except near the magnetic equator.

If the ionosphere is regarded as a layer permitting only horizontal currents it is useful to integrate the conductivity with respect to height, giving

$$\sum_{xx} = \int \sigma_{xx}\,dh \qquad \sum_{xy} = \int \sigma_{xy}\,dh \qquad \sum_{yy} = \int \sigma_{yy}\,dh$$

The variation of these 'layer conductivities' as a function of dip latitude at noon is shown in Fig. 4.18. These values could be substituted into Equation 4.79 to predict the consequences of an applied electric field.

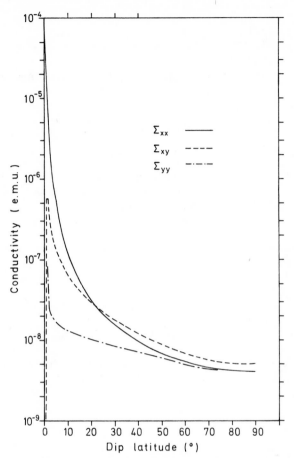

Figure 4.18   Layer conductivities for local noon. (From S.-I. Akasofu and S. Chapman, *Solar-Terrestrial Physics*, Oxford University Press, 1972, after Maeda and Matsumoto, *Rep. Ionos. Space Res. Japan,* **16**, 1, 1962.)

4.5 Further reading

Bauer, S. J. (1973) *Physics of Planetary Ionospheres* (Springer-Verlag).

Ratcliffe, J. A. and Weeks, K. (1960) The Ionosphere, in *Physics of the Upper Atmosphere* (Ed. J. A. Ratcliffe) (Academic Press) Chapter 9.

Rishbeth, H. and Garriott, O. K. (1969) *Introduction to Ionospheric Physics* (Academic Press) Chapters 1, 3, 4.

Whitten, R. C. and Poppoff, I. G. (1971) *Fundamentals of Aeronomy* (Wiley) Chapters 3, 4, 7–9.

Whitten, R. C. and Poppoff, I. G. (1965) *Physics of the Lower Ionosphere* (Prentice-Hall).

LeGalley, D. P. and Rosen, A. (1964) *Space Physics* (Wiley) Chapter 6.

Akasofu, S.-I. and Chapman, S. (1972) *Solar–Terrestrial Physics* (Oxford) Chapter 3.

Matsushita, S. and Campbell, W. H. (1967) *Physics of Geomagnetic Phenomena* (Academic Press) Chapter III-3.

Hess, W. N. and Mead, G. D. (1968) *Introduction to Space Science* (Gordon and Breach) Chapters 2, 3.

Chamberlain, J. W. (1961) *Physics of Aurora and Airglow* (Academic Press) Chapters 9–13.

Anderson, J. G. and Donahue, T. M. (1975) The neutral composition of the stratosphere and mesosphere, *J. atmos. terr. Phys.* **37**, 865.

Special Issue: Fifty years of the ionosphere (1974) *J. atmos. terr. Phys.* **36**, 2069.

Bates, D. R. (1970) Reactions in the ionosphere, *Contemp. Phys.* **11**, 105.

Ferguson, E. E. (1974) Laboratory measurements of ionospheric ion–molecule reaction rates, *Rev. Geophys. Space Phys.* **12**, 703.

Smith, E. V. P. and Gottlieb, D. M. (1974) Solar flux and its variations, *Space Sci. Rev.* **16**, 771.

Hinterregger, H. E. (1976) EUV fluxes in the solar spectrum below 2000 Å, *J. atmos. terr. Phys.* **38**, 791.

Special Issue: Recent advances in the physics and chemistry of the E region (1975) *Rad. Sci.* **10**, 229.

Lemaire, J. and Scherer, M. (1974) Exospheric models of the topside ionosphere, *Space Sci. Rev.* **15**, 591.

Ferguson, E. E. (1971) D-region ion chemistry, *Rev. Geophys. Space Phys.* **9**, 997.

Rees, M. H. and Roble, R. G. (1975) Observations and theory of the formation of stable auroral red arcs, *Rev. Geophys. Space Phys.* **13**, 201.

5
Geographical and Temporal Structure of the Ionosphere

5.1 Introduction

The ideas developed in Chapter 4 provide a set of basic concepts that may be expected to govern the behaviour of the upper atmosphere. Armed with these principles we are now in a position to look at the results of observations. Reference to observational techniques will be minimized, since the main techniques of upper atmosphere study have already been covered in Chapter 3. We note, however, that since it is easier to observe the ionized than the neutral component of the upper atmosphere much of the large body of information now available refers to the ionosphere, and that is where the present discussion will be concentrated.

It has been said that the structure of the ionosphere covers all the scales of space and time that are open to observation; the total picture is therefore a complex one. We start our account by looking in this chapter on the global scale — essentially, that is, at geographical effects and at the diurnal effects due to the Earth's rotation. The emphasis, moreover, is on regularity: on behaviour that tends to repeat from one day to the next.

Not all the regular behaviour (and much less the irregular behaviour) of the upper atmosphere is well understood. Historically, ionospheric observations have been compared with the Chapman theory for the production of an ionospheric layer (Section 4.2.2), and major departures from the theory were called 'anomalies'. Chapman theory excludes ion movement, and some of the classical anomalies have been explained by taking large-scale movements of ionization into account. The causes of some of the others remain in doubt, though in most cases possible causes are known and the problem now is to identify the correct mechanism from among several candidates.

5.2 Diurnal and seasonal variations at middle latitude

5.2.1 E region

The behaviour of the ionospheric E region is close to what would be expected of an α-Chapman layer. On average the critical frequency of the layer, f_0E, varies with the solar zenith angle (χ) as $(\cos \chi)^{1/4}$, implying that the peak electron density, $N_m(E)$, varies as $(\cos \chi)^{1/2}$. There is a certain amount of variability in the value of the exponent. Writing $f_0E \propto (\cos \chi)^n$, values of n range between 0.1 and 0.4.

Since Chapman theory can be applied to the E region, the observed behaviour can be used to determine the recombination coefficient (α). There are three methods:

(i) compare the measured or calculated production rate (q) with the electron density (N) at a given height, $q = \alpha N^2$;

(ii) observe the rate of decay of the layer after sunset, when $q = 0$;
(iii) measure the asymmetry of the diurnal curve with respect to local noon: this is called the *sluggishness* of the ionosphere, the time delay is $\tau = 1/2\alpha N$ and amounts to about 10 min.

These methods give values of α in the range 10^{-7} to 10^{-8} cm^3 s^{-1}.

In fact the E region does not quite disappear at night. The region remains weakly ionized with electron density about 5×10^3 cm^{-3}, compared with 10^5 cm^{-3} by day. The night ionization is irregular with height (Fig. 5.1). One possible cause is the ionization produced by meteors as they burn up in the atmosphere, though there may be other causes at work as well.

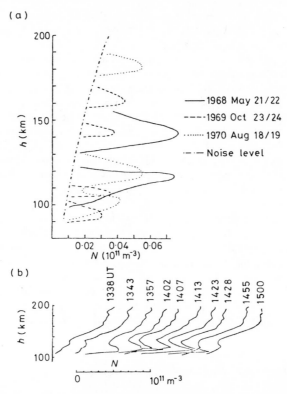

Figure 5.1   Some E layers observed by incoherent scatter. (a) Midnight profiles of $N(h)$ on three different occasions. (b) Growth and decay of a sporadic-E layer by day. The E_s layer appears at height 110 km and lasts for an hour. (After G.N. Taylor, *Rad, Sci.* **9**, 167, 1974. Crown copyright.)

Even during the day the electron density does not always vary smoothly with height, and narrow layers called *sporadic-E* can develop. An example is shown in Fig. 5.1. On an ionogram, sporadic-E appears as an echo at constant height that extends to higher frequency than would be normal for an E-region echo — for instance, to more than 5 MHz. Typical sporadic-E layers are only a few kilometres thick at middle latitudes. Depending on the intensity of the layer, echoes from greater heights may or may not be visible through it. The main cause is thought to be a change of wind speed with height, a wind shear, which in the presence of the geomagnetic field can act to compress the ionization. The mechanism is similar to one acting in the F region that will be discussed in Section 5.2.4. Sporadic-E layers are detected at many latitudes, and several different kinds are recognized. As shown in Fig. 5.2, different diurnal and seasonal patterns are observed at high, middle and low latitude, and each type arises from a different mechanism.

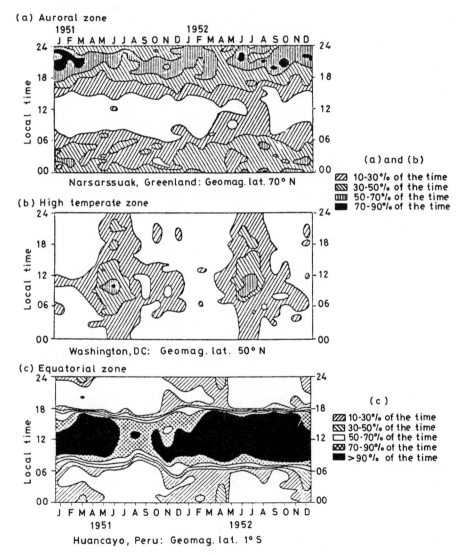

Figure 5.2   Diurnal and seasonal occurrence of three kinds of sporadic-E. (a) The auroral kind maximizes at night but has no seasonal variation. (b) The temperate kind peaks near noon in summer. (c) The equatorial kind occurs mainly by day but shows no seasonal preference. (After E.K. Smith, NBS Circ. 582, 1957.)

5.2.2 F1 region

The F1 layer, often referred to as the *F1 ledge*, is also well behaved. When the layer is present its critical frequency, $f_0 F1$, varies as $(\cos \chi)^{1/4}$, indicating α-Chapman behaviour. However, the seasonal behaviour is rather puzzling at first sight, since the ledge is more pronounced in summer than in winter, but less apparent at sunspot maximum than at sunspot minimum. In fact, F1 is never seen at sunspot maximum in winter. For a correct interpretation one has to compare the transition height, h_t, between α-type and β-type recombination (as discussed in Section 4.2.2) with the height of maximum electron production, h_m. F1 appears only if $h_t > h_m$, and since h_t depends on the electron density (Equation 4.43) the ledge tends to disappear when the electron density is greatest.

5.2.3 D region

The D region also exhibits strong solar control, but the form of diurnal behaviour that is observed depends on the technique used. To a first approximation, very-low-frequency (< 30 kHz) radio waves are reflected in the D region as at a sharp boundary, because the refractive index changes considerably within the distance of one radio wavelength (see Section 3.7.2). As seen by VLF radio waves at steep incidence, the apparent height of reflection (h) varies according to

$$h = h_0 + H \ln \sec \chi \tag{5.1}$$

as predicted by Chapman theory for the height variation of a level of constant electron density. In Equation 5.1, $h_0 \sim 72$ km, and H, which is the scale height of the neutral atmosphere, is about 5 km. At oblique incidence (angles exceeding about $65°$, arising when transmitter and receiver are more than 300 km apart) the diurnal pattern is quite different, with a sharp reduction of reflection height before sunrise, and a fairly rapid recovery after sunset (Fig. 5.3). The reasons for this behaviour are bound up with the complex photochemistry of the D region, which was outlined in Chapter 4 and need not be pursued further here. The part of the ionosphere that behaves in this way is sometimes called a C layer — which serves to remind us that (i) it is the lowest part of the D region and (ii) it is produced mainly by cosmic rays.

Another way of observing the long-term variation of the D region is to measure the absorption of radio waves reflected from the E region (Section 3.7.1). The absorption depends on both the electron density and the collision frequency. The measurement provides an integrated value of absorption rather than the value at a given height (though some height information has come from multi-frequency observations). Writing absorption $\propto (\cos \chi)^n$, it is found that n usually lies in the range 0.7–1.0, somewhat, though not seriously, lower than the values predicted by theory. However, when looking at the seasonal variation of radio absorption a major puzzle appears. During the winter months the radio absorption is greater by a factor of two or three than would be predicted by extrapolation from the summer observations. The absorption is also much more variable from day to day in the winter. This effect, called the *winter anomaly of ionospheric absorption*, again involves the ion chemistry of the D region, since rocket measurements show a large enhancement of electron density between 70 and 80 km on anomalous days. The winter anomaly is a particularly interesting one because of evidence that it is

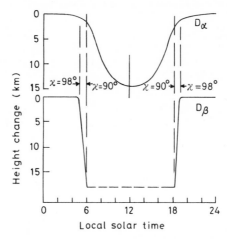

Figure 5.3 Two kinds of diurnal behaviour of the D region observed by VLF radio waves. The regions originally called D_α and D_β are now more usually known as D and C. (After R.N. Bracewell and W.C. Bain, *J. atmos. terr. Phys.* 2, 216, 1952, by permission of Pergamon Press.)

related to other geophysical phenomena, such as the meteorology of the stratosphere and, perhaps, solar activity. These aspects of the D region will be discussed in Chapter 11.

5.2.4 F2 region

Since it is necessary to include the vertical diffusion of plasma in order to understand the existence of the F2 region at all, it is hardly surprising that the behaviour of the region appears to be anomalous in many ways if it is considered only against the Chapman theory. The major classical anomalies of the F2 region at middle latitudes are as follows.

(i)   The diurnal variation may be asymmetrical about noon, with rapid changes at sunrise and slow changes or no change at all at sunset. The daily maximum can occur before or after noon in summer, though it is likely to be near noon in winter.

(ii) The diurnal variation is not repeatable from day to day. Apart from any scientific implication, this makes prediction more difficult, and because of its importance in radio communications the F2 layer is just the one for which accurate predictions would be the most valuable. Fig. 5.4 illustrates points (i) and (ii).

(iii) The seasonal variation is anomalous in several ways. Values of $f_0 F2$ at noon are sometimes greater in winter than in summer – the *seasonal anomaly*, illustrated by Fig. 5.5. If averaged over the whole Earth, to include both hemispheres therefore, electron densities are 20% greater in December than in June, though the changing distance of the Earth from the Sun would account for only 6% difference. This is the *annual anomaly*. The electron content is abnormally high at the equinoxes – the *semi-annual anomaly*.

(iv) A substantial F2 region is maintained at night.

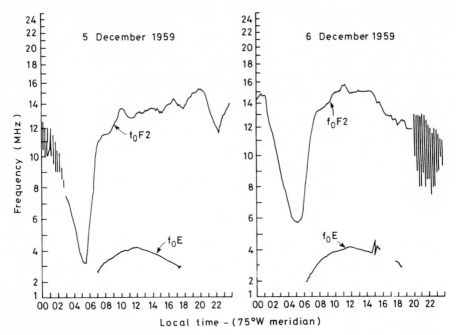

Figure 5.4 Diurnal change of $f_O F2$ on successive days in December 1959 at a low-latitude station, Talara, Peru. Note, by contrast, the regularity of the E layer. (From T.E. Van Zandt and R.W. Knecht, *Space Physics* (Ed, LeGalley and Rosen). Copyright © 1964. Reprinted by permission of John Wiley & Sons, Inc.)

It is not possible or necessary to go into a full discussion of all the complexities of the behaviour of the F2 region, but we should note some of the theoretical ideas to which they have led, and which, taken together, could probably account in principle for most of the phenomena. In most cases, however, new experimental results will be needed before detailed verification of the theories can be made. We will outline here the effects of (1) chemical changes, (2) diurnal heating and cooling, and (3) winds and electric fields.

In the F region, the rate of electron loss depends on the concentration of molecular ions, particularly O_2 and N_2 (see Sections 4.2.2 and 4.2.3), whereas the production rate depends on the concentration of atomic oxygen. The equilibrium electron density therefore depends on the ratio $[O]/[N_2]$ and also on $[O]/[O_2]$. It is thought that the seasonal anomalies of the F2 region are largely due to seasonal changes in the relative concentrations of atomic and molecular species. There appears to be a relative abundance of O_2 in the summer hemisphere due to the global circulation of the atmosphere, and it is observed that the ratio $[O]/[O_2]$ shows a maximum at the equinoxes, consistent with the seasonal variations of $N_m F2$.

The photoelectrons produced by ionization processes are hotter than the neutral atoms from which they were formed. This excess energy is gradually shared with the positive ions, but transfer to the neutral component is less efficient. Consequently the temperature of the plasma exceeds that of the neutral air. By day the electrons are considerably hotter than the ions ($T_e > T_i$), but by night the temperatures are more alike. Diurnal temperature changes are illustrated in Fig. 5.6.

94

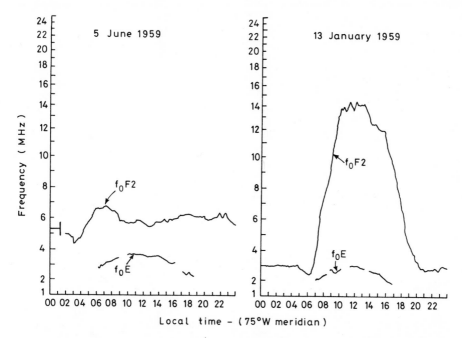

Figure 5.5 Diurnal change of $f_O F2$ in summer and winter at a high-latitude station in the northern hemisphere, Adak, Alaska. The F region is anomalous while the E layer behaves as expected. (From T.E. Van Zandt and R.W. Knecht, *Space Physics* (Ed. LeGalley and Rosen). Copyright © 1964. Reprinted by permission of John Wiley & Sons, Inc.)

The height distribution of the F2 region is governed by the plasma scale height, $H_p = k(T_i + T_e)/m_i g$ (Equation 4.49). When the plasma is hot it moves up to greater altitudes, where it is effectively stored because of the smaller rate of recombination at greater height. When the plasma cools, at night, it moves down again and so helps to maintain the F region against loss by recombination at the lower altitudes. There is no doubt that replenishment from the protonosphere helps to maintain the F region during the night.

Changing temperature can also alter the equilibrium electron density because the rates of reactions controlling the recombination are sensitive to the temperature. For instance the rate coefficient for the reaction

$$O^+ + N_2 \rightarrow NO^+ + N$$

which is the first step of the two-stage recombination process of electron loss in the F region (Section 4.3.2), is very sensitive to the temperature of N_2, increasing by a factor of 20 over the temperature range 1000–4000 °K. Not only does this effect help to explain the persistence of the F region at night, but it also helps to account for the seasonal anomaly whereby $N_m F2$ tends to be larger in winter than in summer.

The third major source of F-region anomalies is the wind in the neutral air at heights around 300 km. It will be seen in Chapter 6 that these winds range between tens and hundreds of m s^{-1}, and in middle latitudes they are directed poleward by day and equatorward by night. By Equation 4.72 a neutral wind, U, exerts a force $F = m\nu U$ on a charged particle of mass m, ν being the collision frequency. Suppose

95

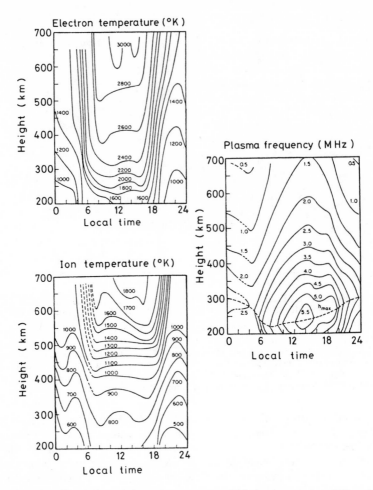

Figure 5.6 Diurnal variation of electron temperature (T_e), ion temperature (T_i), and plasma frequency (f_N) between 200 and 700 km, measured by incoherent scatter at a mid-latitude station, April 1964. (After J.V. Evans, *Planet. Space Sci.* **15**, 1387, 1967.)

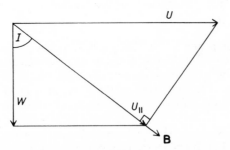

Figure 5.7 Effect on vertical plasma drift of a neutral-air wind blowing horizontally in the magnetic meridian. U = neutral wind; $U_\parallel$ = component of neutral wind along magnetic vector **B**; W = vertical component of $U_\parallel$.

for simplicity that the wind blows horizontally in the magnetic meridian. Then if the magnetic dip angle is I, the neutral air has a velocity component $U_\parallel = U \cos I$ along the magnetic field (see Fig. 5.7). Since the geomagnetic field does not affect the component of ion motion along the field, the ions also move along the magnetic field at velocity $U_\parallel$. Since the horizontal gradient of electron density is much smaller than the vertical gradient, the effect of ion drift on the vertical distribution depends only on the vertical component

$$W = U_\parallel \sin I = \tfrac{1}{2} U \sin 2I \tag{5.2}$$

Note that a given neutral wind has greatest effect on the ionospheric vertical distribution where the magnetic dip angle is 45°. At middle latitude, W is likely to be about 30 m s^{-1} by day and 100 m s^{-1} at night. If the ion drift is downward the maximum of the layer is depressed and, since the loss rate increases downward, the equilibrium electron density is reduced. Exact computations depend on the assumption of a specific model for the processes involved. However it can be shown that by day (in equilibrium conditions) the change of height of the maximum is

$$\Delta h_m \sim WH^2/D \sim W/\beta \tag{5.3}$$

In the equilibrium condition, Equation 4.45 shows that the downward drift speed by diffusion $\approx D/H$ where D is the diffusion coefficient and H the scale height. To this must be added the drift due to wind W, giving a total of $(D/H + W)$. The balance between recombination and diffusion that determines the height of maximum is expressed by $\beta_1 \sim D/H^2$ in the absence of wind. With wind present, this becomes $\beta_2 \sim D/H^2 + W/H$, β_1 and β_2 being the recombination coefficients at the respective heights of maxima, $h_m(1)$ and $h_m(2)$. Since β varies with height as $\exp(-h_m/H)$

$$\frac{(D/H + W)}{D/H} = 1 + WH/D \sim \exp \frac{[h_m(1) - h_m(2)]}{H} \sim 1 + \Delta h_m/H$$

for a small increment of height, Δh_m. Equation 5.3 follows.

If $H = 60$ km, $W = 30$ m s^{-1}, $D = 2 \times 10^6$ m^2 s^{-1} and $\beta = 6 \times 10^{-4}$ s^{-1}, then $\Delta h_m \sim 50$ km. With these parameters, $W \sim D/H$, and therefore drift due to neutral wind is about as important as diffusion in the F region. The effects of neutral winds can go a long way towards explaining the anomalous diurnal behaviour of the F region. By inference, the day-to-day variability is taken to indicate a substantial variability of the winds in the thermosphere, and is evidence for a meteorology of the F region. Seasonal changes in the winds may also contribute to the seasonal anomaly.

When discussing the electrical structure of the upper atmosphere (Section 4.4) we saw that in the F region, where the collision frequencies are much smaller than the gyrofrequencies, an electric field E_x causes the plasma to drift with velocity $v_y = -E_x/B$ (in the coordinate system of Fig. 4.15). In vector notation

$$\mathbf{v} = \frac{\mathbf{E} \times \mathbf{B}}{|\mathbf{B}|^2}$$

The velocity is always perpendicular to the magnetic field. As for plasma motion produced by neutral-air winds, it is the vertical component of the drift that modifies the vertical distribution of ionization. Because of the dip of the magnetic field, the effects of electric fields are much less than the effect of neutral wind at middle latitude. As will be seen in Section 5.4.1, electrodynamic drift is important at the equator.

5.2.5 Variations in the neutral air

We have seen that the electron and ion temperatures vary diurnally, and through collisions the neutral air is also heated so that it, too, undergoes a substantial temperature variation (Fig. 5.8). Correspondingly, the atmospheric density at given height changes by a factor of five between day and night at altitudes of several hundred kilometres, and this is something that can be measured directly, by the satellite drag technique. The density and temperature come to a maximum near 1500 h local time, and a minimum at 0300 h. Calculations have been made of the exospheric temperature distributions, using the known energy input and the thermal conductivity of the exosphere, and there are some discrepancies between the computations and the observations. The density variations are important in providing the driving force for the global wind system (see Section 6.3).

5.3 Effects related to the sunspot cycle

The activity of the Sun goes through an 11-year cycle, which is monitored by the number of sunspots seen on its surface and by the rate of occurrence of solar flares. These visible signs are accompanied by marked changes in the flux of ionizing radiation. Hence the ionospheric production rates depend on sunspot number. The temperature of the upper atmosphere varies by a factor of two during the sunspot cycle and, in consequence, the density at given height varies by a factor of between 10 and 50 (Fig. 5.8).

Observable ionospheric quantities like $f_0 E, f_0 F1, f_0 F2$, and the radio absorption of the D region all increase with the sunspot number, R (defined in Section 7.2.1). In the E layer, the following empirical law holds:

$$f_0 E = 3.3 \left[(1 + 0.008 \, R) \cos \chi \right]^{1/4} \qquad \text{MHz} \qquad (5.4)$$

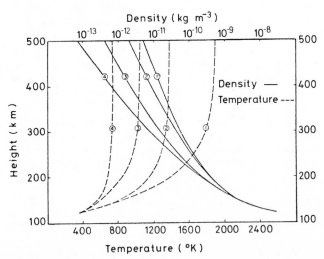

Figure 5.8 Variation of thermosphere temperature and density during the sunspot cycle and from day to night. High solar activity: ① day, ② night. Low solar activity: ③ day, ④ night. (From H. Rishbeth and O.K. Garriott, *Introduction to Ionospheric Physics*, Academic Press, 1969.)

As we saw in Section 5.2.1, the variation with $(\cos \chi)^{1/4}$, χ being the solar zenith angle, identifies the E region as a well behaved α-Chapman layer. The ratio $(f_0 E)^4 / \cos \chi$, known as the *E-layer character figure*, is proportional to the production rate, q. Hence, observations of the E layer can be used to estimate the changes of solar ionizing radiation during a sunspot cycle. Taking $R = 10$ at solar minimum and 150 at solar maximum, Equation 5.4 implies that the E-region production rate varies by a factor of 2 over the sunspot cycle. A similar formula can be given for the F1 layer:

$$f_0 F1 = 4.25 \left[1 + 0.015 R\right) \cos \chi\right]^{1/4} \qquad \text{MHz} \tag{5.5}$$

The F2 layer is not a Chapman layer but, by taking noon values of $f_0 F2$ and averaging over a year to remove seasonal anomalies, it can be seen that here also the ionization density increases strongly with sunspot number as

$$(\overline{f_0 F2}) \propto (1 + 0.02 R)^{1/2} \qquad \text{MHz} \tag{5.6}$$

The essential solar cycle variations of the F region are illustrated in Fig. 5.9. It will be noted that in winter the sunspot number has greatest effect at high latitude, and that at middle latitude and high latitude the effect is smaller in summer than in winter. Fig. 5.9 also shows the seasonal anomaly (Section 5.2.4), which is most marked at sunspot maximum.

The height of the F-region peak increases with sunspot number, At middle latitudes at noon, $h_m F2$ can be below 200 km at sunspot minimum, but above 300 km at sunspot maximum. A similar range of variation affects the night values,

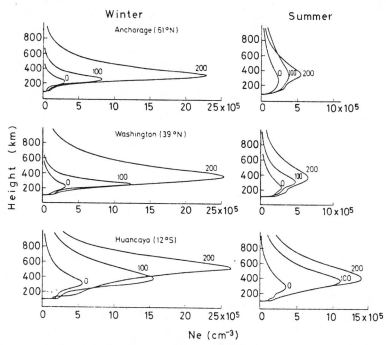

Figure 5.9 Representative electron density profiles at high, middle and low latitude for three levels of solar activity ($R = 0$, 100, 200). These curves are based partly on bottomside ionograms and partly on theory. (After J.W. Wright, *Nature,* **194,** 462, 1962.)

which are all some 100 km higher than those for noon. This solar cycle variation reflects the change of atmospheric temperature and scale height during the sunspot cycle (see Equation 4.58).

5.4 Latitudinal variations

The greatest body of information about the ionosphere has come from observations at middle latitudes, and it is reasonable to take these as the basis when discussing the behaviour of the ionosphere. Moreover, the processes at work in the mid-latitude ionosphere should apply to a greater or lesser extent over a wide range of latitude. However, some striking differences of behaviour are found in high- and low-latitude zones. To a large extent these arise from the orientation of the geomagnetic field, which runs horizontally at the magnetic equator and is nearly vertical at high latitude.

5.4.1 Equatorial ionosphere

The equatorial F region presents an interesting anomaly, usually called the *Appleton anomaly* or just the *equatorial anomaly*. On simple theory the electron density should be a maximum over the equator at the equinox, because the Sun is overhead and so the ionization rate is a maximum. In fact, the electron density falls to a minimum over the magnetic equator and passes through maxima at latitudes $10°$ or $20°$ north and south of the equator. The phenomenon is illustrated in Fig. 5.10,

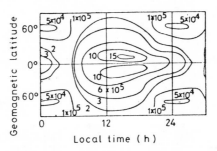

Figure 5.10   The equatorial anomaly as seen in $N_m(\mathrm{F2})$. Contours of $N_m(\mathrm{F2})$ in electrons cm^{-3}. (After D.F. Martyn, *Report of Conference on the Ionosphere*, The Physical Society, 1955, © Institute of Physics.)

as deduced from ionosonde data. The explanation is found in electrodynamic lifting, in which a horizontal electric field, acting in the presence of a horizontal magnetic field, produces a plasma flow in the vertical. The equatorial anomaly can be explained by an electric field of a few hundred μV m^{-1} directed to the east. Given a magnetic field induction of 0.5 G (5×10^{-5} Wb m^{-2}) directed south to north, a vertical drift of 10 m s^{-1} is produced. The electric field is thought to arise in the E layer. As the plasma is lifted to greater altitude it meets field lines that connect to the ionosphere north and south of the equator, down which it may diffuse under gravity. In general the plasma motion is a combination of electrodynamic drift and thermal diffusion, but the net result is depletion over the equator and concentration at latitudes 10° or 20° from the magnetic equator. The phenomenon has been described as a *fountain effect*. This theory has been well substantiated by detailed computations. By combining data from ionosondes and topside sounders, as in Fig. 5.11, it is seen that the separation between the peaks is reduced at greater altitudes, with the concentration tending to lie along a field line.

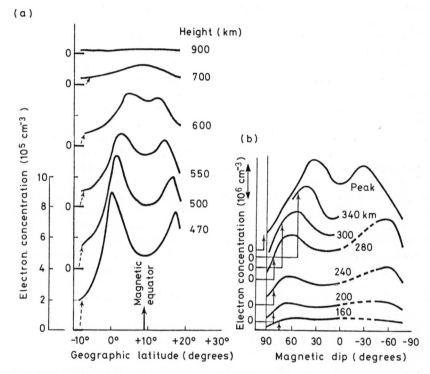

Figure 5.11 Latitudinal variation of electron density across the equatorial anomaly at various altitudes. (a) Above h_{max} (from topside ionograms). (b) Below h_{max} (from bottomside ionograms). (After Croom *et al.*, *Nature*, **184**, 2003, 1959; and Eccles and King, *Proc. Inst. elect. electronics Engr.* **57**, 1012, 1969.)

5.4.2 Polar ionosphere

At high latitude, ion production by solar EUV and X-rays is relatively weak because of the low elevation of the Sun. However, another source of ionization, energetic charged particles, now contributes to the production rate. These energetic particles are associated with the aurora, and, as we shall see in Chapters 7 and 8, they arrive in high-latitude regions because the high-latitude field lines go out into the more distant parts of the magnetosphere where the auroral particles are generated. With charged particles producing more, and ultraviolet less, of the ions, the polar ionosphere behaves rather differently from the mid-latitude regions.

The lower ionosphere, comprising the D and E regions, loses much of its regularity at high latitude, and this is a direct consequence of the ionization produced by auroral particles. Since auroral phenomena occur sporadically, the behaviour at high latitude is irregular. The ionization in the D region is often so intense that radio waves of medium and high frequency, used by ionosondes, are completely absorbed, leading to an ionospheric 'blackout' condition that seriously hinders observation by traditional methods. The presence of spatial structure in the ionization presents the radio observations with further difficulties because of the scattering caused, though this can be turned to advantage in auroral radars operating at higher frequencies.

The F region is also anomalous at high latitude. It appears that the problem of how it is maintained during the long winter night should be capable of explanation by ionization by energetic particles, coupled with the effects of global winds. A key observation is that the diurnal variation of the polar F region, particularly in the Antarctic, is described better by universal time than by local time. Instead of coming to a maximum sometime during the local afternoon, as is the most usual pattern of behaviour at middle latitudes, the high-latitude F region maximizes at 1800 UT in the Arctic and 0600 UT in the Antarctic. These are just the times when the respective geomagnetic poles are closest to the Sun—Earth line. There could well be a contribution from energetic particles to the polar F region (such as the flux entering through the magnetospheric polar clefts, as discussed in Section 7.4.4), but what is certain is that the neutral wind plays an important part in the mechanism. At 1800 UT in the Arctic and 0600 UT in the Antarctic, the daytime neutral wind blows over the dip pole towards the geographic pole. The polar ionization is lifted, and thus, because of the reduction of loss rate with altitude, the peak electron density is maintained at a higher level than would otherwise be the case. The effect is the opposite of that produced equatorward of the dip pole, i.e. at middle latitudes. The neutral wind cannot transport ionization bodily to the polar regions from middle latitude, because that would require the magnetic field lines to be crossed.

There is also a marked latitudinal variation, which in essence is controlled by the shape of the geomagnetic field. The F region is enhanced in the *auroral oval* which, as will be seen in Section 8.2.2, is usually to be found between magnetic latitudes 60° and 80°, and forms an eccentric loop around the magnetic pole. On the equatorward side of the auroral oval the F region is depleted, and then the normal mid-latitude ionosphere appears fairly sharply at about 60° magnetic latitude. The so-called *trough* in F-region electron density, between the mid-latitude ionosphere and the auroral enhancement, is related to the geometrical form of the magnetosphere, for it is connected by the magnetic field lines to a region of the magnetosphere that is relatively short of ionization. These high-latitude features are seen most plainly in topside sounder data (Fig. 5.12). Satellite measurements show that the trough region has rather a complex structure. At noon it is shallow, the electron density dipping by only 20—40%, and it is a matter of some debate just which magnetospheric features (cusp — Section 7.4.4, or plasmapause — Section 7.5.1) are connected to it. The ions, and, particularly, the electrons, are hotter at high latitude, again reflecting the role of energetic particles in creating the ionization.

Within the polar cap, i.e. poleward of the trough, the magnetic field connects to regions deep in the magnetosphere. Instead of returning to a point on the Earth, these lines are effectively 'open', and they provide an escape route for plasma. An

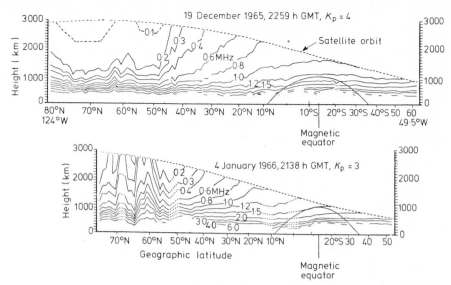

Figure 5.12 Topside sounder sweeps, showing features of the high-latitude F region. The mid-latitude ionosphere ends near 50° N, and the trough and auroral enhancements can be seen at higher latitudes. (After G.L. Nelms and G.E.L. Lockwood, *Space Res.* **VIII,** 604, 1967.)

outflow of light ions, particularly He$^+$, has been observed over the poles. The flow, known as the *polar wind*, is driven by the electric field resulting from the charge separation between electrons and heavy ions under gravity (Section 4.2.3). On closed field lines the force would be balanced against a pressure gradient acting in the opposite direction, but on open lines the ions may escape and so the flow continues. The polar wind plays an important part in the terrestrial helium budget.

The behaviour of the high-altitude ionosphere is closely related to the physics of the magnetosphere and to topics such as the aurora, geomagnetic disturbances, and ionospheric storms, that will be encountered in subsequent chapters. Together they form a group of interrelated problems that are not yet fully solved, and in fact represent one of the most actively studied areas of geophysics at the present time.

5.5 Eclipses

Though not strictly part of the regular behaviour of the ionosphere, the response to an eclipse is — or should be — a well ordered perturbation, since the change in ionizing radiation during the eclipse is regular and predictable. In the D, E and F1 regions, where the electron density is controlled mainly by the photo-ionization and recombination rates, the calculated reduction of production rate provides a means of determining the recombination rate. An example of the D-region effect, as measured by riometers, is shown in Fig. 5.13. If the source of ionizing radiation is uniformly distributed across the solar disc, the variation of flux during the eclipse can be calculated exactly. However, difficulties can arise because some of the EUV and X-rays come from the solar corona and so these fluxes are never totally obscured. Further, if the Sun is active, more of the X-ray emissions come

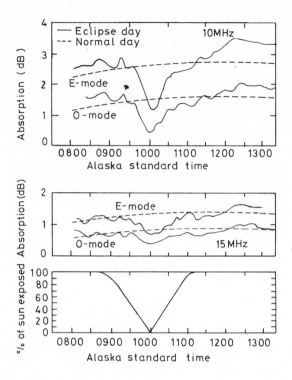

Figure 5.13 Reduction of radio absorption in the lower ionosphere during a total eclipse of the Sun on 20 July 1963. The effect maximizes near the time of totality, but the recovery shows some time delay. (From G.M. Lerfald *et al.*, *Rad. Sci.* **69D**, 939, 1965, copyrighted by American Geophysical Union.)

from localized regions of the solar disc, and these would have to be known in detail from independent observations.

The response of the F2 region to an eclipse was often thought very puzzling when observations were confined to the bottomside of the ionosphere, since $N_m(F2)$ could decrease, increase, or remain unchanged during the eclipse. Methods that looked through the ionosphere, such as beacon satellite measurements of electron content and, particularly, incoherent scatter, showed that the loss of ionization in the lower part of the F region was partly or wholly offset by cooling of the whole region, which allowed ionization to move down from the topside. Thus the peak electron density could be maintained although the electron content decreased. Fig. 5.14 shows an example where the peak electron density increased at totality, though integration of the profile shows that the electron content decreased throughout.

5.6 Summary

The reader who likes his science cut and dried may feel unsettled by the present chapter. Having, in Chapter 4, presented a theoretical basis for the upper atmosphere,

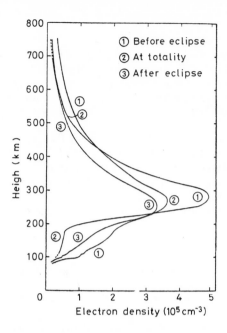

Figure 5.14 F-region changes during the eclipse on 20 July 1963. The curves are composites of incoherent scatter data and bottomside soundings. (After J.V. Evans, *J. geophys. Res.* **70**, 131, 1965, copyrighted by American Geophysical Union.)

we have now seen that some of the observations seem merely to reveal the in-sufficiency of those theories. Yet this contrast makes the very point that the author, as teacher, wishes to bring home to the reader, as student. The interplay of theory and observation or experiment is the essence of the scientific method; and the upper atmosphere, as a scientific study, constantly brings us to the point where existing theories, sound as they are in their own right, are as yet not sufficient to explain all new observed facts as they come along. Difficulties of this kind are exactly what lead to further progress.

As far as the ionosphere is concerned, the classical era, dominated by ideas like those presented in Chapter 4, is behind us. More recent work has led to an appreciation of the importance of ionization movement. And it was during the 1960s, when data from space vehicles were adding greatly to observational detail, and modern computing facilities became generally available to the theorist, that a sound footing was laid for an appreciation of the seemingly anomalous behaviour of the F region. Although now, in the 1970s, many details of the F region still require further study, it seems unlikely that there will be further radical discoveries about the processes governing its regular behaviour at middle and low latitudes. Knowledge of the high-latitude ionosphere, on the other hand, resembles that of the mid-latitude ionosphere some 15 years ago. Many new facts are emerging and important new theoretical concepts have been invented, but there is still much more to be done and it will probably be some time before this disturbed and erratic part of the atmosphere can be said to be properly understood.

5.7 Further reading

Bauer, S. J. (1973) *Physics of Planetary Ionospheres* (Springer-Verlag) Chapter 9.

Keating, G. M. and Levine, J. S. (1972) Response of the neutral atmosphere to variations in solar activity, in *Solar Activity Observations and Predictions* (Ed. McIntosh and Dryer) (MIT Press).

Rishbeth, H. and Garriott, O. K. (1969) *Introduction to Ionospheric Physics* (Academic Press) Chapter 5.

LeGalley, D. P. and Rosen, A. (1964) *Space Physics* (Wiley) Chapter 6.

Ratcliffe, J. A. and Weeks, K. (1960) The Ionosphere, in *Physics of the Upper Atmosphere* (Ed. Ratcliffe) (Academic Press) Chapter 9.

Matsushita, S. and Campbell, W. H. (1967) *Physics of Geomagnetic Phenomena* (Academic Press) Chapters III-3, III-4.

Whitten, R. C. and Poppoff, I. G. (1971) *Fundamentals of Aeronomy* (Wiley) Chapter 5.

Evans, J. V. (1975) High-power radar studies of the ionosphere, *Proc. Inst. elect. electronics Engr.* **63**, 1636.

Whitehead, J. D. (1970) Production and prediction of sporadic E, *Rev. Geophys. Space Phys.* **8**, 65.

Rishbeth, H. (1972) Thermospheric winds and the F-region: a review, *J. atmos. terr. Phys.* **34**, 1.

Rishbeth, H. (1974) Ionospheric dynamics 1946–1970, *J. atmos. terr. Phys.* **36**, 2309.

Murata, H. (1974) Wave motions in the atmosphere and related ionospheric phenomena, *Space Sci. Rev.* **16**, 461.

Beer, T. (1973) The equatorial ionosphere, *Contemp. Phys.* **14**, 319.

Whitten, R. C. and Colin, L. (1974) The ionospheres of Mars and Venus, *Rev. Geophys. Space Phys.* **12**, 166.

6
Winds, Currents, Irregularities and Waves

6.1 Introduction

The structures considered in Chapters 4 and 5 are essentially of large scale. Vertical structure is reckoned in tens and hundreds of kilometres, and the horizontal structures arising from latitudinal and local-time variations are measured in hundreds and thousands of kilometres. Moreover, with some exceptions, these structures tend to be repeatable and, therefore, predictable. In the time scale, we considered mainly variations over periods of days, months and years.

In the present chapter we turn to the smaller structure and the quicker changes. The small-scale structure of the atmosphere tends to be irregular and unpredictable in detail — though predictions of a statistical kind may be possible. The distinction can be illustrated by reference to meteorology, in which the forecaster might predict the average wind speed and typical direction, but it would be a hopeless task to attempt a prediction of the precise wind vector for a stated place and instant of time. Similarly, it would not seem out of the question to attempt predictions of the occurrence of ionospheric irregularities, but to specify complete details would be impossible and, indeed, unnecessary.

Over times shorter than a day the dynamics of the upper atmosphere are concerned with winds and tides, with electric current systems, and with various propagating waves such as acoustic—gravity waves. To a limited extent these topics can be considered separately, but in fact there are also many points where they may interact. Indeed, one of the difficulties of upper atmosphere studies is that so many of the observed phenomena affect others, implying that problems cannot be solved in isolation. Several examples will be noted in this chapter. Because of the limitations imposed by observational techniques, most, though not all, of the observations refer to the ionized atmosphere. Direct information about the movement and the fine structure of the neutral air is more restricted, though it should be remembered that the motions of the ionized and the neutral air are closely related in many parts of the atmosphere.

6.2 Small irregularities

6.2.1 Diffraction of radio waves by small irregularities

Most of the available information about small irregularities in the ionosphere has been obtained by studying the fading of radio waves reflected from, or transmitted through, the ionosphere. In this application the radio signal is scattered coherently, which is to be distinguished from the incoherent scatter discussed in Section 3.7.4.

The signal used in the incoherent scatter (or Thomson scatter) technique is incoherent because it is made up of contributions from numerous irregularities, each smaller than a wavelength of the radio wave. The irregularities being in random motion, there is no defined phase relationship between individual contributions to the total received signal. Coherent scattering is, in a sense, less severe. It appears when the signal is returned from a discrete layer rather than from a volume, and thus there is only one contribution to the total signal from each direction. The signal cancellation of incoherent scattering does not arise, therefore, and the fading is caused by interference between the contributions from different directions if the reflecting layer is not plane and there are differential movements. Coherent scattering is much stronger than incoherent. Echoes from meteor trails (Section 6.3.1), radar aurora (Section 8.2.2), sporadic-E (Section 5.2.1) and spread-F (6.2.3) are examples of coherent scattering from aligned structures. In the present treatment we are concerned with scattering by irregularities in the ionosphere.

Within the regime of coherent scattering, the irregularities come in two varieties — large and small. The distinction is based on the size of the Fresnel zones in relation to the size of the irregularities. The present treatment is concerned with small irregularities.

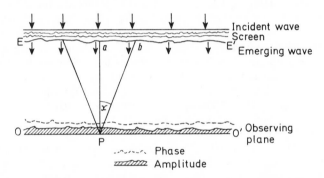

Figure 6.1   Diffraction of a plane radio wave incident on a thin phase-changing screen.

The theory of radio wave diffraction by an irregular propagation medium starts with the concept of a *thin phase-changing screen*. The screen is considered to be infinitely thin, and it is assumed to introduce small phase perturbations along the wavefront of a radio wave passing through it, as in Fig. 6.1. By Huygens' principle, each part of a wavefront is to be considered as the source of secondary wavelets. To obtain the field, i.e. the wave amplitude and phase, at a point P in an observing plane OO′, it is necessary to sum the contributions from each point of the emerging wavefront EE′. Since EE′ is irregular in phase, the field OO′ will also be irregular, and in general both the phase and the amplitude will be affected. When studying ionospheric irregularities by radio propagation methods the general problem is to use the observations at the ground (OO′) to deduce as much as possible about the wavefront emerging from the irregularities, and then to use this information to learn something about the nature of the irregularities themselves. It will be clear from the above that there is not necessarily a one-to-one relation between irregularities in the ionosphere and wavefield irregularities at the ground.

Fortunately, there are some relationships of a statistical nature. The diffraction approach based on Huygens' principle is, of course, valid generally. However, if the irregularities are large enough their effects may be analysed by geometrical optics without introducing the complexity of diffraction theory. Within the diffraction approach, the thin phase screen introducing small phase perturbations (< 1 rad) is the most tractable special case.

To gain further insight we must introduce the concept of the angular spectrum. Just as a wave modulated in time has a frequency spectrum that may be derived by Fourier transformation, so a wave modulated in distance can be expressed as a spectrum in angle. A simple amplitude modulation

$$A(t) = 1 + m \cos(2\pi t/T)$$

where T is the period and m the modulation depth, is equivalent to a frequency spectrum (Fig. 6.2a) composed of a carrier of unit amplitude and two sidebands of amplitude $m/2$ separated from the carrier by frequency $1/T$. Similarly, an amplitude modulation with respect to distance

$$A(x) = 1 + m \cos(2\pi x/d)$$

where d is the spatial period, is equivalent to the angular spectrum of Fig. 6.2b, consisting of an undeviated wave plus two sidewaves at angles $\pm \sin^{-1}(\lambda/d)$ where λ is the wavelength. This will be recognized as the first-order grating formula. The angle between the sidewaves and the undeviated wave is the same whether the initial modulation is of amplitude or phase. (However, additional sidewaves appear at greater angles if the initial modulation depth exceeds 1 rad.)

(a) Frequency spectrum

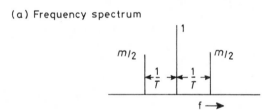

(b) Angle spectrum

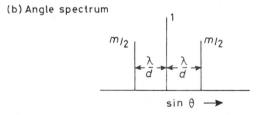

Figure 6.2   (a) Frequency spectrum of an amplitude modulated wave. (b) Angle spectrum of a wave modulated in distance.

If the spatially modulated wavefront contains a spectrum of periodicities, $F(d)$, the equivalent angles also cover a spectrum, $f(\sin\theta)$, where $F(d)$ and $f(\sin\theta)$ are a pair of Fourier transforms. Provided the diffraction screen is shallow, i.e. the modulation imposed on the radio wave is small, and since the angular spectrum received at the ground must be the same as that which left the screen, the spectrum of irregularities formed at the ground will be the same as that at the screen. The comparison must be made using the complex waveform, including both amplitude and phase. Whereas the amplitude of a radio signal is easily measured its phase is not. However, it has been shown that if the observations are made sufficiently far from the screen the statistical properties of the amplitude and the phase irregularities at the ground are the same as each other and as those of the screen; in this case amplitude observations alone will suffice to give the statistical properties of the irregular ionospheric screen.

In nature the irregularities are unlikely to be sinusoidal or to have any regular form that could be described by an analytical function, but will probably look like random noise. This is handled by working out the correlation function.* The form of the correlation function may sometimes be assumed to have a Gaussian form

$$\rho(d) = \exp(-d^2/2d_0^2) \tag{6.1}$$

and in this case the angular power spectrum would also be Gaussian

$$P(\sin \theta) = \exp(-\sin^2 \theta/2 \sin^2 \theta_0) \tag{6.2}$$

where $\sin \theta_0 = \lambda/2\pi d_0$. Thus one has the same kind of relationship as before between angle (θ_0) and irregularity size (d_0), but in terms of quantities expressing the widths of statistical distributions rather than individual features (Fig. 6.3).

To see why the Fresnel zones are important we recall that by definition the first Fresnel zone extends to the point where the distance to the observer exceeds the minimum distance by $\lambda/4$, a difference which produces a phase difference of $90°$. (In Fig. 6.1, if the overhead point is a, and the first Fresnel zone falls at b, then $Pb - Pa = \lambda/4$). If the screen causes phase changes only, the signal at EE' may be sketched as in Fig. 6.4a, where A is the unaffected signal and α_E the perturbation due to the screen. At point P of the observing plane, if the perturbation due to a affects the phase of the total signal, that due to b will affect the amplitude because of the extra quarter wavelength travelled by that element of wavefront. The resulting signal might therefore look like that in Fig. 6.4b. The first contribution to an amplitude perturbation (from b) therefore arrives at angle $\chi = \tan^{-1}(\lambda/2D)^{1/2}$ to the vertical. Assuming small angles, the condition for the appearance of amplitude perturbations is, therefore,

$$\theta_0 \qquad > \chi$$
$$\lambda/2\pi d_0 \quad > (\lambda/2D)^{1/2}$$
$$(\lambda D/2)^{1/2} > \pi d_0 \tag{6.3}$$

Thus a phase screen will produce as much amplitude as phase perturbation in a received signal if the observing plane is sufficiently distant for the first Fresnel zone to contain several irregularities of typical size. When using amplitude observations to deduce properties of an ionospheric phase screen the above limitation must be kept in mind.

*If $a(x)$ are the differences between a quantity $A(x)$ and its mean value $\bar{A}$, and σ^2 is the variance of a, $\overline{(A - \bar{A})^2}$, the bars denoting averages over many values, the correlation function of A over interval y is

$$\rho = \frac{\overline{a(x) \cdot a(x + y)}}{\sigma^2}$$

110

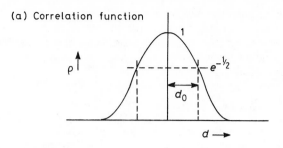

(a) Correlation function

ρ

1

$e^{-1/2}$

d_0

$d \longrightarrow$

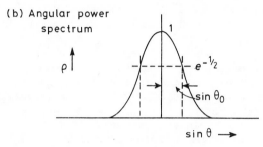

(b) Angular power spectrum

ρ

1

$e^{-1/2}$

$\sin \theta_0$

$\sin \theta \longrightarrow$

Figure 6.3 Correlation function (a) and angular power spectrum (b) for a random diffraction screen.

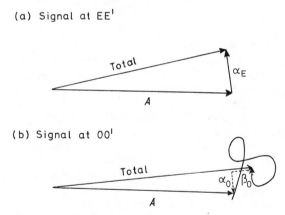

(a) Signal at EE'

Total

A

α_E

(b) Signal at OO'

Total

A

α_0 β_0

Figure 6.4 Development of amplitude and phase perturbations from initial phase perturbations.

In summary, the simple theory of diffraction from a thin sheet of phase-changing irregularities makes it possible to deduce the intensity and spatial size of irregularities in statistical terms from amplitude observations at the ground, but subject to these restrictions:

(i) The modulation must be shallow, < 1 rad. (Otherwise high-order sidewaves are produced which introduce finer structure at the ground. If the modulation is ϕ rad, where $\phi > 1$, the irregularities at the ground are smaller than those in the ionosphere by a factor of ϕ.)

111

(ii) Only phase irregularities smaller than a Fresnel zone can produce amplitude variations at the observing screen.

(iii) Irregularities smaller than one radio wavelength cannot be detected (because the corresponding sidewaves occur at imaginary angles ($\sin \theta > 1$) and these waves are evanescent).

6.2.2 Results of radio observations

Small irregularities in the ionospheric E region are detected and studied by means of reflected radio waves at frequencies of a few MHz. The received signals fade with periods of tens of seconds, and by the use of spaced receivers the correlation distances are found to be a few hundred metres. By the partial reflection technique (Section 3.7.2) irregularities can be observed as low as 70 km. The spatial size is again 100–200 m. The lower height limit is determined by the presence of sufficient ionization to give measureable echoes and does not necessarily indicate the cessation of irregularities. These irregularities of the E and D regions provide a useful method of measuring winds at those altitudes (Section 6.3.1). VLF signals, also reflected in the D region, may fade at night but in this case, no doubt because of the longer radio wavelength, the irregularities appear much larger – about 20 km. Probably these should be classed with the large irregularities (Section 6.5).

The spaced receiver method is also used to study irregularities in the F region. The first approach is like the E-region method, but using a higher radio frequency so that the echo comes from the F region. However, most of the information about F-region irregularities has been obtained by transionospheric propagation where the source, a radio star or a satellite, is outside the ionosphere. The fluctuations that ionospheric irregularities impose on signals passing through the ionosphere are usually called *scintillations*, by analogy with the intensity fluctuations of luminous stars when viewed through a turbulent atmosphere. A vast quantity of information has been gathered about radio scintillation because the fading has an adverse effect on the performance of satellite communication systems.

Morphological studies show that three geographical areas may be defined: polar and equatorial zones where the scintillations are most intense, and the medium latitudes where the fading is weaker and occurs less frequently. In the equatorial zone, within 20° of the geomagnetic equator, the scintillations maximize at night and at the equinoxes. The fading, once begun, usually continues for several hours and is deep, amounting to some 6 dB or more (a factor of two in amplitude) between mean and maximum. The fading is also fairly slow, with $1-8$ fades min^{-1}. Spaced measurements show a typical irregularity size of 100–400 m in the east–west direction, with a considerable (perhaps a factor of 60) elongation north–south – along the geomagnetic field.

Scintillations also maximize by night in the polar zone, situated poleward of 60° geomagnetic latitude. This zone is somewhat offset from the pole, being nearer the equator in the night sector and showing a general, though not exact, correspondence with the auroral oval (Section 8.2.2). These scintillations are more rapid than those at the equator, the period being $1-3$ s in severe fading. The peak-to-peak variations may be 15 dB, or a factor of six in amplitude. When high-latitude scintillations are observed using the transmissions from a geostationary satellite they appear more severe because of the oblique path through the ionosphere.

From some of the more recent satellite radio beacons both phase and amplitude fading can be observed. Since the amplitude fluctuations may be assumed to develop from an initial phase modulation the comparison is one possible way of estimating the height of the irregularities. A more direct method, based on amplitude scintillations only, uses transmissions from an orbiting satellite. The method is illustrated in Fig. 6.5. Since an orbiting satellite moves more quickly than the irregularities, the apparent speed of the irregularities over the ground (V_g), which can be determined by spaced receivers, depends mainly on the satellite velocity (V_s), and on the irregularity height (h_i) as a fraction of the satellite height (h_s)

$$V_g = V_s h_i / (h_s - h_i)$$
$$h_i = h_s / (1 + V_s / V_g) \tag{6.4}$$

Most height determinations fall in the range 200–600 km, with the maximum between 300 and 400 km.

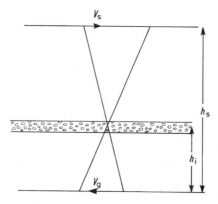

Figure 6.5 Determination of the height of irregularities from the apparent motion of a fading pattern over the ground.

The physical causes of the irregularities are not well established, though attention focuses on the equatorial electrojet as a source of the equatorial irregularities, and on the precipitation of energetic electrons into the auroral ionosphere for the high-latitude zone. Theoretical work has demonstrated mechanisms, based on the different velocities of ions and electrons in an electric field, that can reinforce any gradients that might be present. When such a mechanism is at work the development of irregularities is inevitable.

6.2.3 Other evidence on the nature of small ionospheric irregularities

Although most of the information about irregularities has come by studying the fading of radio waves received at the ground, some illuminating facts have been obtained by other means.

At medium latitude, F-region scintillations tend to occur in patches about 1000 km across, and these patches have been identified with the phenomenon of *spread-F*, long known on ionograms. An example is shown in Fig. 6.6. Here the F-region echo is distributed over a range of heights, instead of being discrete as in an undisturbed ionogram. The occurrence of spread-F (Fig. 6.7) shows maxima at

113

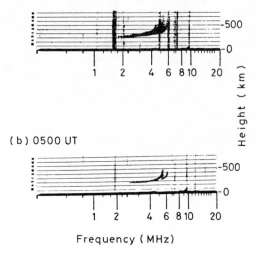

(a) 0300 UT

Height (km)

(b) 0500 UT

Frequency (MHz)

Figure 6.6 Ionograms (a) with and (b) without spread-F. Vertical soundings at Esrange (68.11° N, 20.99° E) 15 March 1970. (From *Kiruna Geophysical Data,* 70/1–3, Kiruna Geophysical Institute, 1970.)

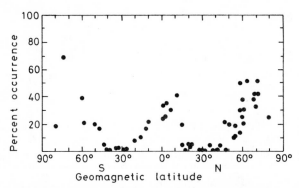

Figure 6.7 Occurrence of spread-F as a function of geomagnetic latitude during August– September 1957. (From T. Shimazaki, *J. Rad. Res. Lab. Japan,* 6, 669, 1959.)

high and low latitudes, and, like scintillations, it also occurs most frequently at night. It seems clear that these two phenomena have a common cause and that spread-F appears on ionograms when the ionosphere is irregular overhead.

Evidence that the ionospheric irregularities are elongated in the direction of the geomagnetic field is the occurrence of ducted echoes on topside ionograms. A topside sounder observes spread-F from above when the region is irregular, but also at such times echoes arrive with such long delays that they can only be explained by assuming that the reflection has occurred in the conjugate hemisphere. The mechanism seems to be that some of the radio energy, propagating upwards from the sounder, becomes trapped between the sheets or columns of enhanced or depleted ionization, which guide it to the other hemisphere. There the signal is reflected in the topside ionosphere and returned, via the aligned ducts, to the sounder. With the radio signal incident obliquely on the duct the variation of electron density need be no greater than 1%. The irregularities appear to be about 0.5 km thick. It is not necessary for a given irregularity to extend the whole distance between hemispheres; a general elongation of all irregularities along the geomagnetic field is sufficient cause.

Some observations of the irregularities have been made directly with satellite-borne plasma probes. Orbiting satellites are good at determining the geographical distribution of a phenomenon, and it has been found that the equatorial scintillation-producing irregularities are more intense at some longitudes than at others (most in the Atlantic sector, least over the Far East). This fact, though suspected, had been difficult to establish by the usual methods because of the limited numbers of ground receiving stations and geostationary satellites.

If scintillations can be measured at the ground while the responsible irregularities are being observed by a direct sensor, then it becomes possible to test the scintillation theory. This is an important aspect, though relatively little has been done along these lines thus far. To take an example, the r.m.s. variation of the electron density of the F region might be observed as 10% of the mean. If the mean is 10^5 cm^{-3} and the ionospheric slab thickness is 200 km, the electron content variation is 2×10^{15} m^{-2}. The phase of a 150 MHz radio wave leaving the ionosphere therefore fluctuates (Equation 3.16) by

$$\frac{1.6 \times 10^3 \times 2 \times 10^{15}}{3 \times 10^8 \times (2 \times 150 \times 10^6)} = 11.3 \text{ rad}$$

The wave is deeply modulated since the phase varies more than 1 rad, and so (Section 6.2.1) the amplitude irregularities produced at the ground should be much smaller than those in the ionosphere (by a factor of 11).

6.3 Winds

6.3.1 Methods of observation

The movement of the neutral air is normally observed by means of a tracer, which might be present naturally or might be injected for the purpose. The tracer serves as an indirect sensor (Section 3.1), which, if observed from a distance, gives information about the medium surrounding it.

Meteor trails provide a good example of a tracer present naturally. As a meteor oblates in the atmosphere, part of the energy goes to ionize the air in a column around the meteor. Most trails are formed in the height range 75–105 km, where they can be tracked by radar as they are blown along in the wind.

Regions of airglow emission (Section 4.3.5) also move with the neutral air, and a Doppler shift is thereby produced. By observing the apparent wavelength of a prominent airglow line it is possible to determine the component of velocity in the line of sight. It has proved practicable to measure winds at F-region altitudes by this technique, using the oxygen line at 6300 Å; the horizontal wind vector is determined by observing in two directions more or less at right angles. The air temperature can be found from the width of the line.

Rocket techniques enable tracers of various kinds to be inserted into the medium. The puff of smoke from a grenade can be followed visually from the ground. Metal foil, or a parachute carrying a reflector, can be tracked by a radar, to give the winds over a range of heights as the target falls. Such methods work well within limited altitude ranges. A parachute will not open above about 60 km, and foil disperses too quickly if released too low down.

Most wind measurements in the thermosphere have been made using chemicals ejected from an ascending rocket. A trail of sodium is luminous at twilight by resonant scattering of sunlight at 5890 Å, and since the trial then shows up against the dark sky its movements can be followed by means of spaced cameras. The irregularities that form in the trail in the minutes after release give evidence of turbulence in the medium, and the irregularities provide necessary reference points for triangulation so that heights and velocities can be determined. Trimethyl aluminium (TMA) is used in a similar manner, but since it reacts with atmospheric atomic oxygen the trail is luminous at night, not just at twilight. The TMA payload is sufficiently robust to be fired from a large gun, and this interesting technique has been used, though not at many sites, to insert TMA as high as 140 km.

A powerful technique, measuring both winds and electric fields, is based on a mixture of barium and strontium. If a barium/strontium cloud is released at twilight the barium ionizes to Ba^+ but the strontium remains neutral. The neutral strontium cloud appears green, and the ionized barium cloud is red. Thus the two clouds can be distinguished, and they are observed to separate after release, the strontium cloud moving with the neutral air but the ionized one moving in response to whatever electric field may be present. The neutral cloud expands uniformly in all directions and retains its spherical shape. The ionized cloud, on the other hand, elongates along the geomagnetic field. At middle latitudes the typical speeds of the red and green clouds are 80 and 40 m s^{-1} respectively, giving a deduced electric field of several mV m^{-1}. The electric fields are much higher at high latitudes.

A propagation method for the determination of neutral winds in the mesosphere and stratosphere is based on the sound emitted from a rocket grenade and received by a microphone array at the ground. The speed of sound depends on the air temperature, as given in Section 3.6.1, but is of course relative to the air. As measured by observers at the ground, e.g. with spaced microphones, the velocity is a vector sum of the wind and the sound velocity. Given a suitable array, therefore, both wind and temperature can be deduced as a function of height.

It will be seen in Section 6.3.2 that the theory of the thermal wind relates the wind profile to the latitude and longitude variations of the temperature profile in the stratosphere and the mesosphere. Winds may thus be computed if temperature data are available, though, since the theory depends on the Coriolis force, it cannot be expected to hold well at low latitudes.

Apart from chemical releases, the most commonly used technique for measuring movements in the upper atmosphere is to observe the fading of radio waves, reflected from or transmitted through the ionosphere, with spaced receivers. If the spaced receivers are not too far apart it is likely that each receiver will see a similar pattern but with a small time displacement, as in Fig. 6.8a. The implication is that a patch of ionospheric irregularities is moving bodily across the observation sites, and it should be possible to deduce the speed and direction of the patch from the time differences and the geometry of the receiving network.

To take a simple case, suppose that the stations form a right-angled triangle, as in Fig. 6.8b. Then time differences t_1 and t_2 give 'apparent velocities' V_x' and

(a)

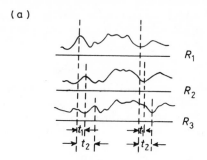

(b)

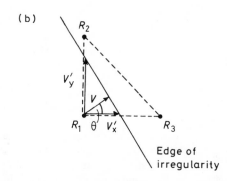

Edge of
irregularity

Figure 6.8 The spaced receiver method for observing the movement of a pattern of irregularities over the ground. (a) Typical fading at stations R_1, R_2 and R_3. (b) Relation between true and apparent velocities.

V_y' in the x- and y-directions

$$V_x' = R_1 R_3/t_2 \qquad V_y' = R_1 R_2/t_1$$

From Fig. 6.8b it is seen that

$$\frac{1}{V^2} = \frac{1}{(V_x')^2} + \frac{1}{(V_y')^2}$$

$$V = \frac{V_x' V_y'}{(V_x')^2 + (V_y')^2} \tag{6.5}$$

and

$$\theta = \tan^{-1}(V_x'/V_y') \tag{6.6}$$

Note that V_x' and V_y' are not velocity components in the usual sense.

The treatment is readily generalized to accept stations in other configurations than the right-angled triangle, but in practice various other complicating factors arise that are not so easily overcome. What may happen is that when the receiving sites are far enough apart to produce measurable time differences it is observed that the fading patterns are no longer identical, implying that the pattern is changing as it moves. In a pattern which is changing but not moving as a whole, the average time difference between the fading at spaced stations is zero. Thus the effect of changes in a moving pattern is to reduce the average time difference, and therefore to increase the apparent velocity. This problem is handled by introducing a conceptual *characteristic velocity*

$$V_c = d_0/t_0 \qquad (6.7)$$

where d_0 is a characteristic distance (the size of the irregularity) and t_0 is a characteristic fading time, both as would be measured by an observer moving with the pattern. Let the true velocity, i.e. the velocity of this observer, be V. Suppose that a stationary observer measures the movement in the direction of the vector **V**. He can in principle determine d_0 correctly, but his measurement of the fading time (t_0') will be less than t_0 because of the movement of the pattern. Thus we can define

$$V_c' = d_0/t_0' \qquad (6.8)$$

It can be shown that V_c and V (the desired quantities) can be obtained from V' and V_c' (the observed quantities) via the relationship

$$V_c'^2 = V_c^2 + V^2 = V'V \qquad (6.9)$$

Anisotropy of the pattern can be handled by further elaboration if the observations are made at four stations.

The foregoing method is beyond reproach if applied to a single irregularity of simple form and all the necessary measurements can be made accurately. In the real ionosphere the irregularities cover a spectrum of sizes, with no guarantee that they all move together, or change at the same rate. To obtain the best average solution the analysis is usually based on correlation coefficients rather than on individual features of the fading pattern; but there remain some basic assumptions about the irregularities that cannot as a rule be justified. The method, nevertheless, has been widely used, since it is relatively convenient, relatively cheap, and often the only one available. When direct comparison with other methods has been possible, the spaced-receiver method has usually been vindicated as a valid one, though perhaps not accurate to the third decimal place! E- and F-region motions can be studied by transmitting from the ground a radio signal to be reflected in the region of interest. F-region measurements are also made by observing the scintillation of a radio star or satellite radio beacon. The technique also works with partial reflections from the D region (Section 3.7.2).

Following our discussion of Sections 4.4.2 and 4.4.3 it will be clear that plasma motions in the E and D regions may usually be taken to indicate the neutral winds, whereas those in the F region are more likely to be a measure of the electric fields.

6.3.2 Prevailing winds in the stratosphere and mesosphere

Wind systems exist in the upper atmosphere just as they do in the troposphere, and knowledge of them has improved considerably in recent years. The winds are weakest near the tropopause and at the mesopause (Fig. 6.9) and it is convenient to consider the stratospheric and mesospheric circulations together. We also distinguish between prevailing winds, which blow continuously in one direction, and tides, which vary with a period related to 24 h.

Air flow in the stratosphere and mesosphere is dominated by prevailing winds flowing zonally, i.e. east–west, and sustained by north–south pressure gradients

118

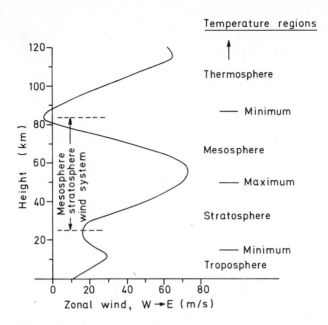

Figure 6.9 General magnitude of zonal winds in the upper atmosphere, illustrated by height profile for 45° N latitude in January. (From COSPAR International Reference Atmosphere, 1972.)

due to the variation of solar heating with latitude. The theory is that of the *thermal wind*, covered in textbooks of meteorology. If the air temperature changes with distance in the south-to-north direction (y) at rate $\partial T/\partial y$, a south-to-north pressure gradient is created at height h, of magnitude

$$\frac{\partial P}{\partial y} = P \frac{h}{HT} \frac{\partial T}{\partial y} \tag{6.10}$$

where H, the scale height, and T, the absolute temperature, are assumed independent of height. This pressure gradient is balanced by the Coriolis force, $\rho u_x f$ associated with an east–west wind u_x. ρ is the air density, and f is the Coriolis parameter $f = 2\Omega \sin \theta$, at latitude θ for terrestrial angular velocity Ω. Thus

$$u_x = -\frac{P}{\rho f} \frac{h}{HT} \frac{\partial T}{\partial y} = -\frac{hg}{fT} \frac{\partial T}{\partial y} \tag{6.11}$$

In a more general case the formula can be proved as follows. Let the west-to-east and south-to-north wind components be u_x and u_y. Then if the pressure gradients are $\partial p/\partial x$ and $\partial p/\partial y$ at a given height, balancing the pressure-gradient and Coriolis forces gives

$$u_x = -\frac{1}{f\rho} \frac{\partial P}{\partial y} \tag{6.12}$$

for geostrophic flow in the zonal direction. For static equilibrium with respect to height

$$\partial P/\partial z = -\rho g \tag{6.13}$$

119

Since $\rho = MP/RT$ from the gas law, Equations 6.12 and 6.13 can be written

$$\frac{\partial P}{\partial y} = -\frac{fMP}{RT}u_x \qquad (6.14)$$

and

$$\frac{\partial P}{\partial z} = -\frac{MP}{RT}g \qquad (6.15)$$

Differentiating Equation 6.14 with respect to z and Equation 6.15 with respect to y, and equating the right-hand sides of the equations:

$$-\frac{fM}{R}\frac{u_x}{T}\frac{\partial P}{\partial z} - \frac{fM}{R}P\frac{\partial}{\partial z}\left(\frac{u_x}{T}\right) = \frac{Mg}{R}\frac{P}{T^2}\frac{\partial T}{\partial y} - \frac{Mg}{RT}\frac{\partial P}{\partial y}$$

The first and last terms of this equation are equal, from Equations 6.14 and 6.15. Thus

$$\frac{\partial}{\partial z}\left(\frac{u_x}{T}\right) = -\frac{g}{fT^2}\frac{\partial T}{\partial y} \qquad (6.16)$$

which integrates to Equation 6.11 if T is independent of height. Considering the component

$$u_y = -\frac{1}{f\rho}\frac{\partial P}{\partial x} \qquad (6.17)$$

gives a second equation

$$\frac{\partial}{\partial z}\left(\frac{u_y}{T}\right) = +\frac{g}{fT^2}\frac{\partial T}{\partial y} \qquad (6.18)$$

Self-consistent distributions of meridional temperature and zonal wind are shown in Fig. 6.10.

6.3.3 Tides

At higher altitudes tides become increasingly apparent. A tide is a perturbation due to an external agent such as the Sun, to which it maintains a fixed relationship while the Earth rotates beneath. A solar tide has a period of 24 h or a submultiple of 24 h (12, 8, 6, etc.), and to an observer on the Earth's surface the perturbation appears to move westward. The tides of the upper atmosphere are generated principally by the daily cycle of heating and cooling, the mechanism being the absorption of solar electromagnetic radiation in the atmosphere, particularly by ozone and water vapour in the stratosphere. The possibility of tidal periods less than 24 h comes about because the heating cycle is more square-wave than sinusoidal and is therefore rich in harmonics. Which periods actually appear is determined by the characteristics of the various possible modes of oscillation of the atmosphere, some of which are depicted in Fig. 6.11. The modes are designated (m, n), where m is the number of cycles per day and there are $(m - n)$ nodes between the poles (not counting those at the poles). To take an example, mode $(2, 4)$ has a period of 12 h and three antinodes, one at the equator and two at middle latitudes. Since the input of solar heat is greatest at or near the equator (depending on the season), modes with an equatorial antinode are generated in preference to those with an equatorial node. Solutions such as those in Fig. 6.11 are called Hough functions.

Since the atmosphere is three dimensional the behaviour of the tidal modes with height must also be taken into account. It is possible for a tidal oscillation

(a)

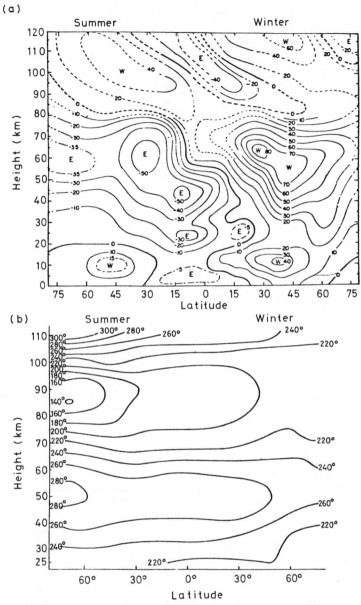

Figure 6.10 Height-latitude cross-sections of (a) zonal wind, and (b) temperature, in stratosphere and mesosphere. (From COSPAR International Reference Atmosphere, 1972.)

to maintain the same phase at all heights, but such a mode is evanescent (Section 2.4.1) and does not propagate. In propagating modes, such as those in Fig. 6.11, the phase varies with height and a vertical wavelength may be defined — that distance over which the phase changes by 360°. If the vertical wavelength is comparable with, or smaller than, the depth of the atmosphere in which most of the

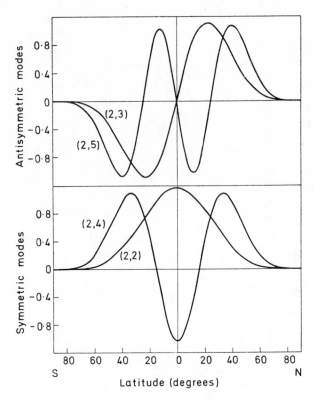

Figure 6.11   Hough functions for the two antisymmetric and symmetric modes of lowest order for the semidiurnal solar tide. (After J.V. Evans, *Proceedings of International Symposium on Solar-Terrestrial Physics,* Boulder, Colorado, 1976.)

heating occurs, that wave will suffer self-interference and as a result will be weak. In the stratosphere and the mesosphere the 24 h tide has a small vertical wavelength and therefore is not the main component; in those regions the 12 h tide dominates.

A propagating tide, like other upward propagating waves, increases in amplitude with height. This happens because, in the absence of dissipation, the energy density of the wave must be constant and independent of height. Thus if ρ is the air density and U is the amplitude of the air velocity associated with the tidal motion, $\rho U^2/2$ is constant and, thus $U \propto \rho^{1/2}$. At ground level the fractional pressure variation is about 10^{-3} and the tidal wind is some 0.05 m s^{-1}. At 100 km the pressure variation is 10% and the corresponding air speed 50 m s^{-1}. The tidal winds are even stronger at 300 km (in the F region).

The best information about tidal motions in the mesosphere comes from the radar meteor technique. Fig. 6.12 clearly shows the semi-diurnal tides at heights between 85 and 100 km, superimposed on a west-to-east prevailing wind. The tidal amplitude and the mean wind are of similar magnitude.

In the thermosphere the 24 h tide is dominant, and is driven largely by the heating due to the absorption of solar EUV at ionospheric levels. The mesospheric

122

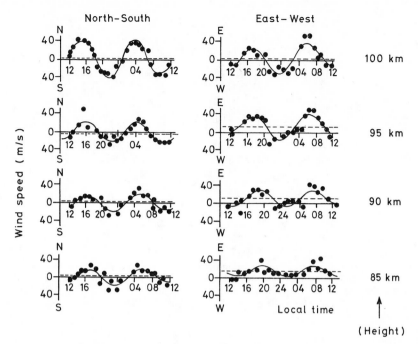

Figure 6.12 Meridional and zonal winds in the mesosphere, according to meteor radar observations. (After J.S. Greenhow and E.L. Neufeld, *Phil. Mag.* **1**, 1157, 1956.)

tide is governed by the need to balance the tide-raising force and the Coriolis force. (Essentially, this tide is a solution of

$$dU/dt = -2\Omega \times U - \nabla\psi \qquad (6.19)$$

where **U** is the air velocity, Ω the Earth's angular rotation, and $\nabla\psi$ represents the force, whatever its nature, that produces the tide.) In the ionosphere another force becomes significant. Due to the collisions between neutral and ionized particles, in a situation where the ions cannot cross geomagnetic field lines, the neutral air experiences a deceleration known as *ion drag*. (This is discussed further in Section 6.4.4.) The ion drag exceeds the Coriolis force in the ionospheric F region, and the tidal equation becomes

$$dU/dt = -\nu_{ni}(U - v_i) - \nabla\psi \qquad (6.20)$$

where ν_{ni} is the rate of collision of a neutral particle with ions and v_i is the ion velocity. Equation 6.20 neglects losses due to viscosity, and in practice these provide another decelerating force. As in the atmosphere very close to the ground, the effect of a large drag force is to make the airflow almost directly down the pressure gradient instead of along the isobars.

The pressure gradients in the thermosphere are related to the temperature distribution, shown in Fig. 6.13. The maximum and minimum occur over the equator, respectively near 1500 and 0400 h LT. The wind computed from this temperature distribution is shown in Fig. 6.14. The wind blows across from the day to the night side of the Earth directly over the poles at high latitude and with

123

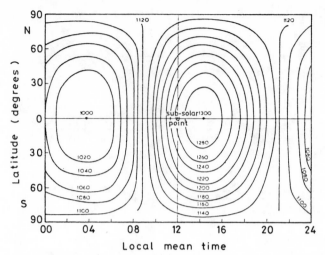

Figure 6.13 Global temperature distribution in the thermosphere (°K) at the equinox in conditions of medium solar activity. No account is taken of high-latitude heat sources due to energetic particles. (After H. Kohl and J.W. King, *J. atmos. terr. Phys.* **29**, 1045, 1967, by permission of Pergamon Press, data from L.G. Jacchia, *Smithsonian Contrib. Astrophys.* **8**, 215, 1965.)

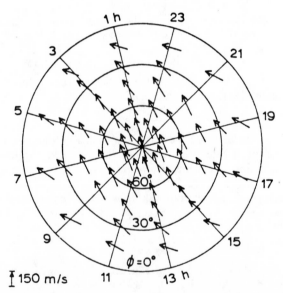

Figure 6.14 Winds at 300 km computed from the temperature distribution of Fig. 6.13. (From H. Kohl and J.W. King, *J. atmos. terr. Phys.* **29**, 1045, 1967, by permission of Pergamon Press.)

westward and eastward components in the morning and evening sectors at middle and low latitudes. The results in Fig. 6.14 are for 300 km, near the maximum of the F region. In theory the wind speeds should be even greater at higher levels if viscosity is not too important, though some observations do not confirm this.

Since the presence of the ionosphere modifies the neutral wind in the thermosphere it is not surprising that the wind has substantial effects on the ionosphere at middle and high latitudes. These points were discussed at some length in Sections 5.2.4 and 5.4.2.

6.4 Electric currents and plasma drifts

6.4.1 Generation

The basic mechanism for generating electric fields and currents in the upper atmosphere is the dynamo action of the horizontal wind system. The neutral molecules collide with ions and electrons, which thereby experience a force in the direction of the neutral wind. The response of the ionized components in the presence of the geomagnetic field was discussed in Section 4.4. In the dynamo region the ion gyrofrequency is smaller than the collision frequency while the electron gyrofrequency is larger. Consequently (Table 4.3) the ions are carried along with the wind but the electrons move across the wind. This relative movement between the ions and the electrons constitutes an electric current, most of the current being carried by the ions. The separation of particles of opposite sign creates horizontal electric fields, and these in turn affect the ion and electron motions. In general the current $\mathbf{J}$, the electric field $\mathbf{E}$, the wind velocity $\mathbf{v}$, and the magnetic induction $\mathbf{B}$, are related by

$$\mathbf{J} = \hat{\sigma}\,(\mathbf{v} \times \mathbf{B} + \mathbf{E}) \tag{6.21}$$

where $\hat{\sigma}$ is the tensor conductivity (Equation 4.76). Equation 6.21 reduces to Ohm's law if $\mathbf{v} = 0$. The first term represents the induced field due to the motion of ions across the magnetic field. If some of the quantities in the equation are known it is possible, in principle, to solve for others. An additional piece of information is that the current must be continuous from place to place (formally, div $\mathbf{J} = 0$), though in another sense this is a complication since it means that the complete global distribution has to be solved for.

There are two approaches to this. $\mathbf{v}$ and $\hat{\sigma}$ are not well known from observations, whereas comprehensive data on $\mathbf{J}$ are available from ground-based magnetometers. A distribution of $\mathbf{J}$ may therefore be taken as a starting point for computing the winds. Alternatively, the computed wind pattern, as seen in Fig. 6.14, may be used to work out the currents and electric fields, the first being verified against the observed current system.

6.4.2 S_q current system

The S_q current system is the pattern related to the solar day (S) in conditions of geomagnetic quiet (q). Some S_q magnetic variations are shown in Fig. 6.15, and distributions of wind and electric field deduced from S_q magnetic variations are given in Fig. 6.16. The winds amount to tens of metres per second, and the fields are a few mV m^{-1}. Both quantities may be expected to vary with height, and the results of Fig. 6.16 are 'effective' distributions only. Naturally, there is a good measure of agreement between this wind distribution and that deduced from tidal theory (Fig. 6.14). According to that model the wind over the polar caps is 400 m s^{-1} at 300 km altitude and 500 m s^{-1} at 700 km. The electric fields at

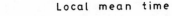

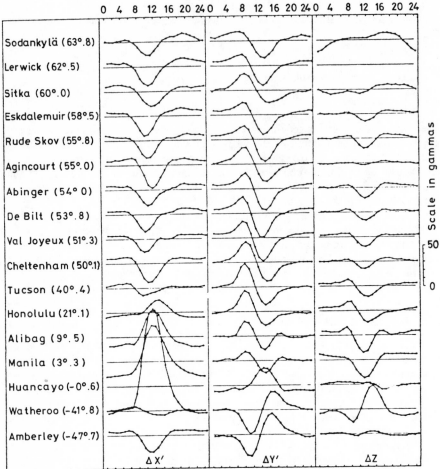

Figure 6.15 S_q magnetic variations at several geomagnetic latitudes under equinoctial conditions. $\Delta X'$, $\Delta Y'$ and ΔZ are the three normal components of the magnetogram. Note the large excursion of $\Delta X'$ at the geomagnetic equator, due to the equatorial electrojet. (From E.H. Vestine, *Physics of the Upper Atmosphere* (Ed. Ratcliffe) Academic Press, 1960.)

these levels will be $10-15$ mV m^{-1}, directed along the dawn-to-dusk meridian. These values are larger than those deduced from the S_q analysis, which will be influenced more by E-region conditions.

6.4.3 F-region drifts

The longitudinal conductivity, σ_0, which governs currents along the geomagnetic field lines, becomes large in the F region (Fig. 4.17), and is larger than the transverse conductivities, σ_1 and σ_2, in both the E and the F regions. If an electric

(a)

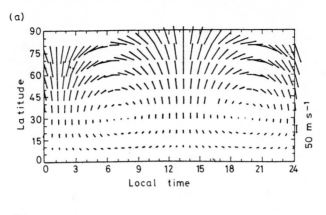

(b)

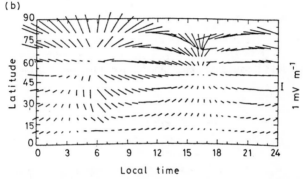

Figure 6.16 Vectors of (a) wind and (b) electric field, deduced from S_q magnetic variations. (After S. Matsushita, *Rad. Sci.* **4**, 771, 1969, copyrighted by American Geophysical Union.)

circuit is envisaged between the E and the F regions, the dynamo being in the E region, it can be seen that the potential drop along the field lines will be small because of their high conductivity. The electric-field pattern of the E region will therefore be reproduced in the F region, where they will drive plasma drifts as discussed in Section 4.4.3. If the E region is a dynamo, the F region is a motor.

F-region drifts can be measured by the spaced-receiver technique, by the drift of barium clouds, and also by incoherent scatter radars (Section 3.7.4). Measurements show the equatorial drifts to be eastward at night and westward by day, the amplitude of the oscillation being between 50 m s^{-1} and about 200 m s^{-1} according to various observations. The amplitude of the east—west component decreases with increasing latitude, while the north—south component shows a 12 h period during daylight hours. However, more measurements are needed before the total picture will be clear. Such observations are important as a test of tidal theory and to indicate which of the various possible modes is being generated in fact. While it appears possible to account fairly well for the day-time drifts on the basis of electric fields communicated from the dynamo region, as discussed above, the night observations agree less well, suggesting that electric fields generated in the F region itself may be important during the night. At high latitude the drifts arise from magnetospheric effects rather than from low altitude winds. This question is discussed in Section 8.1.

6.4.4 Ion drag and air drag

Section 2.2.3 showed how collisions between different species of a gas transfer momentum between them and so lead to a drag force of one on the other. The force on a unit volume of particles of type 1, whose number density is n_1, due to collisions with particles of type 2 is

$$F = n_1 m v_{12} (v_2 - v_1) \qquad (6.22)$$

where v_1 and v_2 are the velocities of types 1 and 2 (assumed to be in the same direction), v_{12} is the collision frequency of a given particle of type 1 with any or all particles of type 2, and m is the mass of a particle (assumed here to be the same for both types).

The effect of ions on the neutral air is called *ion drag*, and the ion drag force per unit volume is

$$F_{ni} = n_n m v_{ni} (v_i - u) \qquad (6.23)$$

where v_i and u are respectively the ion and neutral-air velocities, and the subscripts n and i refer to the neutrals and the ions. The converse effect of neutral particles on the ions is called *air drag*, and that force per unit volume is

$$F_{in} = n_i m v_{in} (u - v_i) \qquad (6.24)$$

Since these forces must be equal and opposite

$$n_n v_{ni} = n_i v_{in}$$

Below the E region the neutral air density and v_{in} are both relatively large, and the ions are carried along in the neutral wind. In the F region the drag forces are smaller, though still significant and with a difference between day and night. The neutral wind effectively blows the ionization along the geomagnetic field (Section 5.2.4) and thereby alters the height of the F-layer maximum by some tens of kilometres. In the thermospheric tide the ion drag force exceeds the Coriolis force, so that the resulting air motion is across the isobars instead of along them (Section 6.3.3). The tidal wind is larger by night than by day because the ion density is smaller at night.

Ion drag is also capable of inducing motion in the neutral air, by communicating to it plasma drift due to electric fields. However, since the ion density is much smaller than the air density, it takes much longer for the ions to set the air in motion than it would for the air to move the ions. The time constant for the acceleration of neutrals by collision with ions is approximately $1/v_{ni} = n_n/n_i v_{in}$. At 300 km the respective values of n_n, n_i, v_{in} and v_{ni} are

$$9.8 \times 10^{20} \text{ cm}^{-3} \quad 10^6 \text{ cm}^{-3} \quad 0.6 \text{ s}^{-1} \quad \text{and} \quad 6.5 \times 10^{-5} \text{ s}^{-1} \qquad \text{by day}$$

and

$$6.6 \times 10^{20} \text{ cm}^{-3} \quad 3.5 \times 10^5 \text{ cm}^{-3} \quad 0.4 \text{ s}^{-1} \quad \text{and} \quad 2 \times 10^{-5} \text{ s}^{-1} \quad \text{by night}$$

The time constant $(1/v_{ni})$ may be as short as 0.5 h per day, but is several hours at night. The day-time plasma drift can therefore be transferred efficiently to the neutral air, but only in the day-time. This effect is likely to be important in polar regions where the electric fields are greatest.

6.5 Propagating atmospheric waves

In addition to the small-scale irregularities, which produce radio scintillation and are associated with a turbulent region of the ionosphere that is either blowing along with the neutral air or moving under electric fields, there is an important class of propagating waves of large scale, whose nature is different. These waves have wavelengths measured in tens and hundreds of kilometres, and periods of minutes and tens of minutes. They travel at speeds related to the speed of sound at the relevant level of the atmosphere, and belong to the family of acoustic–gravity waves.

6.5.1 Theory of acoustic–gravity waves

In an acoustic wave the restoring force is the change of pressure when the medium is compressed. In a gravity wave, the simplest example of which occurs on the surface of an ocean, the restoring force is the action of gravity on the displaced body of fluid. Ocean waves exist at the surface because of the abrupt change of density, but since the atmosphere is stratified, with density a function of altitude, gravity waves may exist within the atmosphere. In general, both the compressional and the gravitational forces should be taken into account, and all the waves properly belong to the acoustic–gravity class. At the high and low frequency ends of the spectrum, however, approximations to pure acoustic and to pure gravity waves can be taken. (It is usual to neglect effects of the Earth's rotation and curvature when dealing with acoustic–gravity waves, though they have to be included for planetary waves and tides, whose wavelength is even greater.)

Whereas most people are familiar with the idea of an acoustic wave, the gravity wave is frequently considered more baffling since it has rather unfamiliar dispersion characteristics. For the simplest solution, which assumes:

small variations of pressure and density;
no loss of energy, i.e. zero viscosity; and
a two-dimensional plane-wave solution of form $\exp j(\omega t - k_x x - k_z z)$, where k_x and k_y are wavenumbers ($k = 2\pi/\lambda$) in the horizontal and vertical directions respectively,

the dispersion relation is

$$\omega^4 - \omega^2 c^2 (k_x^2 + k_z^2) + (\gamma - 1)g^2 k_x^2 + \omega^2 \gamma^2 g^2 /4c^2 = 0 \qquad (6.25)$$

Here, ω is the angular frequency, c the speed of sound, γ the ratio of specific heats of the atmospheric gas and g the acceleration due to gravity. Given the atmospheric properties, Equation 6.25 expresses the permissible relationships between the frequency of the wave and its horizontal and vertical wavelengths. As may be seen from Fig. 6.17, these quantities define the phase behaviour of the wave fully. The phase propagation occurs at elevation angle

$$\theta = \tan^{-1}(\lambda_x/\lambda_z) = \tan^{-1}(k_z/k_x) \qquad (6.26)$$

the wavelength (λ) is given by

$$1/\lambda^2 = 1/\lambda_x^2 + 1/\lambda_z^2 = (k_x^2 + k_z^2)/(2\pi)^2$$

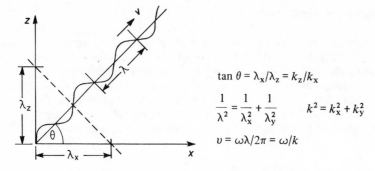

$$\tan \theta = \lambda_x/\lambda_z = k_z/k_x$$

$$\frac{1}{\lambda^2} = \frac{1}{\lambda_x^2} + \frac{1}{\lambda_y^2} \qquad k^2 = k_x^2 + k_y^2$$

$$v = \omega\lambda/2\pi = \omega/k$$

Figure 6.17   Relationships between wavelength, velocity and propagation angle.

and therefore

$$\lambda = 2\pi/(k_x^2 + k_z^2)^{1/2} \tag{6.27}$$

and the phase velocities in directions θ, the horizontal, and the vertical, are respectively

$$\left. \begin{aligned} v &= \omega\lambda/2\pi = \omega/k \\ v_x &= \omega/k_x \\ \text{and} \\ v_z &= \omega/k_z \end{aligned} \right\} \tag{6.28}$$

If $g = 0$, Equation 6.25 becomes

$$c = \frac{\omega}{(k_x^2 + k_z^2)^{1/2}} = \frac{\omega\lambda}{2\pi} \tag{6.29}$$

which is the dispersion relation for a sound wave, in which the phase velocity is independent of direction.

Substituting

$$\omega_a = \gamma g/2c \tag{6.30}$$

and

$$\omega_B = (\gamma - 1)^{1/2} g/c \tag{6.31}$$

into Equation 6.25, and rearranging it, gives

$$k_z^2 = (1 - \omega_a^2/\omega^2)\omega^2/c^2 - k_x^2(1 - \omega_B^2/\omega^2) \tag{6.32}$$

Then, putting $\omega^2 \ll c^2$ (thereby removing the effect of compressibility)

$$k_x^2 = k_z^2(\omega_B^2/\omega^2 - 1) \tag{6.33}$$

which represents a gravity wave. The wave can only propagate if k_x, k_z are real and positive, requiring that $\omega < \omega_B$. From Equation 6.26 it is seen that the angle of propagation is

$$\theta = \tan^{-1}(\omega_B^2/\omega^2 - 1)^{1/2} \tag{6.34}$$

In fact the negative square root should be taken, since the phase propagation is

130

downwards (a point that does not come out of our simplified discussion). In an acoustic wave the velocity is determined by the nature of the medium, so that selection of a frequency fixes the wavelength though not the direction of propagation. Things are different in a gravity wave. Here, selection of the frequency fixes the propagation angle (Equation 6.34) but neither the velocity nor the wavelength. However, v, ω and λ are related through Equations 6.28.

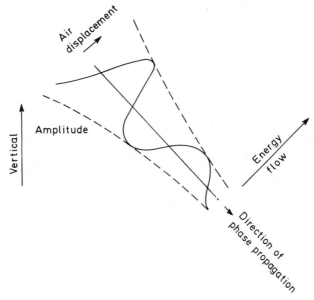

Figure 6.18   A simple gravity wave.

Fig. 6.18 illustrates a simple gravity wave. At the lowest frequencies the air particles move perpendicular to the direction of phase propagation. The energy also travels at right angles to the phase velocity, and the amplitude of the wave increases with altitude (as the square root of the air density, to maintain the energy flux at a constant value). As detected by radio methods the wave in Fig. 6.18 would appear to be moving downwards and from left to right. The source of energy, however, would be on the left and below.

Equation 6.25 includes the full range of acoustic–gravity waves appropriate to our assumptions. In the general case the air particles move in elliptical trajectories, combining the longitudinal motion of an acoustic wave with the transverse motion of a gravity wave. As k_z and k_x must be real, propagating waves are only possible if $\omega > \omega_a$ or $\omega < \omega_B$, these being known as the acoustic and gravity ranges respectively. If $\omega_B < \omega_a$ waves between these frequencies are evanescent. The parameters ω_a and ω_B have physical significance as the *acoustic resonance* and the *buoyancy* or *Brunt-Väisala resonance* respectively. The first is the resonant frequency of the the whole atmosphere in the acoustic mode, when the restoring force is due to compression. The second is the natural resonance of a displaced parcel of air within the atmosphere, the restoring force being buoyancy.

To see the form of the relation for ω_a consider that for a closed organ pipe of length l the resonant wavelength is $\lambda = 4l$, and the resonant frequency is $\omega = \pi c/2l$, where c is the speed of sound. From Equation 4.7, an exponential atmosphere with scale height H would, if brought throughout to the pressure at the base, occupy a thickness H. Substituting H for l, therefore,

$$\omega_a = \pi c/2H \tag{6.35}$$

If the pressure and density of the atmosphere are P and ρ, the speed of sound is $c = (\partial P/\partial \rho)^{1/2} = (\gamma RT/M)^{1/2}$, and the scale height $H = RT/Mg$. Therefore

$$H = c^2/g\gamma$$

and $$\tag{6.36}$$

$$\omega_a = \pi g\gamma/2c$$

An exact treatment leads to Equation 6.30, which differs by a factor of π.

The formula for ω_B is proved as follows.

Consider that a parcel of air, initially in equilibrium with its surroundings at a level where the pressure and density are P_0 and ρ_0, is lifted adiabatically in altitude by Δz to a level where the surrounding pressure and density are $P = P_0 + \Delta P$ and $\rho = \rho_0 + \Delta\rho$. The parcel itself then assumes new values of pressure and density, P' and ρ', and volume V'. The buoyancy force (upward) is $V'g(\rho - \rho') = mg(\rho - \rho')/\rho'$, where m is the mass of the air parcel. Thus the upward acceleration is

$$\ddot{z} = g(\rho - \rho')/\rho' \approx g(\rho - \rho')/\rho_0 \qquad \text{since } (\rho - \rho') \ll \rho_0$$

But

$$\frac{\rho}{\rho_0} = 1 + \frac{1}{\rho_0}\frac{\partial\rho}{\partial z}\cdot \Delta z = 1 - \frac{\Delta z}{H} = 1 - \frac{g\gamma}{c^2}\cdot\Delta z$$

since the surrounding air is in hydrostatic equilibrium with scale height H, and using Equation 6.36; further, since the parcel expands adiabatically,

$$\frac{\rho'}{\rho_0} = 1 + \frac{1}{\rho_0}\cdot\frac{\partial\rho}{\partial P}\,(\text{adiabatic})\cdot\Delta P = 1 + \frac{1}{\rho_0}\cdot\frac{\Delta P}{c^2}$$

Thus

$$\ddot{z} = -g\left(\frac{g\gamma}{c^2}\cdot\Delta z + \frac{1}{\rho_0}\,\frac{\Delta P}{c^2}\right)$$

$$= -g\left(\frac{g\gamma}{c^2} - \frac{g}{c^2}\right)\Delta z \qquad \text{since } \partial P/\partial z = -g\rho$$

$$= -\frac{(\gamma - 1)g^2}{c^2}\cdot\Delta z$$

Since $\ddot{z} \propto -\Delta z$, the motion is simple harmonic and the angular frequency is

$$\omega_B = (\gamma - 1)^{1/2}g/c \tag{6.31}$$

Some values for the acoustic and Brunt frequencies in the atmosphere are shown in Fig. 6.19. At ionospheric levels waves with periods longer than 10–15 min are likely to be gravity waves, and those with periods of only a few minutes are probably acoustic. Acoustic–gravity waves are primarily a phenomenon of the neutral air, but the motions can be communicated to the ionization through collisions (air drag). The response of the ionization depends on altitude: at the lowest levels (collision frequency $\gg$ gyrofrequency) it may be assumed that the ionization will move with the neutral air, but at the higher levels (collision frequency $\ll$ gyrofrequency) ion motion across the geomagnetic field is inhibited

132

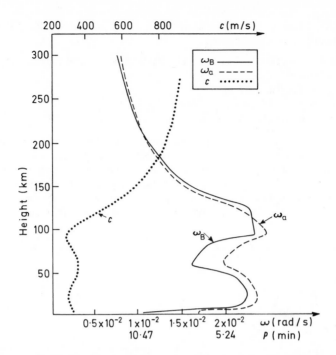

Figure 6.19 Height profiles of the speed of sound (c), the Brunt frequency (ω_B) and the acoustic cut-off frequency (ω_a) in a realistic atmosphere. (After I. Tolstoy and P. Pan, *J. Atmos. Sci.* **27**, 31, 1970.)

and the ionization will respond only to the component of the wave that acts parallel to the field. The ionospheric response is therefore a biased one that does not simply reproduce the motion of the neutral wave, and the interpretation of observations is thereby more difficult.

In addition to the question of the ionospheric response to an acoustic–gravity wave, there are several other complicating factors in the real atmosphere that have been omitted from our first-order discussion. The main ones are:

(i) energy dissipation by air viscosity, particularly at the higher levels where the waves, their amplitude growing with increasing altitude, no longer fit the assumption of relatively small variations;
(ii) reflection at changes of refractive index due to the vertical temperature structure of the atmosphere, which can produce ducted waves;
(iii) velocity changes due to steady winds, which can deviate the waves and again lead to ducting.

Such problems can make it more difficult to identify the sources of waves observed in the atmosphere.

6.5.2 Observations of gravity waves

Despite the bias of the ionospheric response, the most comprehensive information about gravity waves in the upper atmosphere has come from ionospheric observations. The principal methods are:

(a) ionosondes (Section 3.7.1) and, closely related, continuous observations of virtual height at a fixed frequency;
(b) HF Doppler (Section 3.7.1);
(c) electron content measurements (Section 3.7.3);
(d) incoherent scatter (Section 3.7.4).

As seen by these techniques the phenomenon is usually called a *travelling ionospheric disturbance* or TID. These were observed for many years before C. O. Hines explained them as gravity waves in 1961. To the observer they are characterized by a wave-like perturbation that appears at slightly different times at spaced sites, with the obvious implication of a propagating wave motion. The waves are frequently observed at middle and high latitudes and, although individual obser-

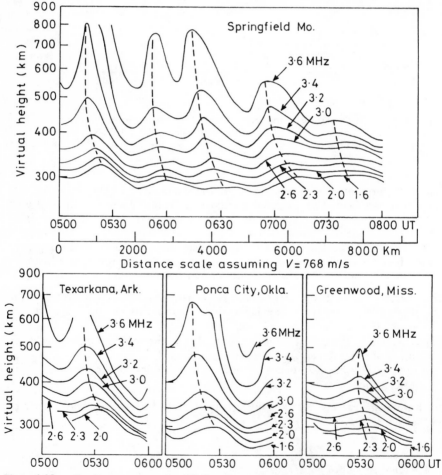

Figure 6.20   Gravity waves seen by HF Doppler at sites several hundred kilometres apart. (After T.M. Georges, *J. atmos. terr. Phys.* **30**, 735, 1968, by permission of Pergamon Press.)

134

vations vary widely as to period, speed and direction, it does appear possible to put them into certain categories.

One group, with periods of 30 min or longer, shows horizontal wavelengths of a few thousand kilometres and horizontal phase velocities in the range 400–700 m s^{-1}. The direction is usually from pole to equator, and the source is thought to be auroral and geomagnetic activity at high latitudes. Theoretical work indicates that the auroral electrojet (Sections 8.2.2 and 8.3.3) is a possible source. For a long-period wave, $\omega \gg \omega_B$, and Equation 6.34 indicates that such a wave propagates almost vertically, the particle motions being nearly horizontal. Thus the horizontal phase velocity is well above the velocity in the direction of propagation. Vertical velocity components can be estimated by HF Doppler if several frequencies are used simultaneously, since higher radio frequencies are reflected from greater altitudes. Fig. 6.20 shows an example. Reflection heights are determined from ionosonde data. Incoherent scatter, though usually confined to a single site, readily provides measurements as a function of height. The technique determines various ion properties such as temperature and velocity that are concerned with properties of the neutral air. Gravity-wave velocities and wavelengths in the vertical direction are readily measured. For a full diagnosis it is best to combine several techniques, so that information about both horizontal and vertical behaviour is obtained on the same wave train.

At rather shorter periods (20–30 min) are waves with horizontal sizes of 100–200 km and speeds of 100–200 m s^{-1}. These are usually known as the *medium-scale* disturbances. Their wavefronts are tilted at about 45° to the vertical, and the coherence distance in the horizontal plane is smaller than for the large-scale events, being a few hundred kilometres against thousands. It is thought that some medium-scale disturbances are generated in meteorological phenomena such as thunderstorms and jetstreams.

Direct evidence for gravity waves in the neutral air is more restricted than the indirect evidence from observations on the ionized medium, but what there is is convincing. The waves are seen in the distortion of meteor trails and vapour trails (Section 6.3.1). Photographs of noctilucent clouds, which occur at 80–90 km, often show a wave-like structure looking very much like waves on the ocean, and which are almost certainly due to gravity waves in the upper atmosphere.

6.5.3 Acoustic waves in the upper atmosphere

Some HF Doppler observations of the ionosphere have shown periodicities of a few minutes, and these are thought due to acoustic waves. The waves propagate upwards at speeds of 500–600 m s^{-1}, and they appear in limited areas not too far from severe weather in the troposphere. Fig. 6.21 shows an example, associated with various severe weather manifestations. Between 0500 and 0700 h UT the period is 3–5 min, which is typical of such observations. The longest period is probably determined by the value of the acoustic resonance in the stratosphere and mesosphere (see Fig. 6.19), while the shortest periods generated tend to be absorbed in the atmosphere and so removed from the spectrum.

The foregoing discussion mainly concerns waves detected in the upper atmosphere that were generated at some lower level, but this rule is not invariable. An exception is waves in the so-called *infrasonic band* of a few seconds to a minute in period, which can be detected at the ground by pressure sensors (microbarographs). One interesting class of infrasonic event is related to the presence of auroral

135

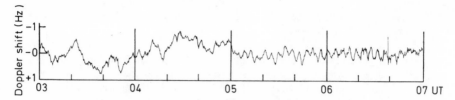

Figure 6.21 HF Doppler record at 4 MHz, 14 June 1967, over Boulder, Colorado, during thunderstorms, hail and tornadoes in the neighbouring state of Nebraska. The rapid fluctuations are due to acoustic waves in the ionosphere. (After A.G. Jean and G.M. Lerfald, personal communication, 1969.)

activity and appears to propagate from the auroral zones. These events are thought to be generated in auroral electrojets through heating and by the electrodynamic force associated with the current. The source is therefore located near altitudes of 110 km.

6.6 Further reading

Chapman, S. and Lindzen, R. S. (1969) *Atmospheric Tides* (Reidel).
Rishbeth, H. (1973) Electromagnetic transport processes in the ionosphere, in *Physics and Chemistry of Upper Atmospheres* (Ed. McCormac) (Reidel).
Akasofu, S.-I. and Chapman, S. (1972) *Solar–Terrestrial Physics* (Oxford) Chapter 4.

Briggs, B. H. (1975) Ionospheric irregularities and radio scintillations, *Contemp. Phys.* **16**, 469.
Aarons, J., Whitney, H. E. and Allen, R. S. (1971) Global morphology of ionospheric scintillations, *Proc. Inst. elect. electronics Engr.* **59**, 159.
Rishbeth, H. (1974) Ionospheric dynamics 1945–1970, *J. atmos. terr. Phys.* **36**, 2309.
Kent, G. S. (1970) Measurement of ionospheric movements, *Rev. Geophys. Space Phys.* **8**, 229.
Lindzen, R. S. and Chapman, S. (1969–70) Atmospheric tides, *Space Sci. Rev.* **10**, 3.
Yeh, K. C. and Liu, C. H. (1976) Motions in the ionosphere, *Sci. Prog.* **63**, 111.
Murata, H. (1974) Wave motions in the atmosphere and related ionospheric phenomena, *Space Sci. Rev.* **16**, 461.
Evans, J. V. (1972) Ionospheric movements measured by incoherent scatter: a review, *J. atmos. terr. Phys.* **34**, 175.
Pudovkin, M. I. (1974) Electric fields and currents in the ionosphere, *Space Sci. Rev.* **16**, 727.
Beer, T. (1972) Atmospheric waves and the ionosphere, *Contemp. Phys.* **13**, 247.
Wilson, C. R. (1975) Infrasonic wave generation by aurora, *J. atmos. terr. Phys.* **37**, 973.
Ratcliffe, J. A. (1956) Some aspects of diffraction theory and their application to the ionosphere, *Rep. Prog. Phys.* **19**, 188.

7
Structure of the Magnetosphere

7.1 Introduction

The magnetosphere is the region where the geomagnetic field dominates other factors in controlling the motion of charged particles. A strict definition is that the energy density of the magnetic field exceeds the energy density of the plasma:

$$\frac{B^2}{2\mu_0} > nkT$$

in MKS units, where B is the magnetic induction, k is Boltzmann's constant, and the plasma has n particles m^{-3} at temperature T. Common usage is somewhat flexible, however. The term *ionosphere* is often used for the ionized region up to 1000 km although the above definition places the topside of the ionosphere within the magnetosphere.

Although there is no physical separation between ionospheric and magnetospheric regions, there are differences of approach and emphasis. Knowledge of the magnetosphere has come mainly from satellite techniques, which map large regions along the orbit and so emphasize the global scale initially. On the other hand, the ionosphere had been studied extensively before the satellite era, by ground-based methods which tend to produce much information about a localized geographical region; the synthesis to a global picture followed subsequently. Of course there are some phenomena (perhaps more than we realize) for which all the plasma must be considered, perhaps with the neutral air as well. The general problem of atmospheric electric fields and currents is an example.

A second characteristic of the magnetosphere is that the difference between undisturbed and disturbed conditions, a distinction that is often useful for the ionosphere, becomes less clear. The magnetosphere exists in the form we observe because of emissions from the Sun, and it responds readily and rapidly to changes in solar activity. Since the activity of the Sun is continually changing, the magnetosphere experiences some degree of disturbance at all times so that terms like 'quiet' and 'disturbed' become merely relative.

The present state of knowledge of the magnetosphere is that, as a result of satellite-borne observations over the last 10–15 years, the basic exploration is well advanced. However, a great deal remains to be done in discovering and understanding the physical processes that control the observed phenomena.

7.2 The Sun as a source of radiation

The Sun is in many respects an average star. It emits 4×10^{33} erg s^{-1} of electromagnetic radiation, irradiating the Earth with about 2 cal cm^{-2} min^{-1}, about half of which reaches the surface. The visible solar surface, the photosphere, approximates to a black body at a temperature of 6400 °K. Above the photosphere are transparent layers: the chromosphere, visible during eclipses, extends some 2000 km above the photosphere and has a temperature up to 50 000 °K; beyond that the corona at 1.5×10^6 °K is observable for more than 10^6 km but in fact has no apparent termination. At the orbit of the Earth the coronal temperature is about 3×10^5 °K. The Sun emits radio waves, X-rays, and energetic particles in addition to visible light.

The Sun rotates with a period that increases with latitude from 25.4 days at the equator to 33 days at latitude 75°. The level of the emissions received at the Earth is modulated by the solar rotation, and for geophysical purposes the rotation period is often taken as 27 days. Apart from the effect of rotation, the intensity of the solar emissions varies with time. Several methods of quantifying the level of solar activity have come into general use, in particular

sunspot counts
solar flare observations
reception of solar radio emissions

7.2.1 Sunspot numbers and the 11-year cycle

Sunspots are observed in 'white light', i.e. with a telescope not using a coloured filter, as darkened areas of the photosphere. They tend to appear in groups and mainly between solar latitudes 5° and 30°. The *sunspot number* is defined as

$$R = k(f + 10g)$$

where f is the total number of spots visible to the observer, g is the number of disturbed regions, i.e. either single spots or groups of spots, and k is a constant for the observatory, related to the sensitivity of the observing equipment. The occurrence of sunspots goes through an 11-year cycle, the *sunspot cycle*. Some recent cycles are shown in Fig. 7.1. In addition to the 11-year variation, sunspot data, which go back over 250 years, show that not all cycles are equally intense, implying a long period modulation of the basic 11-year cycle. The most active cycle on record was cycle number 19, which occurred comparatively recently, the sunspot number reaching 200 during the early months of 1958. The more recent cycle (number 20) that maximized in late 1968 was little more than half as intense. At sunspot minimum, e.g. in 1954 and 1964, the sunspot number dips below 20 and sometimes below 10.

Individual sunspots and groups may both persist for several months and so reappear on successive solar rotations. Sunspot numbers as observed therefore show a strong 27-day periodicity, and for many purposes need to be smoothed to show up changes over the longer term. This may be done by averaging daily values over a month, or monthly values over a year. Recent studies have also indicated quasiperiodic variations within the 11-year cycle, the most marked having a period of about 13 months or 15 solar rotations.

7.2.2 Solar flares

The solar flare, which can be observed in white light as well as through filters, appears as a sudden brightening of a small area of the photosphere that remains

138

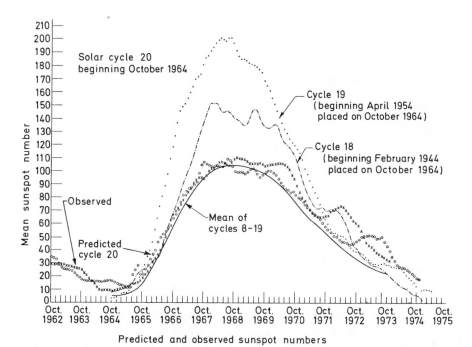

Predicted and observed sunspot numbers

Figure 7.1 Sunspot numbers for recent cycles. Cycles 18 and 19 (chain and dotted lines respectively) have been shifted in time to facilitate comparison. The open circles and crosses show predicted and observed numbers for cycle 20, which was of typical intensity. (From *Solar-Geophysical Data*, June 1974, World Data Center-A, Boulder, Colorado.)

visible for a time between several minutes and several hours. Flares are classified by the area of the brightening when the Sun is viewed through a filter on the $H\alpha$-line (6562 Å) on a scale 1 to 4. In addition, very small flares are designated S (for subflare). Flares tend to occur near sunspots and in the so-called 'plage' areas, which are bright regions seen in Calcium-K or Hydrogen-α light. The rate of flare occurrence is related to the sunspot number by the empirical formula

$$N_f = \alpha (\bar{R} - 10)$$

where N_f is the number of flares per solar rotation (27 days), $\bar{R}$ is the mean sunspot number, and α is a constant, value between 1.5 and 2.

As indices of solar activity, sunspot and flare numbers serve a similar function. In addition, the flare as an event is specially significant with respect to disturbances of the upper atmosphere because of the energy released at the time of the flare, which amounts to some 10^{32} erg. There are three kinds of solar emission which are particularly important in the upper atmosphere, and each has distinct consequences (Table 7.1). These consequences are discussed in other chapters. The solar flare is also important to the geophysicist because it appears to be a solar manifestation of processes, involving magnetic fields and plasmas, that are also important during a substorm in the terrestrial magnetosphere (see Chapter 8). It therefore seems likely that advances in the solution of either of these two problems will also help with the other. The phenomenon of the solar flare is described further in Section 10.1.

Table 7.1   Types of Solar Emission

Emission	Time of transit to Earth	Terrestrial consequences
X-rays	8.3 min	Sudden ionospheric disturbances
Protons, 1–100 MeV	A few hours	Solar proton event
Low energy plasma, velocity ~ 1000 km s^{-1}	1–2 days	Magnetic storm and aurora

7.2.3 Solar radio emissions

The Sun radiates a thermal radio emission at all times, but the level of emission is greatly increased during flares and when active regions are visible. The radio emission during a flare is complex in both time and spectrum, and the details are beyond our present scope. The total emission from the Sun at radio wavelengths is found to be a useful indicator of solar activity, particularly the intensity at a wavelength of 10 cm, and this is routinely reported in 'flux units' of 10^{-22} W m^{-2} Hz^{-1}. This quantity is generally known as the *10-cm flux*.

Scans of the solar disc with narrow-beam antennas are able to map the positions of active regions in some detail, though not, of course, with the resolution of optical methods. Fig. 7.2 shows an assembly of daily east–west scans in which the motion of emitting regions due to solar rotation can be clearly seen. Fig. 7.3 shows solar maps in terms of X-rays, sunspots, magnetic field, radio emissions at two wavelengths and calcium light. By and large the various techniques reveal similar active regions. An obvious advantage of non-optical methods is that they continue to give information from the ground even when the Sun is obscured by cloud.

7.3 The solar wind

7.3.1 History

In 1930, S. Chapman and V. C. A. Ferraro, wishing to explain the disturbance of terrestrial magnetometers after a large solar flare, suggested that a cloud of plasma was ejected from the Sun during a flare. The plasma cloud would reach the Earth about a day later, and there, unable to enter the geomagnetic field, it would be deflected around it. The pressure exerted by the plasma against the geomagnetic field would cause a compression of the field that could be detected by a magnetometer at the ground.

In the Chapman and Ferraro model the plasma clouds were released intermittently. From a study of the tails of comets in 1951, L. Biermann concluded that not all of them could be explained by radiation pressure alone. There must also be pressure from a stream of particles emitted from the Sun continuously. Fig. 7.4 shows a comet with two tails, one due to radiation pressure and the other to solar particles. Biermann deduced a velocity of 500 km s^{-1} for the particle stream.

Theoretical studies by Chapman and by E. N. Parker subsequently showed that, in contrast to the Earth's atmosphere, the solar corona is not in hydrostatic equilibrium, but is expanding continuously, with matter leaving the Sun and flowing out into space. Parker called this flow the *solar wind*.

140

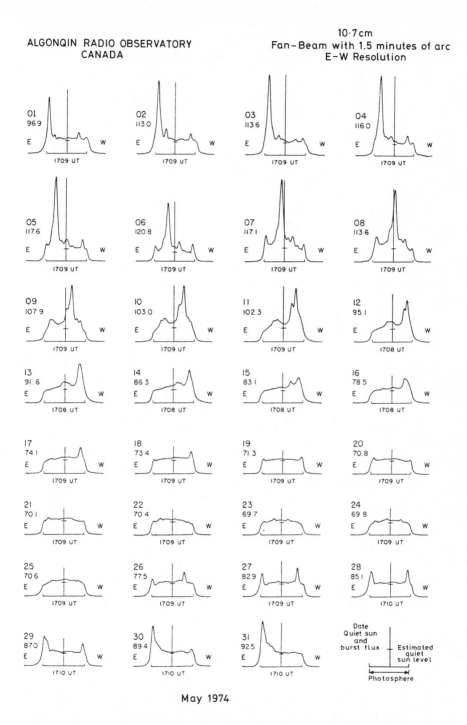

10·7 cm
Fan–Beam with 1.5 minutes of arc
E–W Resolution

May 1974

Figure 7.2 East–west scans of the solar disc at 10.7 cm. (From *Solar-Geophysical Data*, June 1974, World Data Center-A, Boulder, Colorado.)

141

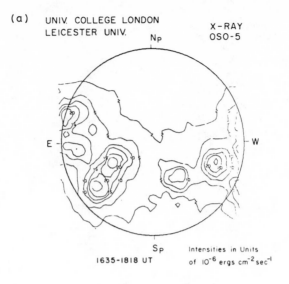

UNIV. COLLEGE LONDON
LEICESTER UNIV.

X-RAY
OSO-5

Np

E

W

Sp
1635-1818 UT

Intensities in Units
of 10^{-6} ergs cm^{-2}sec^{-1}

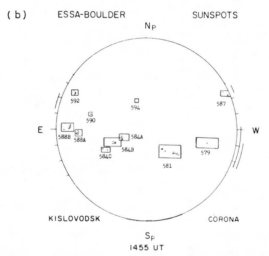

(b)   ESSA-BOULDER   SUNSPOTS

Np

E   W

592   594   587

590

588B 588A   584A

584C   584B   581   579

KISLOVODSK   CORONA

Sp
1455 UT

Figure 7.3 Comparable solar maps for 20 February 1970 in terms of (a) X-rays, (b) sun-spots, (c) magnetic fields, radio emission at (d) 9.1 and (e) 21 cm, and (f) calcium light. (From *Solar-Geophysical Data*, April 1970, World Data Center-A, Boulder, Colorado.)

7.3.2 Theory of the solar wind

From optical measurements it is possible to estimate the electron density in the solar corona at various distances from the Sun. If it is assumed that the corona is moving outward, the velocity can be calculated from the conservation of flux. That is, if n is the particle density at radius r, and v is the velocity, then nr^2v is constant.

142

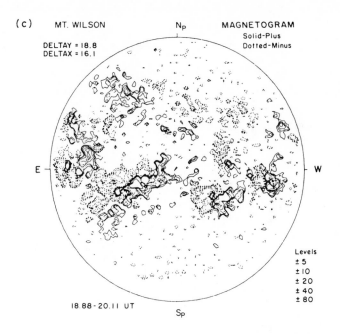

(c)   MT. WILSON N_P MAGNETOGRAM

DELTAY = 18.8   Solid-Plus
DELTAX = 16.1   Dotted-Minus

E   W

Levels
± 5
± 10
± 20
± 40
± 80

18.88 - 20.11 UT S_P

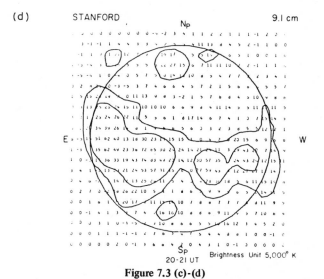

(d)   STANFORD   9.1 cm
                              N_P

E   W

S_P
20-21 UT   Brightness Unit 5,000° K

Figure 7.3 (c)-(d)

The outward acceleration can then be calculated, and compared with the gravitational acceleration towards the Sun. It is found that beyond 15 or 20 solar radii the convective acceleration, acting outward, exceeds the gravitational acceleration acting inward. Matter therefore leaves the Sun.

The simplest treatment assumes a solar corona with one type of particle and a constant temperature throughout. A steady outward flow is assumed, and solar rotation, magnetic fields and viscosity are neglected. By continuity

$$\rho v r^2 = \text{constant} \tag{7.1}$$

143

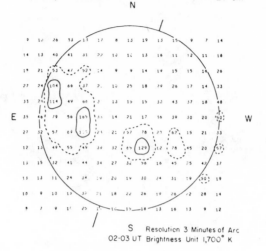

S Resolution 3 Minutes of Arc
02-03 UT Brightness Unit 1,700° K

(f) McMATH-HULBERT CALCIUM REPORT

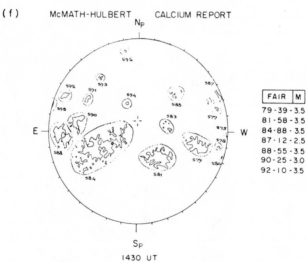

FAIR	M
79 - 39 - 3.5	
81 - 58 - 3.5	
84 - 88 - 3.5	
87 - 12 - 2.5	
88 - 55 - 3.5	
90 - 25 - 3.0	
92 - 10 - 3.5	

1430 UT

Figure 7.3 (e)-(f)

where ρ is the gas density, v is the velocity, and r is the radial distance from the Sun. By equating forces on unit volume of gas

$$\rho v \frac{dv}{dr} + \frac{dP}{dr} + \frac{GM_s\rho}{r^2} = 0 \qquad (7.2)$$

where P is the gas pressure, G is the gravitational constant, and M_s is the solar mass. Also

$$P = (kT/m)\,\rho \qquad (7.3)$$

where k is Boltzmann's constant, T is the absolute temperature, and m is the particle mass. Substituting from Equation 7.3, integrating Equation 7.2, and putting velocity and radial distance in dimensionless form by means of

$$U = \frac{v}{(2kT/m)^{1/2}} \qquad \Pi = r\,\frac{kT/m}{GM_s} \qquad \left(\sim \frac{r}{8R_s}\right)$$

144

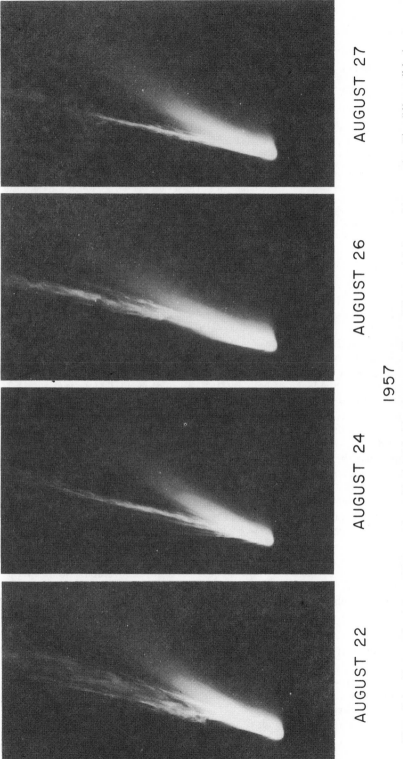

AUGUST 22   AUGUST 24   AUGUST 26   AUGUST 27

1957

Figure 7.4   Photograph of Comet Mrkos, taken with the Schmidt telescope at Mount Wilson and Palomar Observatories. The diffuse tail is due to radiation pressure and the long straight one is caused by the solar wind. (From Hale Observatories.)

Equations 7.1 and 7.2 become

$$\ln \rho = -\ln U - 2 \ln \Pi \tag{7.4}$$

$$U^2 - \ln U = 2 \ln \Pi + 1/\Pi + \text{constant} \tag{7.5}$$

The constant of integration is found as follows. Differentiation of Equation 7.5 gives

$$\left(2U - \frac{1}{U}\right) \frac{dU}{d\Pi} = \frac{1}{\Pi}\left(2 - \frac{1}{\Pi}\right) \tag{7.6}$$

The left-hand side of this equation is zero if $U = 2^{-1/2}$, and since $dU/d\Pi$ cannot be infinite then it follows that $\Pi = \frac{1}{2}$ at that point. These corresponding values of U and Π enable the constant in Equation 7.5 to be evaluated, and thus velocity may be found as a function of distance.

From such equations, Parker in 1958 deduced solar wind velocities at the Earth which ranged between 260 and 1160 km s^{-1} for coronal temperatures between 0.5×10^6 and 4×10^6 °K. The value $U = 2^{-1/2}$ corresponds to a velocity $v = (kT/m)^{1/2}$, which is the acoustic speed, and in the present model this occurs at $r = 4R_s$ where R_s is the solar radius. The corona streams out subsonically to this distance, but supersonically beyond it. Later treatments of the problem have included a varying temperature, viscosity, a magnetic field, and two kinds of particle, protons and electrons.

7.3.3 Observed properties of the solar wind

Space probes confirmed the existence of the solar wind in the early 1960s. Notably, Mariner 2, the first Venus probe, observed that the solar wind blew continuously during the 4 months of the flight.

Measurements of the positive ions using plasma probes show that the flux of particles of energy exceeding 25 eV is between 2×10^8 and 7×10^8 particles cm^{-2} s^{-1}. The relative abundance of helium (He^{2+}) to hydrogen (H$^+$) ions is about 4.5%, and heavier ions make up about 0.5%. The typical particle density is 5 protons cm^{-3} with, for bulk neutrality, a similar density of electrons. The mean mass of solar wind particles is therefore nearly half the mass of a proton, about 10^{-24} g. Fluctuations of density of an order of magnitude occur over times of minutes and hours, and imply that there are irregularities of 10^5 km and larger within the solar wind. The bulk velocity is usually between 200 and 700 km s^{-1} (Fig. 7.5) with a mode at 400 km s^{-1}. Superimposed on the bulk velocity are random components corresponding to a temperature of 10^5 °K, though the temperature is not quite isotropic and the electrons are hotter than the protons.

The importance of the solar wind in solar–terrestrial relations and magnetospheric physics cannot be over stressed. With few exceptions it is the principal medium through which the activity of the Sun is communicated to the environment of the Earth, and as such it is now subject to almost continuous monitoring by spacecraft sensors. The properties of the solar wind plasma that are most important for the magnetosphere are the following:

(a) the wind blows all the time, with velocity generally within the range 200–700 km s^{-1};
(b) the energy is more directed than random ($\Delta E/E \sim 0.01$), and the flow is almost radially away from the Sun; the power flux is about 1/10 of that in the solar EUV spectrum, i.e. 10^{-4} W m^{-2};
(c) it consists mainly of electrons and protons, with a few per cent of α-particles and heavy nuclei;
(d) the wind is gusty with periods of minutes to a few hours;

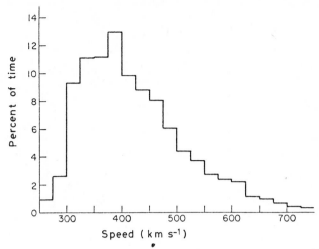

Figure 7.5   Speed of the solar wind. Histogram of measurements between 1962 and 1970. (Reprinted from J.T. Gosling, *Solar Activity Observations and Predictions* (Ed. McIntosh and Dryer) 1972, by permission of The MIT Press, Cambridge, Massachusetts.)

(e) the particle density is usually within the range 3–10 cm^{-3}, the most typical value being 5 cm^{-3}, the proton energy is about 1/2 keV and the electron energy about 1/4 eV.

7.3.4 Interplanetary magnetic field and sector structure

Satellites carrying sensitive magnetometers, beginning with IMP 1 (Interplanetary Monitoring Platform) in 1963, have detected a weak magnetic field in the solar wind. The field strength amounts to only a few gamma (1 gamma = 10^{-5} Gauss). Because of the high electrical conductivity of the solar wind, plasma and magnetic field are 'frozen' together. Unlike the situation in the magnetosphere, the plasma controls the movement in the solar wind because it has the greater energy density. Thus we can write, in MKS units

$$B_s^2/2\mu_0 < nmv^2/2$$

where B_s is the magnetic induction, n is the particle density, m the particle mass and v the solar wind velocity. In the solar wind the particle kinetic energy exceeds the magnetic energy by a factor of eight.

Although the solar wind moves out almost radially from the Sun, the rotation of the Sun gives the magnetic field a spiral form. The effect can be simulated by turning round while watering the garden with a hosepipe. The jet of water follows a spiral path, although the trajectories of individual drops are radials. Thus, the spiral form of the interplanetary magnetic field (IMF) is often said to result from the *garden hose effect*. At the orbit of the Earth the angle between the field lines and the radial is about 45°.

The field can be directed either inward or outward with respect to the Sun, and perhaps the most remarkable result of the early IMP 1 measurements was that

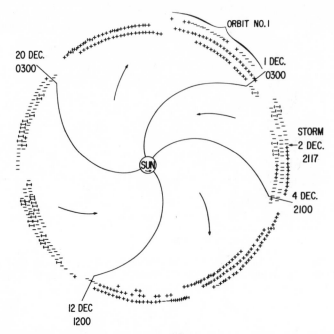

Figure 7.6 Sector structure of the solar wind in late 1963, showing two sectors with the IMF outward and two with it inward. +, field away from Sun. −, field toward Sun. (From J.M. Wilcox and N.F. Ness, *J. geophys. Res.* **70**, 5793, 1965, copyrighted by American Geophysical Union.)

sectors may be identified, the field being in and out in alternate sectors. In 1963 the solar wind had four sectors, as shown in Fig. 7.6, but the structure has evolved since that time; in 1968—69 it contained only two sectors, though four seems to be the more usual number.

Since the solar wind is in fact part of the solar corona, it is to be expected that the sector structure of the solar wind will depend on the level of solar activity and on the magnetic fields that may be observed in the Sun. The relation between interplanetary and solar magnetic fields may be understood with the aid of Fig. 7.7. In region 2, the transverse (or horizontal) components of the solar magnetic field have greater energy density than the solar plasma, and so the plasma is retained by the field. More than 0.6 solar radius above the photosphere the opposite is true and the plasma can move outward, carrying the magnetic field with it. The surface where this transition from region 2 to region 3 occurs is the effective source of the solar wind. Beyond 20 or 30 solar radii the energy density associated with radial flow exceeds both the thermal and the magnetic energies, and this is still the situation at the Earth's orbit (215 solar radii). Models of this kind have been put to the test by predicting, from spacecraft measurements of the interplanetary field near the Earth, what the coronal structure should be during an eclipse. The approach has met with some success, particularly if the Sun was not too disturbed. Observationally, it is also found that the magnitude of the IMF is related to the mean solar field, which is an average over the solar disc of the photospheric magnetic field. A delay of 4.5 days has to be allowed for, since this is the average time taken for the solar wind to reach the Earth.

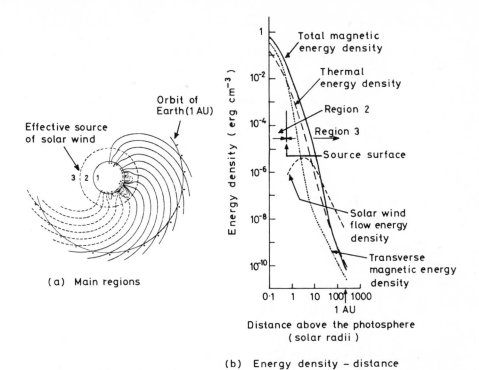

(a) Main regions

(b) Energy density – distance

Figure 7.7 Energy relations in the Sun and solar wind. (a) Main regions (region 1 being within the photosphere). (b) Energy density − distance, showing the effective source of the solar wind where the transverse magnetic energy density falls below the thermal energy of the plasma. (After K.H. Shatten, *Solar Activity Observations and Predictions* (Ed. McIntosh and Dryer) 1972, by permission of The MIT Press, Cambridge, Massachusetts.)

Although secondary to the solar wind plasma in energy, the interplanetary magnetic field plays an essential part in certain aspects of the behaviour of the magnetosphere, as will be seen in Chapter 8.

7.4 The geomagnetic cavity

7.4.1 The geomagnetic field near the Earth

As a first approximation, the geomagnetic field above the Earth's surface can be represented (to an accuracy of about 30% for points within 2 or 3 R_E of the surface) by a dipole field, the poles of the dipole being at geographic latitudes 79° N, 70° W and 79° S, 70° E. The field strength or magnetic induction is given by

$$B(r, \lambda) = \frac{M}{r^3} [1 + 3 \sin^2 \lambda]^{1/2} \tag{7.7}$$

where M is the dipole moment, r is the geocentric radial distance and λ is the magnetic latitude.

149

Although inaccurate, the dipole form is useful for approximate calculations in much of the magnetosphere and some of its properties will be summarized here. The field components in the r and λ directions are

$$B_r = \frac{-2M \sin \lambda}{r^3} \tag{7.8}$$

and

$$B_\lambda = \frac{M \cos \lambda}{r^3} \tag{7.9}$$

Thus

$$B/B_r = r \, d\lambda/dr = -1/(2 \tan \lambda) \tag{7.10}$$

and, by integration of Equation 7.10,

$$r = r_0 \cos^2 \lambda \tag{7.11}$$

which, for a given value of r_0, is the equation of a field line: the path that a single north pole would trace out. The parameter r_0 is the value of r when $\lambda = 0$, i.e. the greatest radial distance of the field line, reached over the magnetic equator.

On the magnetospheric scale it is convenient to express distances in units of the radius of the Earth, R_E. Then, putting $r/R_E = R$,

$$B(R, \lambda) = \frac{0.32}{R^3} [1 + 3 \sin^2 \lambda]^{1/2} \qquad \text{Gauss} \tag{7.12}$$

where 0.32 Gauss is the field strength at the surface of the Earth at the equator. In these terms, the field line equation is

$$R = R_0 \cos^2 \lambda \tag{7.13}$$

where R and R_0 are measured in earth radii. The latitude where the field line through $R = R_0$ intersects the Earth's surface is given by

$$\cos \lambda_E = (R_0)^{-1/2} \tag{7.14}$$

Field lines of dipolar form are illustrated in Fig. 7.8.

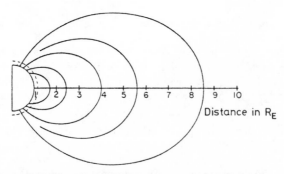

Distance in R_E

Figure 7.8 Dipole field lines. (From D.L. Carpenter and R.L. Smith *Rev. Geophys.* **2**, 415, 1964, copyrighted by American Geophysical Union.)

Improved accuracy is given by the eccentric-dipole approximation, in which the dipole is displaced about 400 km from the centre of the Earth. However, for accurate mathematical work it is necessary to express the field as a series of spherical harmonics, in which the magnetic potential V is given by

$$V = \sum_{n=1}^{\infty} (R_E/R^{n+1}) \sum_{m=0}^{n} (g_n^m \cos m\phi + h_n^m \sin m\phi) p_n^m (\cos \theta) \qquad (7.15)$$

where g_n^m and h_n^m are coefficients, $p_n^m (\cos \theta)$ are associated Legendre functions, and R, θ and ϕ are spherical polar coordinates. The potential distribution, and from it the coefficients, are derived from the observed global distribution of magnetic elements, both at the ground and measured on satellites. A computer program is used to find the potential in the region of interest, the field being obtained as the gradient of the potential. Because of the secular variation of the geomagnetic field, a given set of coefficients relates to a stated epoch.

The form given by Equation 7.15 would be valid at all levels if all the internal contributions, due to the main magnetic field originating within the solid earth, and the external contributions, due to ionospheric and magnetospheric currents, were fully included. However, the external currents are variable and their effects at the Earth's surface are relatively small. It is usual, therefore, to base the expansion on the internal field only. Within this limitation, a modern computation would be accurate to about 0.5% near the Earth's surface, and would be useful out to about $4 R_E$, where in any case the higher-order terms are unimportant relative to the dipole term ($n = 1$). Beyond this distance external contributions due to the distortion of the geomagnetic field by the solar wind become increasingly important.

7.4.2 Formation of the cavity

The geomagnetic cavity, illustrated in Fig. 7.9, is formed because the solar wind cannot enter the geomagnetic field but is swept around it. The non-penetration arises from the high electrical conductivity of the solar plasma, which is 'frozen

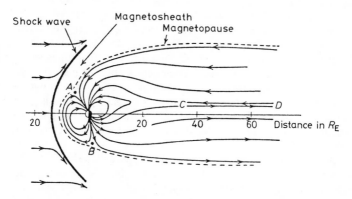

Figure 7.9 A cross-section of the magnetosphere. A and B are neutral points and CD is a neutral sheet. (After J.A. Ratcliffe, *An Introduction to the Ionosphere and Magnetosphere*, Cambridge University Press, 1972.)

151

out' of the geomagnetic field just as the IMF is 'frozen in' to the solar wind. A simple analogy would be a metal plate in the field of a strong laboratory magnet (Fig. 2.5). As soon as the plate is moved, currents are induced in it (Faraday's law of induction) and these currents themselves create a magnetic field in such a direction as to produce a force opposing the original motion. If the plate had infinite conductivity, the slightest movement would produce infinite current and an infinite restoring force. Hence, in the limit the plate could not move. A more rigorous treatment starting from Maxwell's equations of electromagnetism is given in Section 2.3.6.

7.4.3 The magnetopause

To work out the form of the boundary of the geomagnetic cavity it is necessary to find a surface such that at every point the pressure of the solar wind just outside the boundary is balanced by the pressure of the geomagnetic field just inside.

If the plasma contains N particles cm^{-3}, each of mass m g, travelling at velocity v cm s^{-1} at angle ψ to the normal to the boundary, the number striking each cm^2 of the boundary per second is $Nv \cos \psi$. If specular reflection is assumed the rate of change of momentum is $2Nmv^2 \cos^2 \psi$ dyne cm^{-2}, which has to be equated to the magnetic pressure $B^2/8\pi$. Such a calculation readily gives a realistic distance to the boundary on the Earth–Sun line (about $10\,R_E$), though the full computation of the whole surface requires a lengthy iteration procedure.

Fig. 7.9 shows the result of such a computation. The boundary, marking the termination of the magnetosphere, is known as the *magnetopause*. The field within is now distorted, and noteworthy points are that:

(a) the field lines from low latitudes still form closed loops between southern and northern hemispheres, though they may be distorted from the dipole form;

(b) lines emerging from the poles are now swept back, away from the Sun;

(c) there are two high-latitude field lines that intersect the magnetopause, though in fact the intensity vanishes as the boundary is reached, to form neutral points.

Approximate, though useful, solutions can also be obtained by the 'image dipole' method, in which the pressure of the solar wind is replaced by the magnetic pressure due to the image. In this method the image, of moment M_I, is placed parallel to and distance d from the Earth's magnetic dipole of moment M_E, and the fields due to both dipoles are added. The distorted field lines associated with the Earth's dipole are taken to represent the field within the magnetosphere. Those associated with the image have no physical significance, but the boundary between the two sets represents the magnetopause. A satisfactory model is given by $M_I = 28\,M_E$, $d = 40\,R_E$. It should be stressed, however, that the methods described so far are satisfactory only for the 'front' of the magnetosphere, the part towards the Sun.

Spacecraft observations recognize the magnetopause by a sharp drop in magnetic field strength and the appearance of turbulence (which is present in the IMF just outside the magnetopause). Fig. 7.10, which shows the results obtained on one of the first magnetopause crossings, illustrates these points and shows another interesting fact. The smooth solid curve in Fig. 7.10 gives the theoretical variation of field strength with distance if the existence of a boundary is ignored. Note that

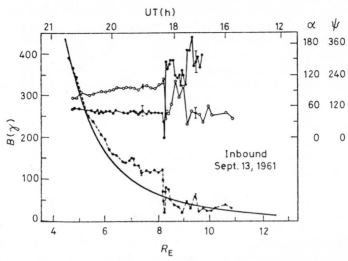

Figure 7.10 The boundary of the magnetosphere, observed at 8.2 R_E by Explorer XII on 13 September 1961. Outside the boundary the field is smaller in magnitude, and is turbulent. (L.J. Cahill and P.G. Amazeen, *J. geophys. Res.* **68**, 1835, 1963, copyrighted by American Geophysical Union.)

the observed field is just twice as intense as the theoretical value at a point immediately inside the boundary. The reason is easily seen in terms of the image method, since the fields due to the Earth and to the image are equal at the boundary.

Generally, good agreement has been obtained between the observed and the computed shapes of the magnetopause. Fig. 7.11 demonstrates this for the ecliptic plane. Measurements such as these are compiled over a period of time, as the orbital

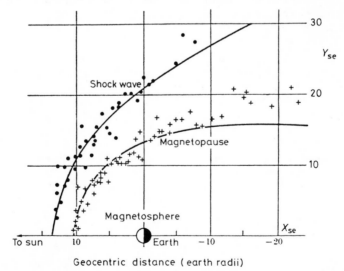

Geocentric distance (earth radii)

Figure 7.11 Comparisons between observed and predicted positions of the magnetopause and of the shock. (After N.F. Ness *et al.*, *J. geophys. Res.* **69**, 3531, 1964, copyrighted by American Geophysical Union.)

153

plane of the satellite rotates with respect to the Earth–Sun line as the Earth moves around the Sun. The result is therefore a long-term avergae. It is not possible to determine the whole magnetospheric surface for a given instant of time. At high latitudes there have been relatively few measurements, and recent observations over the poles show that agreement with theory is less satisfactory in those regions, the magnetosphere being some 20% thicker than predicted by current models.

7.4.4 The polar clefts

Simple models predict that on the surface of the magnetosphere there should be two neutral points at which the total magnetic field is zero. These points are the terminations of field lines intersecting the Earth's surface at latitudes near $\pm 78°$ at local noon. Lines at lower latitudes form closed loops into the other hemisphere, while those at higher latitude are swept back away from the Sun. In the simple model all the field at the surface of the magnetosphere converges into the two points where the neutral points project to the Earth's surface. These are therefore very interesting regions, where one might expect particles from the magnetopause, and even solar-wind plasma, to be guided along the field into the atmosphere.

At latitudes between 70 and 80°, low-orbit satellites have found particles with energy and flux characteristics similar to those of particles outside the magnetosphere, and these regions have therefore been associated with direct entry. It appears from the spatial distribution, which typically covers some 5° of latitude, that they enter not at a point but over a finite area. Therefore the neutral points are now considered rather as 'cusps' or 'clefts', with the form like that shown in Fig. 7.12. The clefts vary in size but are typically several degrees of latitude by 8 h of local time. In these regions various ionospheric effects are observed, such as enhancements of electron density and temperature, and a greater degree of irregularity, and these almost certainly result from an influx of particles with energies of a few hundred electron volts from the cleft.

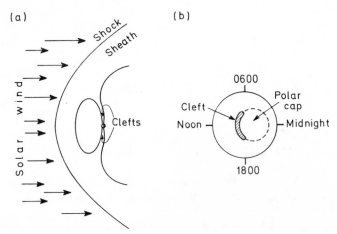

Figure 7.12   The magnetospheric clefts. (a) View in noon–midnight meridian plane. (b) View of ionosphere from above north pole. (After V.M. Vasyliunas, *EOS*, 55, 60, 1974, copyrighted by American Geophysical Union.)

The existence of the clefts raises important questions regarding the structure of the magnetosphere and its dynamics, some of which will be touched in Chapter 8.

7.4.5 The shock and the sheath

We now come to an interesting feature of Figs. 7.9, 7.11 and 7.12, which though not strictly a part of the geomagnetic cavity arises directly from its presence in the solar wind. Tenuous though it may be by any ordinary standard, the magnetosphere is a relatively solid object in comparison with the plasma that comprises the solar wind. As mentioned in Section 7.3.2, the solar wind is 'supersonic' at the orbit of the Earth. That is, its velocity exceeds that at which waves, whether acoustic or hydromagnetic, can propagate through it. In the solar wind the Alfvén speed, $v_A = B/(4\pi\rho)^{1/2}$, where B is the field strength in Gauss and ρ is the particle density in g cm^{-3}, is about 50 km s^{-1}. Thus the solar wind has an *Alfvén Mach number* of about 8.

The situation may therefore be likened to that of a solid object flying through the air supersonically, and, as with a bullet or a supersonic aircraft, a shock wave is formed. The existence and position of the shock were predicted theoretically in the early 1960s and subsequent observations confirmed the predictions. As may be seen in Fig. 7.11, the theoretical and observed positions agree. At the subsolar point the shock is usually $2-3\,R_E$ ahead of the magnetopause, and the separation increases away from the subsolar point.

A shock wave is a discontinuity of the medium, produced when information about an approaching perturbation is not transmitted ahead into the medium. The shock wave in front of the magnetosphere is recognized by the change of particle properties. The solar wind is here deflected so as to flow around the magnetosphere. Its velocity is reduced to about 250 km s^{-1}, and the kinetic energy is dissipated as thermal energy so that the temperature is increased to 5×10^6 °K, which is 5−10 times hotter than the solar wind.

The region between the shock and the magnetopause has properties distinct from either the solar wind or the magnetosphere, though the plasma and magnetic field observed there are plainly of solar rather than terrestrial origin. This region is called the *magnetosheath*. Particles observed beneath the polar clefts have properties appropriate to the magnetosheath rather than to the solar wind proper.

7.4.6 The tail of the magnetosphere

Having learned that the magnetosphere has a pause, two clefts, a shock and a sheath, the reader will not be surprised to hear that it also has a tail. In fact the magnetospheric tail, or simply *magnetotail*, is perhaps the most remarkable feature of the magnetosphere and one that is probably the source of phenomena that are visible from the ground and are by no means negligible in some human activities.

Fig. 7.13 shows the magnetic field components parallel to and below the XY-plane in the solar−magnetospheric coordinate system (a system in which the X-axis points towards the Sun, and the XY-plane includes the geomagnetic dipole axis). In a dipole field these components would point radially away from the Earth, but the observations show that the field tends to lie parallel to the solar

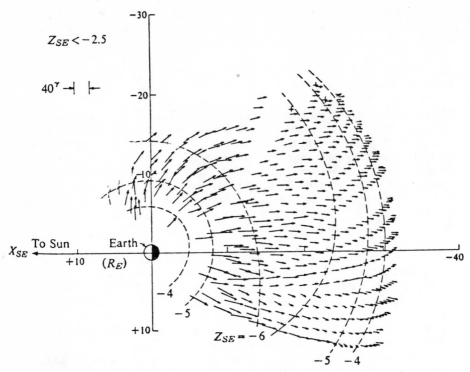

Figure 7.13 Magnetic field vectors in planes parallel to and below the solar-magnetospheric *XY*-plane, observed 27 November 1963 to 31 May 1964 by IMP-1. (After N.F. Ness, *J. geophys. Res.* **70**, 2989, 1965, copyrighted by American Geophysical Union.)

wind downstream from the Earth. Beyond about $10\,R_E$, the field is deformed into a tail, whose boundary with the solar wind is roughly cylindrical, the field pointing approximately towards the Earth in the northern half and away in the southern half. The length of the tail is not certain. It extends more than $80\,R_E$ and may have been observed at $1000\,R_E$, but an attempt to detect it at $3300\,R_E$ proved negative. Fig. 7.9 illustrates the form of the tail in the *XZ*-plane of the solar–magnetospheric system, the plane including the magnetic poles. The strength of the tail field is about 20 gamma (20×10^{-5} Gauss), though much weaker than this in the region where it reverses direction.

The cause of the magnetotail has been the subject of some debate. One suggestion is that momentum is transferred from the solar wind to the outer part of the magnetosphere, as in the viscous interaction that occurs when one fluid flows past another. The difficulty met by this approach is to find the physical process equivalent to viscosity when the fluids involved are collisionless. The cause that is now thought to be more likely than viscous interaction invokes the idea that lines of the geomagnetic field can at times become connected to lines of the interplanetary magnetic field. Since the latter are being swept past the Earth in the solar wind, the geomagnetic lines tend to be carried along as well, to accumulate on the down-wind side of the Earth.

156

To simulate the magnetotail in a model magnetosphere it is necessary to add an electric current, which flows across the tail from the dawn side of the magnetosphere to the dusk side. This may be taken to imply the existence of a large-scale electric field across the tail. These ideas will be explored in more detail when the circulation of the magnetosphere is considered in Chapter 8.

7.5 Plasma in the magnetosphere

7.5.1 The plasmasphere

The ionized particles in the ionosphere have temperatures up to several thousand degrees Kelvin, and electron energies are therefore several tenths of an electron volt. ($E = 3kT/2$ where $k = 1.38 \times 10^{-16}$ erg $^\circ$K, and 1 eV $\equiv 1.6 \times 10^{-12}$ erg. Hence, to take an example, 2000 $^\circ$K $\equiv$ 0.26 eV.) The particle density is typically 10^4 cm^{-3} at altitude 1000 km, falling off with increasing height. The fraction of total gas that is ionized increases with height and for present purposes we assume the magnetospheric gas to be fully ionized.

There seems to be no good reason why the hydrostatic equilibrium that holds in the topside ionosphere (Section 4.3.3) should be maintained at greater altitudes. Collisions are infrequent and therefore, as for the exosphere, the gas laws cannot be used. However, observational evidence suggests that the distribution is not far from exponential provided an appropriately large scale height (of several thousand kilometres) is taken and the distributions are worked out along the field lines rather than in vertical columns. As for the neutral gas, gravity cannot retain the plasma; in the present case the geomagnetic field provides the mechanism for retention. This will be discussed further in Section 7.6.1.

The region of plasma of ionospheric character is limited, nevertheless, to a volume extending to several earth radii which is known as the *plasmasphere*. Its outer boundary is the *plasmapause*.

Electron densities in the magnetosphere are determined by two main techniques. They can be studied from the ground by analysing the characteristics of *whistlers*, the VLF radio signals that arise in lightning flashes and travel between hemispheres by propagating along the geomagnetic field in the whistler mode. This method is powerful because, being ground-based, it can be used continuously at a given place to study changes with time in a selected part of the magnetosphere. The other major approach is direct sensing by satellite-borne instruments. The great merits of satellite measurements are that readings are interpreted with a minimum of assumptions, and for the ability to measure parameters like temperature and composition as well as concentration.

The basic theory of whistler propagation was given in Section 3.7.5. Here we will discuss its diagnostic application to the magnetosphere. If the travel time of a whistler is displayed against frequency, it is found that there is one frequency where the travel time is a minimum. This is in fact a characteristic of all whistlers, but those which show it plainly enough to be suitable for this analysis are called *nose whistlers*. The frequency corresponding to minimum travel time indicates which field line the whistler has travelled along, and the time taken gives the smallest electron density encountered along that field line.

The group refractive index for a signal travelling in the whistler mode is

$$\mu_g = f_N f_B / [2f^{1/2} (f_B - f)^{3/2}]$$ (3.27) (7.16)

157

where f is the wave frequency, f_N is the plasma frequency and f_B is the gyrofrequency. There-fore the total time of propagation between hemispheres is

$$t = \frac{1}{c} \int \mu_g \, ds = \frac{1}{2c} \int \frac{f_N f_B \, ds}{f^{1/2} (f_B - f)^{3/2}} \qquad (7.17)$$

where c is the velocity of light and the integration is along the field. For a given value of f_N, the travel time will be large if f is small or if $f \sim f_B$. Between these values of f is one for which the travel time is a minimum, and that is the *nose frequency*. Fig. 7.14 shows the dispersion curve for one particular form of variation of f_N along the field line. The ratio of the nose frequency to the minimum gyrofrequency along the path depends on the form of the model taken for the electron density, but is typically close to 0.42 for distances around $4 R_E$. Thus the nose frequency indicates how far into the magnetosphere the whistler travelled. It can also be seen from Equation 7.17 that for given values of f_N and f, the largest contribution to the travel time comes from the regions most distant from the Earth's surface, because (i) the factor $(f_B - f)$ is then smallest, and (ii) relatively more of the path lies at the greater distances because of the shape of the field lines. Hence it is possible to determine the electron density in the region where the field line along which the whistler travelled crossed the equatorial plane.

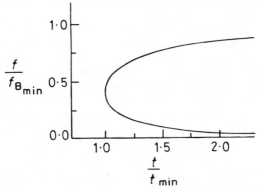

Figure 7.14   Theoretical normalized dispersion curve for a nose whistler. (From D.L. Carpenter and R.L. Smith, *Rev. geophys.* **2**, 415, 1964, copyrighted by American Geophysical Union.)

By this means, analysis of a number of nose whistlers can produce a plot of electron density against distance, as shown in Fig. 7.15. The remarkable feature of such plots is that they often show a sharp drop of electron density, by a factor of ten or more, at about $4 R_E$ geocentric distance. The drop occurs over a relatively small distance, less than $0.5 R_E$, and this is the plasmapause, sometimes called the *knee*. If traced along the geomagnetic field to the ionosphere, the plasmapause is found to correspond approximately with the position of the high-latitude trough (Section 5.4.2). Thus there can be little doubt that the plasmasphere occupies a doughnut-shaped region in the inner part of the magnetosphere where the field lines are not too distorted from the dipolar form, as illustrated in Fig. 7.16. The further significance of the plasmasphere and its dynamic behaviour will be considered in Section 8.1.5.

7.5.2 Low-energy plasma and the plasma sheet

Satellite instruments not only confirm that the electron density is smaller beyond the plasmapause, but also show that the outer plasma is much hotter than that in

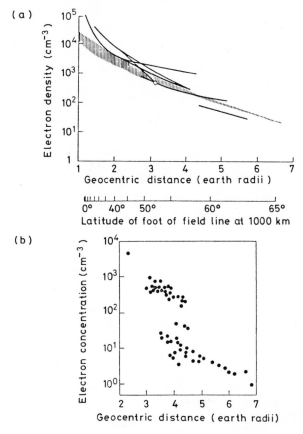

(a)

Electron density (cm^{-3})

Geocentric distance (earth radii)

0°   40° 50°   60°   65°
Latitude of foot of field line at 1000 km

(b)

Electron concentration (cm^{-3})

Geocentric distance (earth radii)

Figure 7.15 Electron density in the equatorial plane as a function of geocentric distance, determined from whistler studies. (a) A compilation of early measurements, showing typical values out to 4 or 5 R_E. (From Carpenter and Smith, *Rev. Geophys.* **2**, 415, 1964. copyrighted by American Geophysical Union.) (b) Detail, showing the knee near 4 R_E. (From J.A. Ratcliffe, *An Introduction to the Ionosphere and Magnetosphere,* Cambridge University Press, 1972, after D.L. Carpenter, *J. geophys. Res.* **71**, 693, 1966, copyrighted by American Geophysical Union.)

the plasmasphere. Energies range over 10^2 to 10^4 eV, much higher than those of the 'thermal' population within the plasmasphere. Nevertheless, the particles are considered 'low energy' by comparison with the even higher energies of the Van Allen particles that also inhabit this outer region but had the advantage of having been discovered first.

The distribution of the outer plasma is sketched in Fig. 7.17. It extends down the magnetotail, where it forms a thin plasma sheet dividing the two lobes where the field is directed in opposite directions. The tail field is small in the centre of the tail, and a pressure balance is maintained between the magnetic pressure of the tail field outside the sheet, $B_T^2/8\pi$ (in cgs), and the pressure due to particles within the sheet, nkT. The particle density is typically 0.5 cm^{-3} for both electrons and protons, i.e. 1 particle cm^{-3} in all, and the average particle energies are 0.6 keV for electrons and 5 keV for the protons. The total energy density is about 3 keV cm^{-3}. The plasma sheet is several earth radii thick and has an identifiable inner

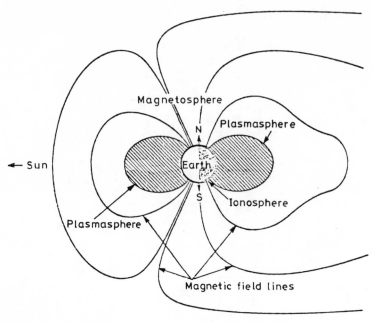

Figure 7.16 Location of the plasmasphere within the magnetosphere. (After P.M. Banks, *Space Res.* **XII**, 1053, 1972.)

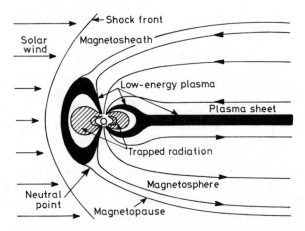

Figure 7.17 Distribution of the low-energy plasma in relation to other magnetospheric features. (After B.J. O'Brien, *Solar-Terrestrial Physics* (Ed. King and Newman) Academic Press, 1967.)

edge some 7 earth radii from the Earth at local midnight. However both the thickness and the position of the inner edge are variable.

The presence of the plasma sheet between the two magnetotail lobes, where the magnetic fields lie in opposite directions, makes a very interesting physical situation. Such a configuration represents a store of energy. If the two lobes could come into contact mutual annihilation should occur, enabling the field to return to a dipole form, the excess energy being communicated to the plasma. To understand how the tail is maintained it is necessary to consider the dynamics of the magnetosphere (Section 8.1).

7.6 Van Allen particles

To complete this account of the structure of the magnetosphere we consider the particles of higher energy, generally known as *the trapped radiation* or *the Van Allen particles*. These were the first major discovery in the exploration of the Earth's more distant environment by satellites. Although not themselves an essential element in the formation and maintenance of the magnetosphere, they do introduce important aspects of the relationships between particles and fields, notably in the mechanism by which charged particles are trapped in the geomagnetic field. We shall therefore deal with the principles of trapping, and with the morphology of trapped particles as revealed by the observations.

7.6.1 Trapping theory and adiabatic invariants

We saw in Section 2.3.2 that a charged particle with mass m, and charge e, gyrates in a magnetic field of induction B at an angular frequency $\omega_B = eB/m$ (in MKS units). ω_B is the gyrofrequency. The radius of gyration is $r_B = mv/Be$, where v is the velocity of the particle. To take an example, for particles of energy 1 MeV at altitude 2000 km above the Earth's equator the periods are 7 μs for electrons and 4 ms for protons, the radii of gyration being 300 m and 10 km respectively.

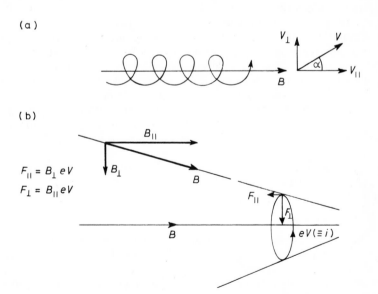

Figure 7.18   Spiral motion of a trapped particle and forces near a mirror point. (a) Motion of a positive particle along the geomagnetic field. (b) Retarding force on equivalent current loop.

If a particle in the magnetosphere has velocity $v_\perp$ normal to the direction of the geomagnetic field and velocity $v_\parallel$ parallel to it, the trajectory is a spiral (see Fig. 7.18). If no work is being done on or by the particle the magnetic flux through the orbit (Φ_m) is constant. (If $d\Phi_m/dt \neq 0$ an electric field would be generated and the particle would be accelerated or retarded.) Hence

$$\Phi_m = B \times \pi r_B^2 = 2\pi m E_\perp /e^2 B = \text{constant}$$

where

$$E_\perp = mV_\perp^2/2 \tag{7.19}$$

the kinetic energy associated with the transverse velocity. Hence, $E_\perp/B = \text{constant}$. But $E_\perp/B$ is the magnetic moment of the current loop represented by the gyrating particle:

$$\mu = \text{current} \times \text{area of loop} = \frac{eV_\perp}{2\pi r_B} \times \pi r_B^2 = mV_\perp^2/2B = E_\perp/B \tag{7.20}$$

This gives us the *first adiabatic invariant* of charged particle motion in a magnetic field: the magnetic moment is constant. This holds provided the magnetic field does not change significantly within a gyration period.

If the total velocity of the particle is V

$$V_\perp = V \sin \alpha \tag{7.21}$$

where α is the angle between the velocity vector and the magentic field, i.e. the *pitch angle*. Hence, $E_\perp/B = E \sin^2 \alpha/B = \text{constant}$, where the total energy, E, is also constant. Therefore $\sin^2 \alpha \propto B$. As a particle moves from the equator towards the pole in a dipole field it encounters increasing field strength and so the pitch angle increases. Eventually $\alpha = 90°$, and here the forward motion stops. This is the *mirror point* at which the particle is reflected back towards the other pole. By this mechanism the particle is trapped in the geomagnetic field.

The reason why the particle turns around at the mirror point may be seen by considering the forces on the equivalent current loop. As illustrated in Fig. 7.18, the convergence of the magnetic field introduces a small perpendicular component which tends to push the loop towards the region of weaker field. The cause of mirroring can also be thought of in terms of the forces on a small magnetic dipole in a converging magnetic field. The magnetic dipole equivalent to the current loop would be aligned with the geomagnetic field and directed so as to oppose that . field. In a non-uniform field the net force is such as to push the dipole towards weaker field.

For a given particle the location of the mirror point is determined by the pitch angle (α_0) as the particle crosses the magnetic equator (where the field is weakest on that field line)

$$B_0/B_M = \sin^2 \alpha_0/1 \tag{7.22}$$

B_0 and B_M being the magnetic induction at equator and mirror point respectively. Since the total energy does not change, $E = E_\perp + E_\parallel$, and $V_\parallel^2 = V^2 - V_\perp^2 = V^2 \cos^2 \alpha$. The parallel energy falls to zero at the mirror point, and in energy terms the trapping process represents an alternation of kinetic energy between the perpendicular and parallel components, the total remaining constant.

The time taken for a particle to travel between mirror points is often known as

162

the *bounce time* and in the case of particles crossing 2000 km above the equator is 0.1 s for electrons and 2 s for protons of energy 1 MeV.

In addition to gyrating about the field line and oscillating between hemispheres, a trapped particle also drifts around the Earth. There are two causes of longitude drift. First, the particle finds itself in a weaker magnetic field when in that part of its gyration furthest from the Earth. The radius of gyration therefore changes during the gyration and in consequence a lateral displacement is introduced, as illustrated in Fig. 7.19. The second cause is the curvature of the geomagnetic field, which introduces a centrifugal force to which the particle responds by drifting in a direction normal to both the field and the force. Given such a force, the mechanism is similar to that described in Section 4.4 by which an upper-atmosphere wind causes a current to flow at right angles to the wind if collisions are negligible.

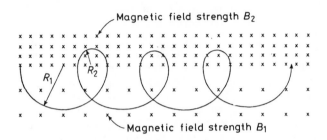

Magnetic field strength B_2

$(B_2 > B_1;$ therefore, $R_1 > R_2.)$

Figure 7.19 Longitude drift caused by a gradient in geomagnetic field intensity. (After W.N. Hess, *Radiation Belt and Magnetosphere*, used by permission of Ginn and Co., Copyright, 1968 by Xerox Corporation.)

A detailed treatment of gradient and curvature drifts is beyond our present scope, but it will be noted that in a dipole magnetic field the two effects are of similar magnitude for particles mirroring near the surface of the Earth, and that each causes electrons to drift eastward and protons to drift westward. For particles of flat pitch angle, mirroring near the equator, curvature drift would be negligible. Hence the longitudinal drift rate depends on the mirror point. For particles of 1 MeV, 2000 km above the equator, an electron completes one circuit of the Earth in about 50 min, and a proton takes about 30 min.

More exactly the period is given by

$$T_R = 172.4 \times \frac{1 + \xi}{\xi(2 + \xi)} \times \frac{m_e}{m} \times \frac{1}{R_0} \times \frac{G}{F} \quad \text{min} \tag{7.23}$$

where $\xi = (1 - \beta^2)^{1/2} - 1$, β being the ratio of particle velocity to the speed of light, m/m_e is the mass of the particle in electron masses, R_0 is the geocentric distance to the equatorial crossing of the particle measured in earth radii, and G/F is a factor increasing from 1.0 for a particle mirroring at the equator to 1.5 for one mirroring at the pole. For non-relativistic particles, $\beta^2 \ll 1$, and the formula reduces to

$$T_R = \frac{733}{E} \times \frac{1}{R_0} \times \frac{G}{F} \quad \text{h} \tag{7.24}$$

In a non-dipole field, the path followed by the drift is determined from the *second adiabatic invariant*, alternatively known as the *integral invariant*, which

163

says that the integral of the parallel momentum over one complete bounce between mirror points is constant:

$$J = 2 \int_{l_1}^{l_2} m \, v_\parallel \, \mathrm{d}l = 2mv \int_{l_1}^{l_2} (1 - B/B_\mathrm{M})^{1/2} \, \mathrm{d}l = \text{constant} \qquad (7.25)$$

Here l is the distance measured along the field line between mirror points at which the induction is B_M. The quantity

$$I = \int_{l_1}^{l_2} (1 - B/B_\mathrm{M})^{1/2} \, \mathrm{d}l \qquad (7.26)$$

is a property of the field configuration and defines a surface on which a given particle remains in the absence of external perturbations. The second invariant holds provided the field does not change appreciably within the time of one bounce. Since the first invariant defines the field at mirror points, the first and second taken together define the loci of mirror points for a drifting particle. It has been shown that the flux of particles perpendicular to the field will be the same at all points having the same values of B and I, and thus these variables are often used in mapping observations of trapped particles in space.

An important development from the second invariant is the *L parameter* introduced by C. E. McIlwain in 1961. L has the dimension of a length and is usually measured in earth radii. It is analogous to the distance to the equatorial crossing of a field line in a dipole field (R_0 in Section 7.4.1), to which it reduces if the field is indeed dipolar. Along most geomagnetic field lines, L is constant to about 1%, and so it provides a convenient way of identifying field lines even though they might not be strictly dipolar. By analogy with a dipole field, the *invariant latitude* may be defined:

$$\Lambda = \cos^{-1} (1/L)^{1/2} \qquad (7.27)$$

Compare this with Equation 7.14.

A complication that arises in an asymmetrical field like the Earth is the phenomenon of *shell splitting*. This comes about because the drift path depends on the pitch angle as well as the *L*-value of the field line. If particles are introduced in the midnight meridian, those mirroring at the equator ($\alpha_0 = 90°$) drift so as to follow a constant field intensity. Since the magnetic field is squashed in by the solar wind on the day side of the Earth, increasing the field at given distance, these particles will move further from the Earth in the day sector. However, particles mirroring at high latitude (α_0 small) will, in following a shell of constant I, be more influenced by the length of the field line and these will not move out so far. Thus, the spatial distribution of particles tends to be smeared out. Because of shell splitting, particles from some regions are unable to complete a full circuit of the Earth but are lost beyond the magnetopause or into the tail. These regions may be called *pseudo-trapping* regions.

The *third adiabatic invariant* or *flux invariant* says that the total magnetic flux enclosed by the drift orbit is constant. From the flux invariant it is possible to compute the effects of very slow changes of magnetospheric structure on particle orbits. The third invariant is violated if changes occur in times less than the time for one circuit of the Earth — which happens not infrequently. The third invariant is important in relation to magnetic storms, where conditions may change rapidly at the beginning and return to normal more slowly.

The basic concepts of particle trapping, to which the reader has been introduced in this section, were developed for the Van Allen particles, but their importance goes beyond that with applications to much discussion of particle and field phenomena in the magnetosphere.

7.6.2 Observations of Van Allen particles

The discovery of energetic trapped particles in the magnetosphere, made by J. A. Van Allen's group at the University of Iowa, is generally regarded as the most important of the early satellite era. The first indication came from a Geiger counter carried on the first satellite of the United States, Explorer I, launched on 31 January 1958. The observation was in fact rather peculiar. At times the Geiger counting rate was very high and at others it fell below the threshold expected from the background of cosmic rays. This latter point was correctly interpreted as a malfunction brought about by an exceptionally high flux of particles. Explorer III and Sputnik III also detected them. The Geiger counter gave little information about the nature of the particles and they were at first thought to be electrons in the energy range 30–150 keV. Subsequent satellites with better instruments showed that they were actually protons of energy exceeding 30 MeV.

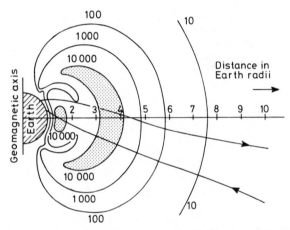

Figure 7.20   Van Allen's first map of the radiation belts. The contours are counting rates of the Geiger counter. (After J.A. Van Allen and Frank, *J. geophys. Res.* **64**, 1683, 1959, copyrighted by American Geophysical Union.)

At the end of 1958, the probe Pioneer III went out 107 400 km into space and found that the particle distribution had a double structure, as in Fig. 7.20. From the early measurements it was thought that there were two distinct belts of particles, the inner belt centred at 1.5 earth radii and the outer belt centred at 3.5 earth radii. Subsequent work has refined this picture. The most energetic protons are indeed concentrated near $1.5 R_E$, and the more energetic electrons, with energies exceeding 1.6 MeV, occur in the region formerly described as the outer belt. At lower energies, however, the distributions occupy larger volumes of space and extend further from the Earth, as shown in Fig. 7.21. The proton spectrum covers 100 keV–100 MeV, the maximum occurring at greater distance at lower energy. An outer limit is set by the distortion of the geomagnetic field. Beyond about $3 R_E$ the flux varies greatly with the level of geophysical activity.

The outer boundary varies with the time of day. Fig. 7.22 illustrates the latitude of the outer boundary of electrons whose energy exceeds 40 keV. It is seen that the trapped radiation extends to a higher latitude by day than by night. Since the

165

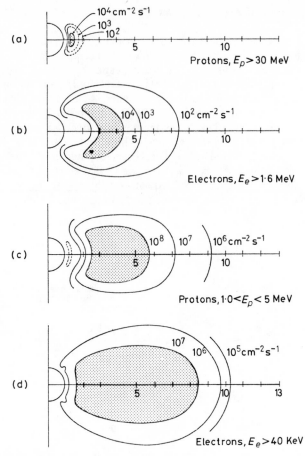

Figure 7.21 Distributions of trapped protons and electrons of various energies. The more energetic particles maximize closer to the Earth. The contours are particle fluxes in $cm^{-2}\,s^{-1}$ (After W. N. Hess, *Radiation Belt and Magnetosphere*, used by permission of Ginn and Co., Copyright, 1968, by Xerox Corporation.)

trapping mechanism requires closed field lines between northern and southern hemispheres, the trapped radiation provides a means for studying the extent of the closed field region. The asymmetry of Fig. 7.22 relates to the asymmetry of the magnetosphere in the solar wind.

7.6.3 Production and loss mechanisms

Since the trapped particles cannot, by definition, leave the trapping region, neither can charged particles of the same energy enter it. Therefore we must look for production mechanisms which involve the creation of the particles *in situ* or their introduction through the violation of adiabatic invariants. Given that there are production mechanisms, there must also be loss mechanisms.

In the inner regions the source is probably the decay of neutrons that have themselves been produced by the arrival of cosmic rays. Each cosmic ray proton of

166

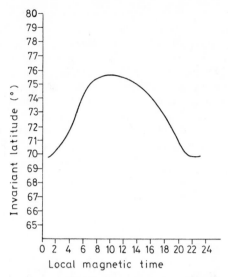

Figure 7.22 Invariant latitude of the trapping boundary of 40 keV electrons; median during October 1968 – April 1969. (After D.E. Page and M.L. Shaw, *Space Res.* **XII**, 1471, 1972.)

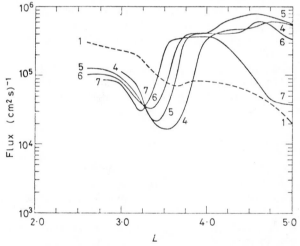

Figure 7.23 Flux of trapped electrons ($E \geqslant 1.6$ MeV) on days after a magnetic storm, showing that electrons are drifting towards the Earth during this period: 1, 7 December 1962; 4, 20 December 1962; 5, 23 December 1962; 6, 29 December 1962; 7, 8 January 1963. (After L.A. Frank *et al.*, *J. geophys. Res.* **69**, 2171, 1964, copyrighted by American Geophysical Union.)

energy 5 BeV produces about seven neutrons, each of which decays to a proton, an electron and an antineutrino. Calculation shows this source to be sufficient for the inner zone. In this region losses involve an interaction with the atmosphere near the mirror points, with retardation and charge exchange for protons, and scattering into the loss cone for electrons after which they lose energy by ionizing atmospheric gases.

167

The processes operating in the outer regions are somewhat more problematical, though there is no doubt that they work by violating an adiabatic invariant through varying electric or magnetic fields or by an interaction with electromagnetic waves. These processes involve storm phenomena. For example, Fig. 7.23 shows the distributions of trapped electrons on successive days, and provides direct evidence for the filling of the outer zone after a magnetic storm. Storm phenomena will be considered in later chapters. However we may recall here an early suggestion that the trapped radiation was the source of the aurora. Further studies of the energy required showed that this could not be so, and current thinking has, if anything, reversed the roles. The aurora and the trapped radiation are closely associated, but in a sense the second can be considered a byproduct of the first.

7.7 Further reading

Akasofu, S.-I. and Chapman, S. (1972) *Solar–Terrestrial Physics* (Oxford) Chapters 1, 5.

McIntosh, P. S. and Dryer, M. (1972) *Solar Activity Observations and Predictions* (MIT Press) particularly the articles by Evans, Dodson and Hedeman, Schatten, Dryer and Cuperman.

LeGalley, D. P. and Rosen, A. (1964) *Space Physics* (Wiley) Chapters 9–12, 14.

Hess, W. N. and Mead, G. D. (1968) *Introduction to Space Science* (Gordon and Breach) Chapters 4, 8, 9.

Matsushita, S. and Campbell, W. H. (1967) *Physics of Geomagnetic Phenomena* (Academic Press) Chapter V-3.

Hess, W. N. (1968) *Radiation Belt and Magnetosphere* (Blaisdell).

Vette, J. I. (1970) Summary of particle populations in the magnetosphere, in *Particles and Fields in the Magnetosphere* (Ed. McMormac) (Reidel).

Roederer, J. G. (1970) *Dynamics of Geomagnetically Trapped Radiation* (Springer-Verlag).

Dungey, J. W. (1976) The magnetosphere, *Sci. Prog.* **63**, 315.

Tandberg-Hanssen, E. (1973) Solar activity, *Rev. Geophys. Space Phys.* **11**, 469.

Hundhausen, A. J. (1970) Composition and dynamics of the solar wind plasma, *Rev. Geophys. Space Phys.* **8**, 729.

Neugebauer, M. (1976) Large-scale and solar-cycle variations of the solar wind, *Space Sci. Rev.* **17**, 221.

Schatten, K. H. (1971) Large-scale properties of the interplanetary magnetic field, *Rev. Geophys. Space Phys.* **9**, 773.

Roederer, J. G. (1969) Quantitative models of the magnetosphere, *Rev. Geophys.* **7**, 77.

Walker, R. J. (1976) An evaluation of recent quantitative magnetospheric magnetic field models, *Rev. Geophys. Space Phys.* **14**, 411.

Kavanagh, L. D., Schardt, A. W. and Roelof, E. C. (1970) Solar wind and solar energetic particles: properties and interactions, *Rev. Geophys. Space Phys.* **8**, 389.

Kovalevsky, J. V. (1971–72) The interplanetary medium and its interaction with the Earth's magnetosphere, *Space Sci. Rev.* **12**, 187.

Willis, D. M. (1971) Structure of the magnetosphere, *Rev. Geophys. Space Phys.* **9**, 953.

Ness, N. F. (1969) The geomagnetic tail, *Rev. Geophys.* **7**, 97.

Carpenter, D. L. and Smith, R. L. (1964) Whistler measurements of electron density in the magnetosphere, *Rev. Geophys.* **2**, 415.

Gringauz, K. I. (1969) Low-energy plasma in the Earth's magnetosphere, *Rev. Geophys.* **7**, 339.

Special Issue: The Physics of the Plasmapause (1976) *J. atmos. terr. Phys.* **38**, 1039.

Van Allen, J. A. (1969) Charged particles in the magnetosphere, *Rev. Geophys.* **7**, 233.

Schultz, M. (1976) Geomagnetically trapped radiation, *Space Sci. Rev.* **17**, 481.

White, R. S. (1973) High-energy proton radiation belt, *Rev. Geophys. Space Phys.* **11**, 595.

Kennel, C. F. (1973) Magnetospheres of the planets, *Space Sci. Rev.* **14**, 511.

8
The Dynamical Magnetosphere and the Substorm

8.1 Circulation of the magnetosphere

8.1.1 The Axford and Hines model

As we have seen, a tolerable approximation to the form of the magnetosphere can be obtained by assuming that the solar wind exerts a pressure against the boundary but otherwise flows smoothly past. Such an assumption must, however, be questionable. Most objects moving at high speed through a fluid experience some friction, and the same would be expected of the magnetosphere as it travels at super-Alfvénic speed through the solar plasma. The frictional force would provide a means for energy to be transferred from the solar wind into the magnetosphere, and much interest has centred, therefore, on the question of a dynamical interaction across the magnetopause.

In 1961, W. I. Axford and C. O. Hines suggested that momentum is transferred to the outer layers of the magnetosphere by the flow of the solar wind past the boundary. The mechanism was physically unspecified but was considered equivalent to friction. The difficulty with friction is that it depends on collisions between gas particles, and with a mean free path of 10^7 km in the solar wind collisions are much too infrequent to have much effect. It was thought that some irregularity or turbulence generated at the boundary, possibly involving a wave–particle interaction (Section 9.4), might fill the same role.

It was supposed that the transfer of momentum across the magnetopause produces a general circulation of the magnetosphere, in which the magnetospheric plasma and magnetic field move away from the Sun near the boundary and back towards the Sun in the centre of the magnetotail as sketched in Fig. 8.1a. If the rotation of the Earth is taken into account the pattern is distorted as in Fig. 8.1b. The concept of magnetospheric circulation does not rest solely on the existence of a plausible mechanism, because it follows as a deduction from the analysis of one of the ionospheric current systems, the S_q^p system, whose existence is derived from observations with medium- and high-latitude magnetometers. The general form of S_q^p is a current flowing over the poles towards the Sun, with a return current in middle latitudes (Fig. 8.2). As we saw in Section 4.4.3, in the dynamo region of the E region the electron gyrofrequency exceeds the collision frequency, which exceeds the ion gyrofrequency ($\omega_e > \nu_e$; $\omega_i < \nu_i$). Therefore the ions move with the neutral air but the electrons move with the geomagnetic field. Circulation of the magnetosphere will carry field lines over the poles and produce a current like S_q^p.

If an electric field $\mathbf{E}$ is applied across a magnetoplasma containing magnetic induction $\mathbf{B}$, the magnetoplasma moves at velocity

$$\mathbf{v} = \mathbf{E} \times \mathbf{B}/|\mathbf{B}|^2 \tag{8.1}$$

(a)   (b)

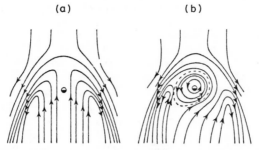

Figure 8.1 Circulation of the magnetosphere in the equatorial plane: (a) due to friction at the magnetopause; (b) including the effect of the Earth's rotation. (After A. Nishida, *J. geophys. Res.* **71**, 5669, 1966, copyrighted by American Geophysical Union.)

Sun

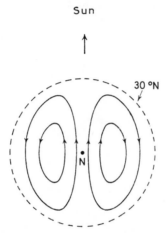

Figure 8.2 S_q^p current system due to the motion of the feet of magnetospheric field lines over the pole. (From J.A. Ratcliffe, *An Introduction to the Ionosphere and Magnetosphere*, Cambridge University Press, 1972.)

(see also Sections 4.4.3 and 5.2.4). Conversely, it may be shown that if magneto-plasma moves at **v** with respect to a stationary observer, the observer will measure an electric field

$$\mathbf{E} = \mathbf{v} \times \mathbf{B} \qquad (8.2)$$

Motion of magnetoplasma and an electric field are in fact equivalent when the conductivity is high as in the magnetosphere. (This equivalence can be used to clarify the idea of a moving magnetic field, that meets the objection that since field lines are merely a physicist's fiction how can one be labelled so that its motion may be observed? The solution is to say that the motion of a magnetoplasma is the same as that of an observer moving within it in such a way that he detects zero electric field.) The circulation of the magnetosphere is therefore equivalent to an electric field. The field is directed across the tail from the dawn side to the dusk side, with a magnitude of about 0.3 mV m^{-1}. The total potential difference across the magnetosphere is thus about 40 kV. The further significance of the cross-tail electric field will be considered in the ensuing sections.

8.1.2 Reconnection with the interplanetary field

To appreciate the current ideas about the driving force of the magnetospheric circulation one must go to a suggestion by J. W. Dungey in 1961. Fig. 8.3 depicts the geomagnetic field in polar section added to an interplanetary field directed either northward or southward. In the second case neutral points are formed in the equatorial plane, and some interplanetary field lines are connected to lines of the geomagnetic field. In the first case the neutral points are on the axis of the dipole and there is no connection between the two fields. In fact the IMF tends to lie in the solar—ecliptic plane, orientated at the 'garden hose' angle (Fig. 7.7). However, the field usually has some component out of the ecliptic plane, and this component can connect with the geomagnetic field when directed southward.

The IMF is frozen into the solar wind and carried along with it. The geomagnetic field lines connected to the IMF are also carried along, being dragged over the polar regions as in Fig. 8.3c from the day side to the night side. The velocity of field over the poles can be estimated from Equation 8.1. Taking the radius of the polar caps

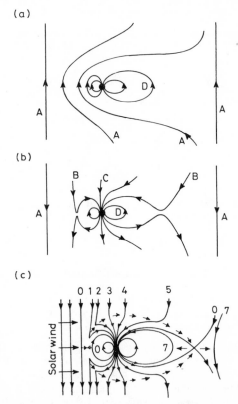

Figure 8.3 Terrestrial and solar-wind fields in polar section: (a) with northward IMF, giving a closed magnetosphere; (b) with southward IMF, giving an open magnetosphere; (c) showing the circulation due to the flow of the solar wind. ((a) and (b) after C.T. Russell, *Critical Problems of Magnetospheric Physics*, 1972, with permission of the National Academy of Sciences, Washington, DC, after J.W. Dungey; (c) after R.H. Levy *et al.*, *Am. Inst. Aeronaut. Astronaut. J.* **2**, 2065, 1964.)

(from which field lines go to the solar wind) as 15° of latitude or 1.7×10^6 m, and the total magnetospheric potential difference as 60 kV, the polar cap electric field (E_p) is 20 mV m^{-1}. Since $B_p = 0.55$ $g = 5.5 \times 10^{-5}$ Wb m^{-2},

$$|v| = |E_p|/|B_p| \approx 400 \text{ m s}^{-1}$$

In the tail the broken lines reconnect to form closed field lines and these move in towards the Earth. Fig. 8.1 and 8.3 are drawn for different sections of the magnetosphere, but the three-dimensional pattern of circulation is of course the same whether friction or an electric field is taken as the driving force.

Note also that the magnetosphere shown in Fig. 8.3a is closed overall, with no direct connection to the solar wind. Version (b) is 'open', since such connections exist. Evidence for solar wind events within the magnetosphere is therefore evidence for (b) and an open magnetosphere.

8.1.3 The magnetosphere as MHD generator

If the magnetosphere is driven by an electric field one might well wonder what might be the source of such a large-scale electric field. The generator is in fact the interplanetary magnetic field blown across the magnetosphere with the solar wind. The system is like a laboratory magneto-hydrodynamic (MHD) generator, in which a jet of plasma is forced across a static magnetic field and an electric potential is developed by dynamo action. If the magnetosphere were static, the solar wind plasma would blow across the field lines of type C in Fig. 8.3b, and the generator so formed would produce an electric field across the magnetosphere. The magnitude of this electric field is, of course, just enough to produce the circulation which keeps the merged field lines moving with the solar wind.

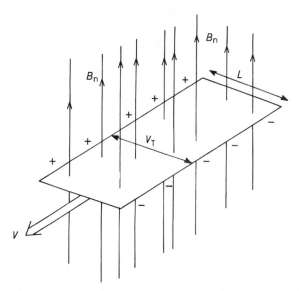

Figure 8.4   The magnetosphere as an MHD generator. A conducting plate of width L, moving at speed v through magnetic induction B_n, generates potential difference V_T. A jet of plasma serves as the conducting plate. (After S.-I. Akasofu, *Chapman Memorial Lecture, 1973*.)

173

The MHD generator is shown in Fig. 8.4. The total potential drop across the magnetotail is of magnitude

$$V_T = vLB_n \qquad (8.3)$$

where v is the solar wind speed, L the width of the tail, and B_n the magnetic induction normal to the magnetotail boundary. To estimate B_n we say that the total magnetic flux leaving the polar caps (approximately the regions enclosed by the auroral ovals — Section 8.2.2) ultimately connects with the IMF over the whole surface of the magentotail. With a polar cap of radius R_p and magnetic induction B_p, a tail of radius R_T and length S_T

$$B_n = \pi R_p^2 B_p / 2\pi R_T S_T \qquad (8.4)$$

Taking $R_p = 1.7 \times 10^6$ m, $B_p = 5.5 \times 10^{-5}$ Wb m^{-2}, $R_T = 20 R_E = 1.3 \times 10^8$ m, $S_T = 200 R_E = 3.2 \times 10^9$ m, then $B_n = 4.5 \times 10^{-10}$ Wb m$^{-2} = 0.45\,\gamma$. If the solar wind speed is 500 km s^{-1}, $V_T = 58$ kV. While such a calculation assumes values that are not well known, e.g. S_T, it does in fact give a result of about the correct magnitude according to the evidence available. The success of this approach does not rule out the possibility that some of the driving force might be friction from the solar wind itself, as in the original Axford and Hines concept.

8.1.4 Field merging in the neutral sheet

In the dynamic magnetosphere the plasma sheet is the region where open field lines reconnect and move in towards the Earth. Satellite observations show a small N–S component across the plasma sheet, indicating that the field lines are not strictly parallel (as in Fig. 7.9) but are rather extended dipole lines. Such a configuration cannot be static, since the tension in the field lines will produce a net force towards the Earth. The resulting depletion of field from the tail is made up by lines arriving over the poles and moving in through the lobes towards the neutral sheet. These motions are what would be expected in the cross-tail electric field, E_T, as may be verified qualitatively by the left-hand rule.

The physical mechanism of merging in the tail is not certain, but, as will be seen later, it is a most significant question in solar–terrestrial physics. Merging is not likely to occur uniformly along the tail, but probably takes place in limited regions where the current sheet is locally less intense. The situation in such a region

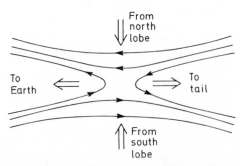

Figure 8.5 X-type neutral point in the magnetotail. The configuration is dynamic, with magnetoplasma arriving from the lobes and moving earthward and tailward subsequently.

would be as in Fig. 8.5, and is known as an X-type neutral point. If the induction normal to the sheet is 2γ, about 10% of the tail field, the inward velocity of field lines towards the Earth is about 100 km s^{-1} (calculated as the $\mathbf{E} \times \mathbf{B}$ drift in an electric field of 0.3 mV m^{-1}). Plasma trapped in the shortening tubes of force is energized by Fermi acceleration (Section 2.3.7). If the field lines shorten by 10% the energy is increased by 20%. It is more likely that merging in the tail is intermittent and catastrophic, and that this phenomenon is the source of the substorm and the aurora, to be discussed in Section 8.2.

8.1.5 Dynamics of the plasmasphere

Continuing the equivalence between moving magnetoplasma and an electric field, it can be seen that a dipole magnetic field rotating about the axis of the dipole produces to a stationary observer an electric field of magnitude $r\omega B$ at radial distance r if the angular frequency is ω. For rotation of the Earth's field, the electric field is directed inward. The inner part of the magnetosphere co-rotates with the Earth (Fig. 8.1b). This is the region where the co-rotation electric field is relatively large, and it is identified with the plasmasphere. In the outer part of the magnetosphere, which takes part in the general circulation, the cross-tail field is larger than the co-rotation field. The location of the plasmapause can be estimated by equating the cross-tail and co-rotation fields:

$$E_T = \frac{B_E}{L^3} \times LR_E\omega$$

where L is the L-value, R_E the radius of the Earth, and B_E the geomagnetic induction at the surface of the Earth at the equator. Putting in numerical values

$$E_T = 14.4/L^2 \qquad \text{mV m}^{-1} \tag{8.5}$$

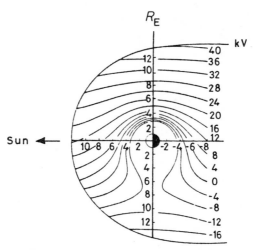

Figure 8.6 Calculated distribution of equipotentials about the Earth, derived by adding the co-rotation field to a cross-tail field of 2kV/R_E. (From R.J. Walker and M.G. Kivelson, *J. geophys. Res.* **80**, 2074, 1975, copyrighted by American Geophysical Union.)

(a)

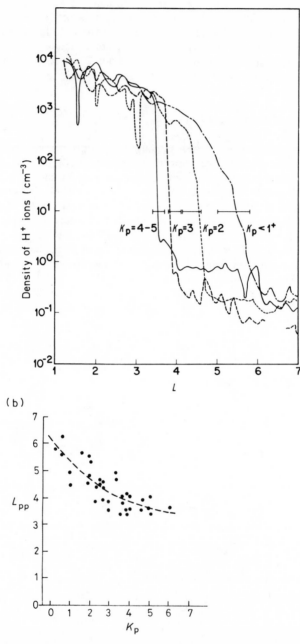

(b)

Figure 8.7 Data on the plasmapause position: (a) satellite observations of ion density, showing the plasmapause at several K_p levels; (b) relation between plasmapause distance, L_{pp}, and K_p. (After C.R. Chappell *et al., J. geophys. Res.* **75**, 50, 1970, copyrighted by American Geophysical Union.)

It will be seen that a plasmapause at $L = 4$ corresponds to $E_T \sim 1\ \text{mV m}^{-1}$. A computed distribution of the electric equipotentials around the Earth is shown in Fig. 8.6. The terrestrial magnetosphere is an interesting one in that the co-rotation and the cross-tail fields are of similar magnitude. Jupiter's magnetosphere is dominated by rotation and Mars's by the solar wind.

Since the plasmapause distance depends on the magnitude of the cross-tail electric field, it is to be expected that the size of the plasmasphere will vary with the solar wind and will show a correlation with other quantities that depend on the solar wind. Observations show the plasmapause moving in and out as the level of geophysical activity changes. It usually occurs between L-values 3 and 6, though it has been seen as close to the Earth as $L = 2$, i.e. only one earth radius above the surface. According to whistler observations (Sections 3.7.5 and 7.5.1), the geocentric distance to the plasmapause in earth radii (L_{pp}) during the post-midnight hours is related to the greatest value of K_p (the 3-h index of magnetic activity, described in Section 8.2.2) recorded during the previous 12 h ($\hat{K}_p$) by the empirical law

$$L_{pp} = 5.7 - 0.47 \hat{K}_p \tag{8.6}$$

Some actual measurements of plasmapause position by a satellite are shown in Fig. 8.7, and these are related to the K_p value at the time of the observation. Justification for a delayed effect, as implied by Equation 8.6, comes from the nature of the probable mechanism. The loss of plasma when geophysical activity increases is probably due to an increase in the magnetospheric electric field, which alters the magnetospheric flow pattern in such a way that the outer layers of the plasmasphere are peeled off, as suggested in Fig. 8.8. Direct evidence for peeling is the observation of dense plasma regions in the afternoon sector which appear to have become separated from the main body of the plasmasphere. The detached plasma is lost into the magnetosphere, and eventually into the solar wind. (There might also be some draining of plasma into the ionosphere during the storm.)

As activity subsides the protonosphere gradually refills with ionization that diffuses up from the ionosphere. The whistler measurements indicate that this is rather a slow process. The rate of filling is determined by the speed of diffusion of protons (formed in the thermosphere by the charge exchange reaction between neutral hydrogen atoms and oxygen ions) along the magnetic lines of force into a large, initially vacant, volume. The volume defined by a tube of force depends on the fourth power of the L-value of that tube, so that it takes much longer to fill up tubes originating at a high magnetic latitude. Observations of the rate of filling are shown in Fig. 8.9. Since active periods may recur every few days, there will be times when the outer tubes are never refilled before being depleted again by the next storm.

8.2 The substorm and auroral phenomena

8.2.1 Luminous aurora

The aurora is the most readily observed consequence of the dynamic behaviour of the magnetosphere. For thousands of years men must have seen the northern and southern lights in the night sky, the aurora borealis and aurora australis, which therefore must surely qualify for a place among the oldest known geophysical phenomena. The term *aurora borealis* dates from 1621, and detailed reports of

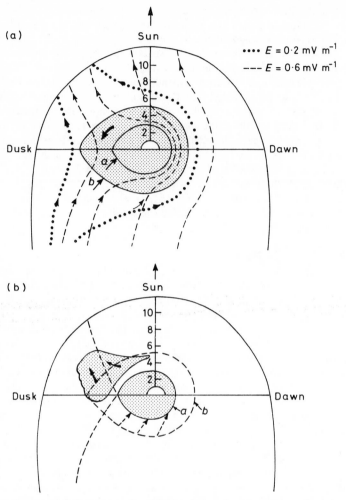

Figure 8.8 Detaching of plasma due to changing flow patterns during a magnetic storm: (a) flow patterns for cross-tail fields 0.2 and 0.6 mV m^{-1}; (b) resulting detached plasma. (After C.R. Chappell *et al.*, *J. geophys. Res.* **76**, 7632, 1971, copyrighted by American Geophysical Union.)

auroral displays date from 1716. Many original, even fanciful, explanations were devised, from electric discharges and 'elastic fluids' to the reflection of sunlight from polar snow or ice crystals in the polar atmosphere. Considering the antiquity of observations it comes perhaps as a surprise to realize how recent is any real progress towards an understanding of the phenomenon. Only in the early 1950s was it proved that the immediate source of the auroral light is the excitation of atmospheric gases by energetic particles; and not until rockets were first fired into aurora, in 1958, was it shown that energetic electrons are the major source. Such observations led immediately to another question — Where do these particles come from? — and progress towards the solution of that problem is even more recent. It depends, first, on a fairly detailed knowledge of the magnetospheric structure, such

178

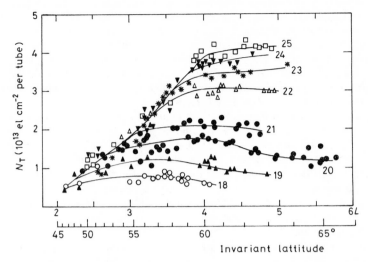

Figure 8.9 Refilling of the plasmasphere after a storm, 18–25 June 1965. The measurements are of the electron content between conjugate points as a function of L-value, by the whistler technique. The content is almost independent of L while the tube is filling, whereas the content of full tubes increases strongly with L. (After C.G. Park, *J. geophys. Res.* **79**, 165, 1974, copyrighted by American Geophysical Union.)

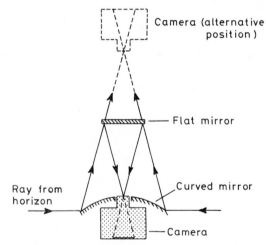

Figure 8.10 Arrangement of an all-sky camera. Alternative positions are shown for the camera, depending on whether the flat mirror is used. (From M. Gadsden, *Antarctica* (Ed. Hatherton) Methuen, 1965.)

as has emerged only during the last decade; and, second, it requires a detailed understanding of physical processes in the magnetosphere and that is not yet available. Consequently, it would not be true to say that the aurora is now fully comprehended, though excellent progress has been made in a remarkably short period of time. In this section we consider the aurora first as a phenomenon in its own right, and then in relation to the magnetosphere.

Classical studies of luminous aurora followed two lines: morphology and spectroscopy. Morphological studies were to map out the occurrence of the aurora in space and time, and to investigate the structure of individual aurorae in detail. Auroral spectroscopy was concerned more with the photochemical processes producing the light – this aspect of auroral studies has affinities with the subject of airglow (Section 4.3.5). Morphological and spectroscopic studies went along largely independent of each other before 1950.

The aurora is a highly structured phenomenon. Some features are as thin as 100 m, and time variations can be as rapid as 10 s^{-1}. One of the most basic instruments is the all-sky camera (Fig. 8.10), first used during the 1950s, which at regular intervals produces a photograph of the night sky from horizon to horizon. The structure is classified according to its general appearance, as in Table 8.1. When structure is present the height of the luminosity may be determined by triangulation. Between 1911 and 1943. C. Störmer made 12 000 determinations by means of spaced cameras (not all-sky cameras), and found that the low borders of auroral forms occur around 100–110 km. The luminosity is often concentrated into a band 10–20 km deep, though this varies with the auroral form (Fig. 8.11).

Table 8.1   Classification of Auroral Forms

Forms without ray structure

Homogeneous ray structure: a luminous arch stretching across the sky in a magnetically east–west direction; the lower edge is sharper than the upper, and there is no perceptible ray structure

Homogeneous band: somewhat like an arc but less uniform, and generally showing motions along its length; the band may be twisted into horseshoe bends

Pulsating arc: part or all of the arc pulsates

Diffuse surface: an amorphous glow without distinct boundary, or isolated patches resembling clouds

Pulsating surface: a diffuse surface that pulsates

Feeble glow: auroral light seen near the horizon, so that the actual form is not observed

Forms with ray structure

Rayed arc: a homogeneous arc broken up into vertical striations

Rayed band: a band made up of numerous vertical striations

Drapery: a band made up of long rays, giving the appearance of a curtain; the curtain may be folded

Rays: ray-like structures, appearing singly or in bundles separated from other forms

Corona: a rayed aurora seen near the magnetic zenith, giving the appearance of a fan or a dome with the rays converging on one point

Flaming aurora: waves of light moving rapidly upward over an auroral form

---

Aurorae are most likely to be seen in a narrow zone centred at about 67° magnetic latitude, the frequency falling off both equatorward and poleward of the auroral zone (Fig. 8.12). As will be seen from this figure, an aurora may be seen at some time every night at the centre of the zone under favourable observing conditions.

Auroral spectroscopy is carried out with a grating spectrograph, the image being recorded by a high speed camera, a TV system, a photomultiplier, photocell or infrared converter. Interference filters or Fabry–Perot monochromators may be set up to study particular lines of the spectrum. Observations have been extended into the ultraviolet by instruments carried on rockets or satellites. The auroral spectrum is complex. A bright aurora appears green or red to the eye, these colours

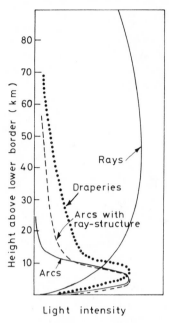

Figure 8.11 Profiles of auroral luminosity along various forms. (From B. Hultqvist, *Physics of Geomagnetic Phenomena* (Ed. Matsushita and Campbell) Academic Press, 1967.)

being due to the atomic oxygen emissions at 5577 and 6300 Å respectively. (Refer to Section 4.3.5 and Fig. 4.14.) The 3914 Å line of singly ionized molecular nitrogen is present in the violet, and the overall effect is usually greyish yellow in an aurora of average intensity. Some rays appear red at high altitude but green lower down. This is due to quenching of the ^{1}D state of atomic oxygen, as in airglow. Such aurorae are known as *Type A*. Some aurorae, known as *Type B*, have a red lower border; these are at exceptionally low altitude and the red light probably comes from molecular oxygen emissions, the green 5577 being quenched.

The interpretation of auroral lines is by no means a simple problem, since they are in general 'forbidden' according to the spectroscopists' terminology. This does not mean that the responsible transitions cannot occur at all, but that, because they defy the usual selection rules, they are relatively improbable. The green line was not identified until 1925, following the accurate measurement of its wavelength 2 years previously. The spectroscopy of the early 1950s was particularly significant in the development of auroral concepts, for in 1950 C. W. Gartlein identified the H-α line in the aurora and in 1951 A. B. Meinel discovered a Doppler shift in the line. These observations showed beyond doubt that at least part of the luminous auroral phenomenon is produced by the arrival of energetic protons from a higher level of the atmosphere. (This accounts for the so-called *proton aurora*. It was several years later, when rockets were fired into auroral displays, that the greater role of energetic electrons was discovered.) Thus, for the first time, the way was cleared for a general acceptance that aurorae are caused by an influx of energetic particles to the atmosphere.

The intensity of an aurora may be classified on a scale of I to IV, as in Table 8.2. The table also shows the standards of comparison that a visual observer might use,

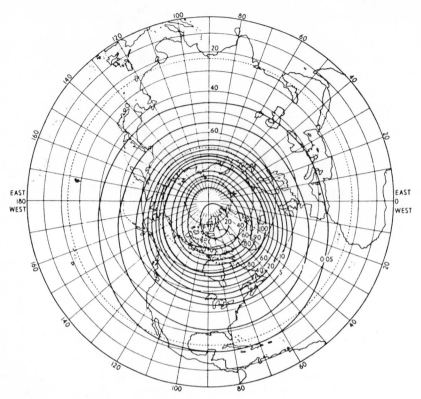

Figure 8.12 Northern auroral zone, based on the percentage of good observing nights when aurora may be seen. (From D.R. Bates, *Physics of the Upper Atmosphere* (Ed. Ratcliffe) Academic Press, 1960.)

Table 8.2   Intensity Classification of the Aurora

Intensity	Equivalent to	Kilorayleighs	Energy deposition (erg cm^{-2} s^{-1})
I	Milky Way	1	3
II	Thin moonlit cirrus	10	30
III	Moonlit cumulus	100	300
IV	Full moonlight	1000	3000

the equivalent intensity in kilorayleighs (defined in Section 4.3.5) and the approximate rate of energy deposition at such times. For exact measurements of auroral intensity one would use a photometer, either directed at some fixed point in the sky or scanned so that the spatial distribution of intensity is determined as well. Neither scanning photometers nor cameras can register the most rapid fluctuations in auroral light, and TV techniques, both monochrome and colour, have been applied to auroral photography in recent years. Apart from their scientific value, some of the colour observations, when re-run as cine film, include sequences of great visual beauty and help to bring to a wider audience the sense of wonder and fascination that seems to grip many of those who witness and study the aurora.

182

8.2.2 Modern auroral phenomena

The first half of the 20th century may be considered as the age of development of auroral studies, when valid scientific investigations were paving the way for the explosive growth of knowledge that was shortly to follow. Modern studies really began with the International Geophysical Year of 1957–58, which built up world-wide interest in geophysical problems and fostered international cooperation in solving them. The IGY also brought a great expansion of scientific work in the Antarctic continent, and coincided with the first scientific applications of rocket and satellite technology. As a result, not only were many more details found concerning the distribution, occurrence and mechanisms of the luminous aurora, but also, because of the variety of observational techniques that were available by this time, the wide range of the auroral phenomena came to be appreciated.

When energetic particles enter the atmosphere there are several observable consequences. Each of the following is either a consequence of energetic particles arriving in the atmosphere or is closely connected with the processes of particle precipitation:

(a) luminous aurora, as discussed in the previous section;
(b) radar aurora, in which radio echoes are received from ionization in the region of the aurora;
(c) auroral radio absorption, the absorption of radio waves in the excess ionization produced by auroral particles that penetrate into the lower ionosphere;
(d) auroral X-rays, detectable from high-altitude balloons;
(e) magnetic disturbances due to enhanced electric currents in the auroral E region, detected by ground-based magnetometers;
(f) very-low-frequency radio emissions, generated in the magnetosphere and propagating to the ground in the whistler mode (Section 3.7.5);
(g) magnetic micropulsations, essentially the fine structure of magnetic disturbances, some components being generated in the magnetosphere.

Common to all auroral phenomena is the nature of their association with solar activity. In Chapter 10 we shall discuss disturbances which show one-to-one association with individual solar flares. No such direct association is found for auroral phenomena, but there is a general and non-specific association in that aurorae occur more often when the Sun is more active. From the 1930s the term 'M-region' was used to describe a solar region that caused aurorae and magnetic storms. The M-region was a hypothesis which has become less necessary as more has come to be known about solar–terrestrial relations, though it retains some use as a descriptive term.

Like the luminous aurora, the auroral phenomena tend to be highly structured in both space and time, with a concentration into high-latitude zones. The concept of the auroral zones has evolved considerably in recent years. The classical picture of auroral occurrence (Fig. 8.12), though correct as far as it goes, is nevertheless insufficient and somewhat misleading. In 1963 Y. I. Feldstein, analysing IGY data, pointed out that at a given time the locus of the aurora is oval rather than circular, the magnetic latitude of the maximum being near 67° at local midnight but near 77° at local noon. The incidence of the aurora in the oval also depends on local time, with maxima near midnight and again on the day side (Fig. 8.13). Since observational conditions favour the night sector, the classical auroral zone is actually the locus of

183

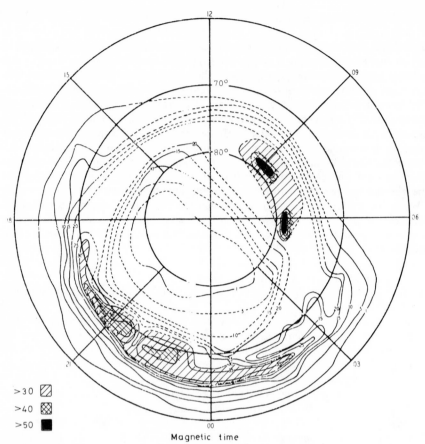

>30 ◻
>40 ▨
>50 ■

Magnetic time

Figure 8.13 Iso-auroral diagram for the IQSY (1964–65), showing contours of equal incidence of all auroral forms as seen from Alaska. (From W.J. Stringer and A.E. Belon, *J. geophys. Res.* **72**, 4415, 1967, copyrighted by American Geophysical Union.)

the midnight aurora as the Earth rotates beneath the oval, which (to a first approximation) remains in a fixed orientation with respect to the Sun.

For many years, studies of the luminous aurora concentrated on the discrete forms, since they are spectacular and their fine and active structure makes them more readily observed against the background light of the sky. In the early 1960s it was reported that a diffuse auroral glow is usually present at the same time as the discrete aurora. This component, originally called the *mantle aurora* but more recently the *diffuse aurora*, contributes at least as much light, in total, as the discrete forms, though it is more difficult to observe from the ground because its intensity per unit area is low. Striking auroral pictures have recently been obtained from photometers carried on orbiting satellites, which look down on the aurora from above. These observations tend to be dominated by the diffuse aurora, which is confirmed to be lacking in fine structure. Discrete forms may occur within the diffuse glow or poleward of it, but they are not seen on the equatorward side. On the evidence available the diffuse aurora seems to be associated with the plasma sheet (see Section 7.5.2 and Fig. 7.17).

184

The satellite pictures also confirm directly the shape of the auroral oval and give the interesting new result that the oval is more than a statistical distribution. It may be seen as a virtually continuous band of light encircling the pole, consistent with the idea that the aurora is always present to some extent.

When energetic electrons are stopped in the atmospheric gas, one electron–ion pair is produced for every 35 eV of energy dissipated. Taking into account the effective recombination coefficient (about 10^{-7} at 90–100 km) it can be shown that the electron density can be increased to 5×10^6 cm^{-3} during auroral activity. According to Equation 3.7, such an electron density will reflect radio waves of frequency below 20 MHz. Experience shows that auroral ionization may be detected also at VHF (> 30 MHz) and UHF (> 300 MHz), when the echoes are partial reflections from irregularities in the auroral ionization. Such irregularities tend to be field aligned, and the echoes are aspect sensitive, being received most strongly when the radar lies in a plane normal to the geomagnetic field line. The irregularities form because of instabilities in the electrojet, and their motion may, through the theory, be interpreted to give the drift speed of the electrons. The heights of radar aurorae are similar to those of luminous aurorae. The echoes generally occur in the evening, midnight and morning sectors (Fig. 8.14).

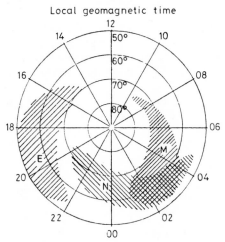

Figure 8.14   Occurrence of radar aurora. (From R.S. Unwin, *J. atmos. terr. Phys.* **28**, 1167, 1966, by permission of Pergamon Press.)

The enhanced ionization also absorbs radio waves (Equation 3.10). Ionosondes operating at a few megahertz are often 'blacked out', i.e. show no echoes, during auroral disturbances. A riometer (Section 3.7.3), operating perhaps at 30 MHz, can measure the attenuation of cosmic radio noise at such times, and the amount of the absorption is typically a few decibels, values over 10 dB being rather rare. On an active day the auroral absorption might look like Fig. 8.15. The statistics of auroral absorption indicate an occurrence zone more circular than oval. The largest events tend to appear several hours before midnight, and again before noon, but if the absorption is simply averaged over many days the maximum is seen between 0800 and 1000 h local time. Auroral absorption is produced lower in the atmosphere

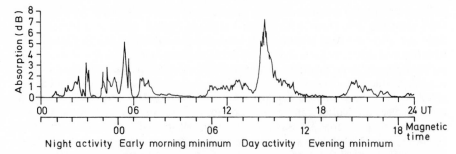

Figure 8.15 Auroral radio absorption on one day, seen on 15 October 1963 at Byrd Station, Antarctica, with a 30 MHz riometer. The descriptions below the axis refer to the typical behaviour; the evening minimum was not respected on the day shown! (After J.K. Hargreaves, *Proc. Inst. elect. electronics Engr.* **57**, 1348, 1969.)

than the luminosity, and the largest contributions probably arise near 90 km by night and 75 km by day. Most riometers operate with a wide-beam antenna, and thus cannot detect fine structure like that in the luminosity. However, such evidence as is available suggests that most of the absorption is in large patches extending over several hundred kilometers, corresponding, possibly, with the diffuse auroral light rather than the discrete forms.

There are some striking similarities between auroral radio absorption and the auroral X-rays that may be detected from high altitude balloons. Both phenomena are observed at greatest intensity during the local morning at a latitude some 10° lower than that of the auroral oval. The events observed by a riometer and an X-ray

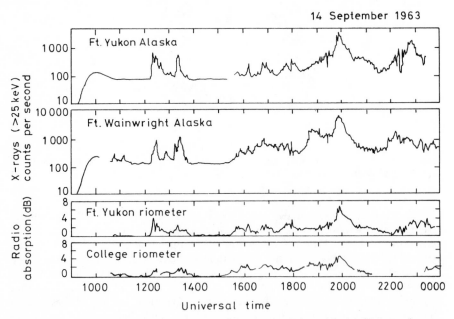

Figure 8.16 Auroral X-rays observed on balloons over Alaska, with simultaneous riometer measurements. (From J.R. Barcus and R.R. Brown, *J. geophys. Res.* **71**, 825, 1966, copyrighted by American Geophysical Union.)

detector often have similar time structures. The first observations of auroral X-rays were made in the mid-1950s using *rockoons* — small rockets launched from balloons — but in most of the recent work the detectors are carried directly on a balloon at an altitude near 30 km. Individual balloons stay aloft for several hours to several days, and thus time continuity is achieved although the balloon drifts away from the launching site on the prevailing wind. Some space–time resolution can be attained by launching several balloons at intervals. The X-rays detected have energies exceeding 30 keV and arise from the bremsstrahlung mechanism (Section 8.2.4) as the primary particles are stopped in the atmosphere. Fig. 8.16 indicates the auroral X-radiation measured on balloon flights over Alaska. Note the general similarity of structure between the X-rays and the auroral radio absorption measured simultaneously. However, the X-rays do at times show rapid pulsations that are not seen with a riometer.

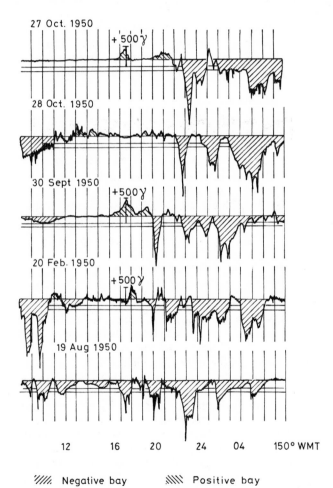

Figure 8.17   Negative and positive bays seen on magnetograms at College, Alaska. The time zone is that of the 150° west meridian. (After M. Sugiura and S. Chapman, *Abh. Akad. Wiss. Göttingen Math. Phys. K1 Spec. Issue* **4**, 53, 1960.)

The enhancement of E-region ionization during auroral activity increases the conductivity and, hence, the E-region electric currents. Through their magnetic effects these perturb ground-based magnetometers, giving a series of *bays* (Fig. 8.17) on the magnetograms. The bays can amount to 1000 γ in the auroral zone. The perturbation tends to be positive just before midnight and negative after, suggesting eastward and westward currents respectively. The changeover point is sometimes called the *Harang discontinuity*. Apart from their use in studying the currents flowing in individual disturbances, magnetic records provide the basic disturbance indices which are compiled and published routinely to give, in a standard form, a semiquantitative measurement of the level of geophysical activity. Several indices are in use, but the most common are K_p and A_p. These are based on the range of variation over 3 h periods of the day in the records from a number of selected magnetic observatories. After some local weighting, and averaging, the K_p index is obtained for each 3-h interval of the UT day on a scale ranging from 0 for 'very quiet' to 9 for 'very disturbed'. A_p is a daily index, derived from the same basic data but converted to a scale related to the deviation of a magnetometer trace. A_p is essentially linear, whereas K_p is approximately logarithmic in relation to the magnetic effect. A K_p value can be expressed as an equivalent 3-h a_p, related as in Table 8.3. K_p is often presented as a Bartels *musical diagram* (Fig. 8.18).

Table 8.3   Equivalent Values of Magnetic Indices

K_p	0	1	2	3	4	5	6	7	8	9
a_p	0	3	7	15	27	48	80	140	240	400

Plotted thus against the day of the solar rotation it shows up the recurrence tendency of the M-regions.

One of the first results of direct solar wind measurements was that the velocity (v) of the solar wind is related to terrestrial magnetic disturbances by the empirical law

$$v(\text{km s}^{-1}) = (8.44 \pm 0.74) \sum K_p + (330 \pm 17) \tag{8.7}$$

where ΣK_p is the sum of the 3-h K_p indices taken over 1 day. This observation showed that the solar wind is the agent which communicates M-region behaviour to the Earth. It also brought a practical consequence, that a basic property of the solar wind could now be estimated from a readily available geophysical index. (The general problem of monitoring the space environment is discussed in Section 12.4.)

Some of the radio noises detected from the magnetosphere do not show the dispersion characteristic of whistlers. Such noises may occur repeatedly in the same form, and this, coupled with their alternate appearance at geomagnetically conjugate points, indicates that the electromagnetic emission is not merely being reflected like a whistler but is actually being regenerated in the magnetosphere. The energy originates with a bunch of charged particles mirroring back and forth along the magnetic field lines, and wave–particle interactions (Chapter 9) transfer some of this energy into electromagnetic waves. The electromagnetic noises from the magnetosphere cover a wide range of frequencies, from below 1 kHz to as high as 100 kHz. While some emissions are discrete, i.e. show a well defined frequency–time characteristic, others are unstructured and sound like random noise when

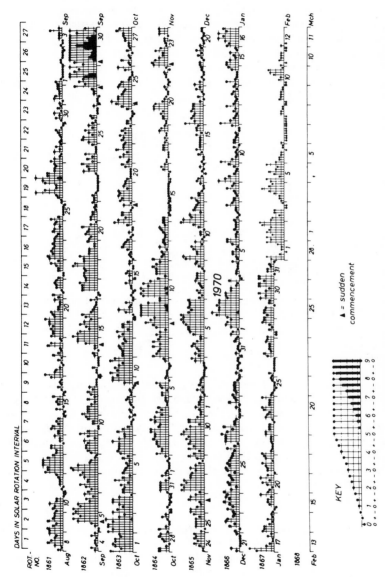

Figure 8.18 A Bartels 'musical diagram', displaying K_p as a function of solar rotation. (From *Solar-Geophysical Data*, January 1970, World Data Center-A, Boulder, Colorado.)

189

played back through a loudspeaker. The terms *chorus* and *hiss* are often used to describe different kinds of emission, the first being derived from a resemblance to birdsong at dawn.

At even lower frequencies, corresponding to periods of seconds or minutes, the emissions show up on magnetometers, when they are known as micropulsations. In this band the waves are hydromagnetic (Section 2.4.6) rather than electromagnetic. Some magnetospheric emissions are connected with auroral particle precipitation and some are not. Their principle significance now is the evidence they provide on wave–particle interactions in the magnetospheric plasma; they also have a practical role in diagnosing the state of the magnetosphere from the ground.

The range of the auroral phenomena and some links between them are indicated in general terms in Fig. 8.19.

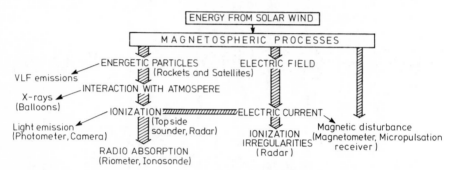

Figure 8.19   Some links between auroral phenomena. Techniques of observation are shown in brackets. (From J.K. Hargreaves, *Proc. Inst. elect. electronics Engr.* **57**, 1348, 1969.)

8.2.3 *Direct observations of auroral particles*

Many direct observations have been made of the incoming auroral electrons and protons by means of rocket-borne detectors. A rocket flight, being of short duration, is most valuable in providing a spot measurement of the energy flux, the energy spectrum and the pitch-angle distribution. The first such flight, in 1960, showed a mono-energetic flux of electrons of energy 5 keV, but this is not typical. More usually the spectra cover a broad band of energies to 100 keV and beyond.

It is sometimes possible to represent the spectral distribution by a simple form such as the exponential

$$N(E) = N_0 \exp(-E/E_0) \tag{8.8}$$

which would mean that $N(E)\,dE$ particles arrived per unit cross-sectional area per second over the energy interval E to $E + dE$. The characteristic energy, E_0, might be 5–10 keV at the lower energies and 10–40 keV at the higher energies. It is not usual for a single characteristic energy to describe the spectrum over the full range. A representative spectrum is shown in Fig. 8.20. Mono-energetic components are present on some occasions, when they appear superimposed on a background spectrum of normal form.

Satellites also monitor flux, energy and pitch angle, but particularly they provide information about the spatial distribution. The division of the day side auroral

190

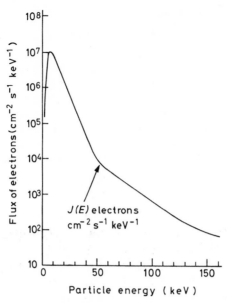

Figure 8.20 A typical spectrum of auroral electrons. (After W.N. Hess, *Radiation Belt and Magnetosphere,* used by permission of Ginn and Co., Copyright, 1968, by Xerox Corporation.)

region into two zones can be distinguished with orbiting satellites, and the rapid latitude coverage is convenient for integrating the total energy input during an auroral disturbance. Such estimates suggest that some 0.1–0.2% of the solar-wind energy intercepted by the magnetosphere appears as auroral particles.

A particularly interesting type of precipitation event observed from satellites is the *inverted-V*, from the appearance of the spectrum as the satellite moves through the precipitation region. In an inverted-V event the spectrum is peaked, the peak having highest energy in the centre of the region, as in Fig. 8.21a. These events have been identified with auroral arcs, and they seem to be connected with the monochromatic fluxes reported from rocket flights. There is evidence that the pitch angle is peaked along the geomagnetic field, which is consistent with acceleration by an electric field of form like that shown in Fig. 8.21b. It will be seen that such a distribution of electric potential implies that the vertical electric field will be greatest in the centre of the structure, and therefore the particles will undergo greatest acceleration in the centre of the arc.

8.2.4 *Effects of energetic electrons in the atmosphere*

The luminous aurora is produced by primary electrons in the energy range 1–10 keV, and some other auroral phenomena, such as X-rays and radio absorption, are caused by electrons with energies between 10 or 20 and 100 or 200 keV. In order to explain these phenomena it is necessary to study the interaction of energetic electrons 1–100 keV with air at low pressure. The relation of aurora to the primary particle fluxes is also important in a practical sense, since it may be considered part of the business of calibrating ground-based and balloon-based auroral sensors so

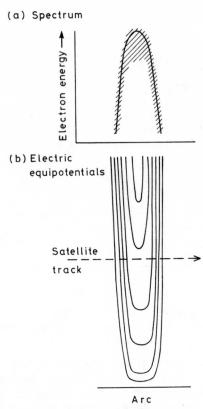

(a) Spectrum

Electron energy →

(b) Electric equipotentials

Satellite track

Arc

Figure 8.21 An 'inverted-V' event: (a) electron energy; (b) electric equipotentials to explain the observed energy structure. (After D.A. Gurnett, *Critical Problems of Magnetospheric Physics*, 1972, with permission of the National Academy of Sciences, Washington, DC.)

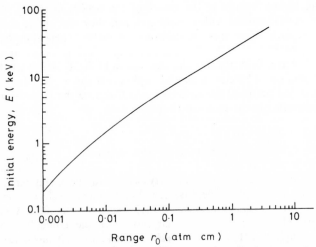

Figure 8.22 Range of electrons in air, according to various measurements and computations. (After M.H. Rees, *Planet. Space Sci.* **12**, 722, 1964.)

that all available techniques can be effectively brought into action in auroral investigations.

The range of energetic electrons in air is known from laboratory measurements. Fig. 8.22 summarizes measurements of the range r_0 (in atm. cm) as a function of the initial electron energy, E. The range is determined principally by the loss of energy due to excitation and ionization as the electrons collide with atmospheric atoms. A reduction of gas pressure is accompanied by a proportional increase in the distance travelled, since the number of gas particles per unit cross-section along the total path is constant. An auroral particle entering the atmosphere vertically from above travels into a medium of increasing density. The particle therefore penetrates to an altitude, h_p, such that the product of pressure and distance, integrated above h_p, is equal to the range r_0 in Fig. 8.22.

That is

$$\int_{h_p}^{\infty} P(h)\,dh = r_0$$

P being in atm. and h in cm. By the gas laws the range, which is the distance the electron would travel in air at 1 atm. pressure, is also given by

$$\int_{h_p}^{\infty} \rho(h)\,dh/\rho'$$

where ρ' is the air density at 1 atm. It depends, therefore, on the mass of gas in a column of unit cross-section down to altitude h_p. The mass is called the *atmospheric depth* at the penetration altitude, R. At any other altitude we call the atmospheric depth z. (It is readily shown that $z = P/g$, where P is the pressure and g is the gravitational acceleration.)

The efficiency of ion production depends on the speed of the incident electron at the level considered, and thus on the fraction of the atmospheric depth traversed, z/R. The efficiency $\lambda(z/R)$ also depends on the pitch-angle distribution of the electrons, since particles of large pitch angle travel along a flatter spiral than those with small pitch angle. If the incident particles are mono-directional, λ is a maximum at $z/R = 0.4$. For broader pitch-angle distributions λ maximizes at a smaller value of z/R such as 0.1 or 0.2.

If an incident electron has energy E, all this energy will be dissipated over the range r_0, and the average ion production rate will be $E/r_0 \Delta E$, where ΔE is the energy required (35 eV) for each ionization. At some point along the path where the density is ρ, the production rate is

$$\frac{E}{r \Delta E} \lambda \left(\frac{z}{R} \right)$$

where r is the range corresponding to density ρ. However, $r = R/\rho$, where R depends on the energy of the incident electron but not on the air density. Hence, since $\rho \propto n$, the number density, the rate of ion-pair production at depth z can be written

$$q_z = \frac{E}{\Delta E r_0} \cdot \lambda \left[\frac{z}{R} \right] \cdot \frac{n(z)}{n(R)} \tag{8.9}$$

A full computation, taking into account the height variation of density and pressure in the atmosphere, gives a family of ionization rate curves such as those in Fig. 8.23. This diagram is for a mono-energetic flux. Integration with respect to energy is required to give the effect of a realistic spectrum. Electrons entering the atmosphere with energies between 1 and 10 keV produce most of their ionization at heights down to 110 km. To ionize at 90 km or below the initial energy must exceed 30 keV.

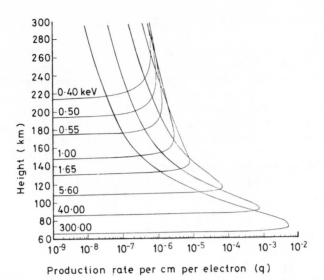

Figure 8.23 Computed production rate profiles. (After M.H. Rees, *Planet. Space Sci.* **11**, 1209, 1963.)

The amount of light emitted during auroral excitation depends on the photo-chemical processes concerned in producing a particular emission. To take an example, one prominent line of the auroral spectrum is that due to ionized molecular nitrogen, N_2^+, at 3914 Å. At this wavelength approximately one quantum of energy is emitted for every 50 ion pairs produced. The intensity of 3914 Å radiation is therefore proportional to the ionization rate calculated as above. Observations of the height profile of auroral luminosity (Fig. 8.11) may therefore be related to the incoming flux and energy spectrum of the auroral electrons.

The increased electron density of the lower ionosphere during auroral activity may be calculated from the production rate (Equation 8.9) and the effective re-combination coefficient (α_e): $N_e = (q/\alpha_e)^{1/2}$, a recombination type of process being assumed at this relatively low altitude. The height profile of α_e is less well known in the D region than is the production rate profile, but the best estimates agree with rocket measurements that auroral ionization densities can reach 10^6 cm^{-3}, far exceeding that due to solar EUV at this latitude.

A small amount of the energy is converted to X-rays by the bremsstrahlung process, which depends on the rapid deceleration of a fast electron on collision with a particle of the atmospheric gas. The X-rays penetrate deeper into the atmo-sphere and may be observed by detectors carried on balloons at 30 or 40 km altitude. The computation of the bremsstrahlung X-ray flux is fairly complicated. An electron of energy E can produce photons of energy E or less, and the required cross-sections for this process have been worked out. The radiation is emitted over a range of angles, and for non-relativistic electrons ($v \ll c$) the intensity varies according to

$$f(\theta) = \sin^2 \theta \tag{8.10}$$

where θ is the angle between the direction of the primary electron and the emitted

194

X-ray photon. For a realistic pitch-angle distribution of electrons the X-rays are emitted over a wide range of angles.

The X-rays produce further ionization when they, in turn, are stopped in the atmosphere. This ionization rate is several factors of ten smaller than that due to the primary electrons, but it occurs at a much lower altitude, the peak being as low as 50 km. The effective recombination rate is also larger at these levels, so that the resulting electron density is small compared with typical ionospheric values. Nevertheless, bremsstrahlung X-rays are the major source for the lower D region during auroral precipitation, exceeding the direct ionization rate below 70 or 80 km.

Table 8.4   Direct and X-ray Ion Production

E (keV)	Height of maximum production (km)		Maximum production rate (ion pairs/cm-electron)	
	Direct	X-ray	Direct	X-ray
3	126	88	2.5×10^{-5}	5.9×10^{-10}
10	108	70	1.4×10^{-4}	1.3×10^{-8}
30	94	48	5.6×10^{-4}	2.3×10^{-7}
100	84	37	1.9×10^{-3}	1.3×10^{-5}

Table 8.4 compares the heights and the maximum production rates due to direct and bremsstrahlung ionization for several initial electron energies.

8.2.5 The auroral substorm

It has been realized since the time of K. Birkeland in the opening years of the 20th century that auroral activity tends to occur in bursts. To a visual observer the aurora appears to be active for periods of about an hour, interspersed with quiescent periods of 2–3 h. Each of the active periods is a *substorm*, and since about 1964, when S.-I. Akasofu published extensive studies based on all-sky camera pictures, the substorm has become the central unifying concept of the auroral phenomena.

Akasofu's representation of the luminous aurora is shown in Fig. 8.24. The sequence of events begins with a quiet arc that brightens and moves towards the pole. Active auroral forms appear on the equatorward side, so that the oval is now much wider than before in the midnight sector. At the same time active aurorae move eastward towards the morning sector and travel westward towards evening. After 30 min to an hour the night sector recovers and the substorm as a whole dies away. This ballet of the discrete aurora (as summarized in Fig. 8.24) is unlikely to be the whole story since it is based on only one of the auroral phenomena. However, the representation is important in demonstrating the substorm not only as an intermittent occurrence but also as a dynamic one. Similar behaviour has been sought in other aspects of the aurora, and Fig. 8.25 shows a generalization that recognizes the two auroral ovals. Again we see the substorm beginning at midnight near 67° magnetic latitude, and subsequently spreading in both latitude and longitude.

The concept of the substorm, originally formulated from ground-based observations, has been elegantly confirmed by auroral observations from space. It is found that the amount of brightening in the auroral oval, i.e. the extent of the

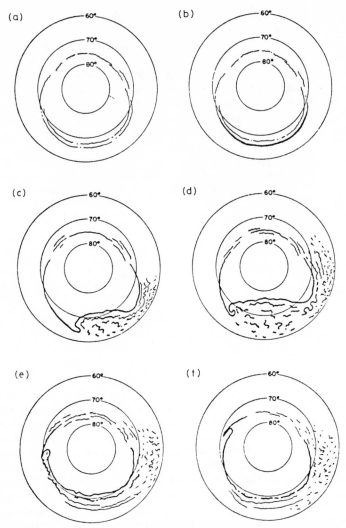

Figure 8.24 The substorm in the luminous aurora. (a) $T = 0$; (b) $T = 0-5$ min; (c) $T = 5-10$ min; (d) $T = 10-30$ min; (e) $T = 30$ min $- 1$ h; (f) $T = 1-2$ h. (From S.-I. Akasofu, *Polar and Magnetospheric Substorms, Astrophysics and Space Science Library, Vol. II*, Reidel, 1968.)

active aurora, is simply related to the area contained within the auroral oval as a whole. Akasofu has expressed the relationship in the formula

$$A_B = 0.05 (A - A_0)^2 \qquad (8.11)$$

where A_B is the bright area, A_0 is the area within the oval at quiet times, and A is the contained area during a substorm. Regarding the oval as the boundary between open and closed field lines, it will be seen that Equation 8.11 relates the rate of energy dissipation during a substorm to the excess energy of the magnetosphere above a threshold or *ground state*.

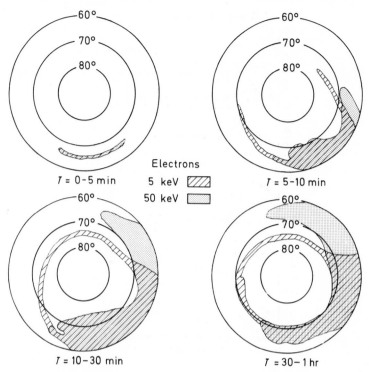

Figure 8.25 Typical development of electron precipitation in a substorm. Note that the two zones are distinct on the day side. (From S.-I. Akasofu, *Polar and Magnetospheric Substorms, Astrophysics and Space Science Library, Vol. II,* Reidel, 1968.)

8.2.6 The substorm in the magnetosphere

Substorm effects can be detected directly by satellite-borne detectors within the magnetosphere, and it is in this context of the magnetosphere as a whole that the substorm phenomenon is coming to be understood. Satellites at geostationary distance observe an increase of energetic electron fluxes during substorms — which is perhaps not surprising, since auroral-zone field lines pass close to the geostationary distance $(6.6 R_E)$ if the field is not too distorted. Even more significant is the appearance of energetic particles at various distances down the tail. At 20 or 30 R_E the particle burst appears to be moving away from the Earth, and this activity occurs some tens of minutes after the substorm is seen in the aurora, indicating that the source is within 20 R_E. The plasma sheet also undergoes characteristic changes, becoming thinner just before the substorm begins (the *growth phase*) and thicker again during the substorm. The configuration of the tail field at three phases of the substorm is sketched in Fig. 8.26. In the *expansion phase*, when the aurora is most active, the field is consistent with the intermittent merging at a point, as indicated in Section 8.1.4.

Fig. 8.27 follows the history of a selected field line: connecting with the IMF, being convected over the poles as an open line, reconnecting in the tail, and returning to a nearly dipole form in the return earthward flow. If R_M and R_T are the rates

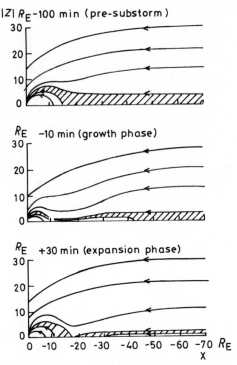

Figure 8.26 Possible magnetotail structure at three phases of the substorm. The time is reckoned from the visible onset in the auroral oval. The growth phase shows the plasma sheet thinning in a localized region, with field lines being stretched down the tail. In the expansion phase a neutral line has formed at $15-18\,R_E$. (After A. Nishida and N. Nagayama, *J. geophys. Res.* **78**, 3782, 1973, copyrighted by American Geophysical Union.)

of field connection at the front of the magnetosphere and in the tail respectively, T is the rate of transport of open magnetic flux into the tail and r is the rate at which closed flux is returned from the tail to the closed magnetosphere, $R_M = T = R_T = r$ implies a state of steady magnetospheric circulation (Section 8.1). If averages are taken over a long enough time it must still be true that

$$\bar{R}_M = \bar{T} = \bar{R}_T = \bar{r} \tag{8.12}$$

Intermittent merging will cause an imbalance. When the IMF turns southward R_M increases (Section 8.1.2) and for a while $R_M > R_T$; since more open field is being produced than is being removed, the auroral oval, as the boundary between open and closed field, moves towards the equator so that more open field can be contained. During a substorm $R_T > R_M$, and there is a net loss of open field. This probably accounts for the shrinking of the oval as the brightening arcs move poleward at the beginning of the substorm.

8.3 Magnetospheric current systems

8.3.1 The cross-tail current

It was pointed out in Chapter 7 that the extension of the magnetosphere down wind of the Sun can be considered due to an electric current flowing in a sheet

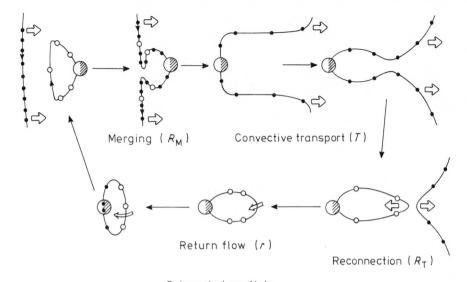

Merging (R_M)   Convective transport (T)

Return flow (r)

Reconnection (R_T)

• Solar wind particle
o Plasma sheet or Van Allen particle

Figure 8.27   The history of a field line. (After S.-I. Akasofu, *Chapman Memorial Lecture,* 1973.)

from the dawn side of the magnetosphere to the dusk side. The circuit is completed over the surface of the tail, so that the tail is confined within a pair of current solenoids as in Fig. 8.28. If the current density over the surface is i_T A m^{-1} (the plasma sheet density being $2i_T$), the magnetic induction along the tail is

$$B_T = \mu_0 i_T \tag{8.13}$$

Since $B_T \sim 20\,\gamma$, $i_T \sim 2 \times 10^{-8}$ (Wb m^{-2})/$4\pi \times 10^{-7}$ (H m^{-1}) $= 1.6 \times 10^{-2}$ A m^{-1}. The plasma sheet current density is therefore $\sim 3 \times 10^{-2}$ A m^{-1}. The total tail current is about 10^8 A, so that if the cross-tail potential difference is 60 kV the power taken from the solar wind amounts to 6×10^9 kW.

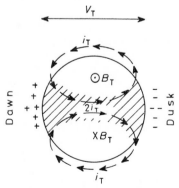

Figure 8.28   The electric currents across and around the magnetotail, seen from the Earth. (After L. Svalgaard, NASA Report SP-366, 1975.)

The current carriers in the plasma sheet are the electrons and protons that arrive from the lobes by drifting in the crossed electric (E_T) and magnetic (B_T) fields. If there are n carriers per unit volume of each sign and the width of the tail is $2R_T$, carriers reach the plasma sheet at a rate $(2n)(2R_T)E_T/B_T$ per unit length of the tail. In a steady state, carriers are removed again, at the same rate, from the dawn and dusk extremities of the sheet. The sheet current per unit length must therefore be

$$2i_T = 4nR_T E_T e/B_T \tag{8.14}$$

From Equations 8.13 and 8.14,

$$n = B_T^2/2R_T E_T e\mu_0 \tag{8.15}$$

Taking $B_T = 20\,\gamma = 2 \times 10^{-8}$ Wb m^{-2}, $R_T = 15\,R_E = 10^5$ km, $E_T = 3 \times 10^{-4}$ V m^{-1}, $e = 1.6 \times 10^{-19}$ C, $\mu_0 = 4\pi \times 10^{-7}$ H m^{-1}, the plasma density in the lobes of the tail is estimated as $n = 3 \times 10^4$ m^{-3}, which is of the correct general magnitude. This treatment neglects the component of tail field normal to the sheet, but this is small relative to the parallel component.

8.3.2 Birkeland currents

For many years it was assumed that ionospheric electric currents flowed only horizontally. Nevertheless, the conductivity is much greater along the magnetic field than across it (see Section 4.4.4) at the higher levels, and thus the possibility of currents along the field cannot be lightly dismissed. The first suggestion of such currents was made by Birkeland in 1908, and experimental evidence for their existence has been produced in recent years. Observations of the electron and proton fluxes entering the upper atmosphere show that the fluxes are unequal, leaving a net electric current density of about 3×10^{-7} A m^{-2}.

Most of the evidence, however, has come from magnetometers carried on satellites since 1966. When crossing the auroral ovals these show magnetic perturbations such as that in Fig. 8.29a. Here, a westward magnetic field was detected over a limited distance of 2 or 3 degrees of latitude (200–300 km), whose explanation requires a double sheet current of some 5×10^{-7} A m^{-2} directed upward at the lower latitude and downward at the higher latitude. These currents, which are plainly Birkeland currents, have been mapped in some detail and they are found to correspond with the auroral oval (Fig. 8.29b), the components into and out of the ionosphere changing places between the evening and the morning sectors. The typical current density is about 1 μA m^{-2}, values being rather greater on the high latitude side than in the lower latitude component. Thus there is a net flow of current into the ionosphere in the morning, and a net flow out in the evening. The currents vary with geomagnetic activity, to which they probably contribute directly. There is probably a good deal of space and time structure that does not appear in statistical analyses like Fig. 8.29b. The total current fed into and out of the ionosphere by the Birkeland currents is some 3.5×10^6 A. The currents are therefore very large and they form an important link between the magnetosphere and the ionosphere.

8.3.3 Electrojets

The ionospheric currents associated with substorms have traditionally been deduced from ground-based magnetograms, such as those in Fig. 8.17. Negative excursions

200

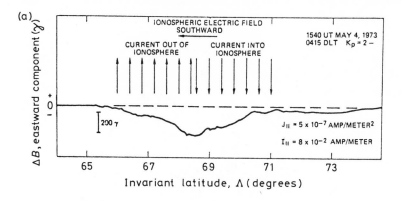

(a)

IONOSPHERIC ELECTRIC FIELD
SOUTHWARD

CURRENT OUT OF
IONOSPHERE

CURRENT INTO
IONOSPHERE

1540 UT MAY 4, 1973
0415 DLT $K_p = 2 -$

ΔB, eastward component (γ)

200 γ

$J_{\parallel} = 5 \times 10^{-7}$ AMP/METER^2

$I_{\parallel} = 8 \times 10^{-2}$ AMP/METER

Invariant latitude, Λ (degrees)

(b)

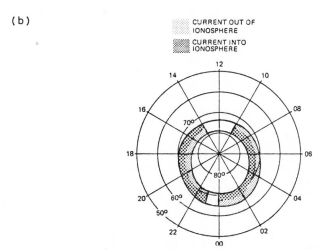

CURRENT OUT OF
IONOSPHERE

CURRENT INTO
IONOSPHERE

Figure 8.29   Birkeland currents: (a) observation by satellite-borne magnetometer, indicating the existence of a parallel electric current; (b) statistical distribution of Birkeland currents around the auroral oval. (From Zmuda and Armstrong, *J. geophys. Res.* **79**, 4611, 1974, copyrighted by American Geophysical Union.)

of the trace, which tend to occur after midnight, are interpreted as due to a westward current in the ionosphere; and positive bays are taken to imply eastward currents. The total flow pattern obtained from magnetic observations has evolved significantly over the years as the coverage of high latitudes has improved. Modern versions usually look rather like Fig. 8.30, with one major current circulation offset from the pole. Such currents, the most intense parts of which would be called *auroral electrojets*, are of course intermittent; more important, they are hypothetical! The analysis that produces such results neglects the possibility of vertical currents — indeed, there is no way of deducing the vertical current component from purely ground-based observations — and so the result is an *equivalent current system* rather than the true one. The essential equivalence between vertical and horizontal currents detected by their magnetic effects below the ionosphere is illustrated by the current in Fig. 8.31. In (a), two Birkeland currents are connected by an electrojet. (a) is equivalent to (b) plus (c), where (c) is the classical model

201

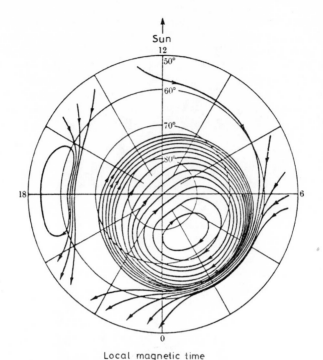

Local magnetic time

Figure 8.30   Equivalent current system of a magnetic substorm. The concentrations of current lines in the early morning and near 1800 would appear as electrojets. (From S.-I. Akasofu and S. Chapman, *Solar-Terrestrial Physics,* Oxford University Press, 1972.)

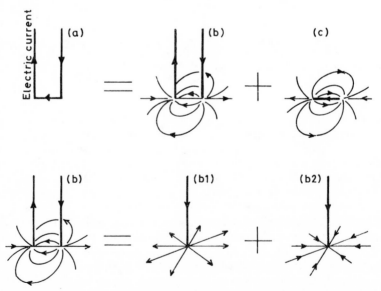

Figure 8.31   The equivalence of different current systems to a magnetometer on the ground. (From N. Fukushima, *Rep. Ionosphere Space Res. Japan,* **23**, 209, 1969.)

of an electrojet in which it was assumed that the return currents were distributed poleward and equatorward of the jet. (b) may be divided in (b1) and (b2), each comprising a vertical current and a spreading current in the ionosphere. It can be shown that neither (b1) nor (b2) produces any magnetic effect beneath the ionosphere. Therefore (a) and (c) look the same to a magnetometer on the ground. On the other hand, (b1) and (b2) do produce magnetic effects above the ionosphere, and therefore vertical currents can be detected from rockets or satellites.

8.3.4 Ring currents

The fourth kind of current that we shall consider can be detected by ground-based magnetometers at the lower latitudes. During a period of high geophysical activity, when high-latitude magnetometers show bays, an instrument at middle or equatorial latitude is likely to show a pattern like that in Fig. 8.32. The pattern represents the classical magnetic storm, consisting of three phases:

(i) an increase of magnetic field lasting only a few hours;
(ii) a large decrease in the H component, the perturbation building up to a maximum in about 1 day; and
(iii) a slow recovery to normal over several days.

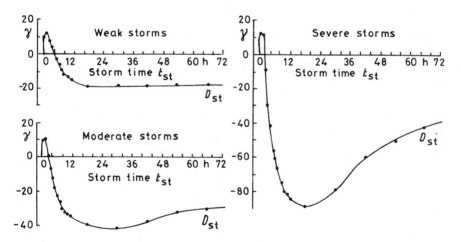

Figure 8.32   Representative magnetic deviations during magnetic storms. (After M. Sugiura and S. Chapman, *Abh. Akad. Wiss. Göttingen Math. Phys. Kl. Spec. Issue*, 4, 53, 1960.)

The first, or *initial phase*, is caused by the compression of the magnetosphere by the arrival of a burst of solar plasma at the magnetopause. The second or *main phase* is what concerns us here, since it is explained by an electric current encircling the Earth at a distance of several earth radii — the ring current.

The magnetic effect $H(\gamma)$ at the centre of a ring current $I(\mathrm{A})$ of radius $r\,(\mathrm{km})$ is $\Delta H = 2\pi I/10r$. To reduce the geomagnetic field the current must flow clockwise as seen by an observer looking down on the north pole. Currents induced in the ground increase the effect, and a factor of 3/2 is sometimes used to account for this. Thus

$$\Delta H = 3\pi I/10r \sim I/r \qquad (8.16)$$

If $I = 10^6$ A and $r = 4.5\ R_E = 3 \times 10^4$ km, $H \approx 30\ \gamma$. The currents associated with typical magnetic storms are $10^6 - 10^7$ A.

A clockwise ring current is produced by the drift of energetic particles in the geomagnetic field (Section 7.6.1), protons drifting to the west and electrons to the east. There is, of course, no way of discovering the distance of the ring current, or the nature and energy of the particles, from observations at the ground; but space observations place the current at $4-6R_E$ and have shown that it is due, not to the energetic Van Allen particles, but to the lower energy particles, mainly the protons. Fig. 8.33 shows an example where the energy band $20-120$ keV contributed most of the current.

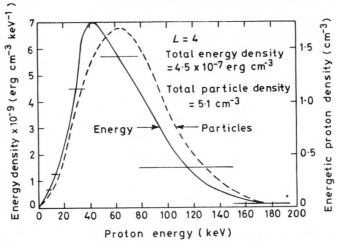

Figure 8.33   Energy and number spectra of protons in the ring current at $L = 4$ during a magnetic storm. (From M.H. Rees and R.G. Roble, *Rev. geophys. Space Phys,* **13**, 201, 1975, copyrighted by American Geophysical Union.)

According to Equation 7.24, the longitudinal drift rate is proportional to the particle energy. Since each proton carries the same charge ($+e$), the total current is proportional to the product of particle density and particle energy integrated over all energies. Thus the total current is proportional to the total energy of the trapped particle population; which in the circumstances of the terrestrial magnetosphere is in effect the total energy of the low energy protons (E). Combining these ideas with Equation 8.16 gives

$$\Delta H(\gamma) = -2.6 \times 10^{21}\ E\ (\text{erg}) \qquad (8.17)$$

The energy density of the ring current protons can exceed that of the geomagnetic field (2×10^{-7} erg cm^{-3} at $5R_E$), and when this happens the field is distorted. This is called the inflation of the geomagnetic field.

The formation of the mid-latitude stable red (SAR) arc (Section 4.3.5) has been attributed to processes in the ring current region, on the basis that instabilities form within the ring current whenever it moves close enough to the Earth to encounter the plasmapause. It is thought that these instabilities feed energy down to the ionosphere to heat the gas by several thousand degrees and so produce the 6300 Å emission. (See Chapter 9 for an introductory discussion of instabilities.)

8.3.5 Currents during a substorm

The magnetospheric currents across the tail, along field lines, in the auroral ionosphere and in the trapped particle belt are linked by the substorm concept. When the tail collapses at a neutral point the cross-tail current must be locally reduced (otherwise the collapse could not occur), and this current becomes diverted along field lines into the auroral ionosphere where the circuit is completed by an electrojet (Fig. 8.34). The growth of the electrojet comes at a time when the conductivity of the auroral E region is enhanced by particle precipitation. Some of the particles accelerated by the collapsing field become trapped to feed the Van Allen belt and also to produce the ring current. It is likely that some of the ring current is partial and does not encircle the Earth. As the particles drift away from the injection region in the midnight sector some will be lost from the magnetosphere and so the westward ring current will taper off towards the day side. Since all the current must be accounted for, it is thought that this electric circuit is completed via Birkeland currents to the ionosphere and an eastward current in the ionosphere.

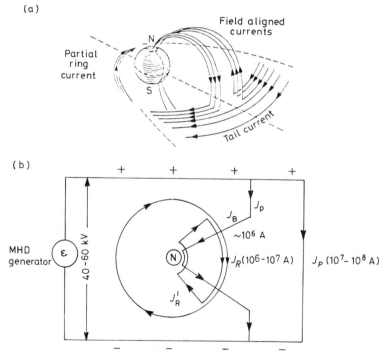

Figure 8.34 Current systems during a substorm: (a) pictorial representation (From L. Svalgaard, *NASA Report* SP-366, 1975); (b) schematic diagram (After W.J. Heikkila, *J. geophys. Res.* **79**, 2496, 1974, copyrighted by American Geophysical Union.)

It can readily be shown that the magnetotail does contain enough energy to maintain a substorm. If the electrojet is 5×10^5 A and the cross-tail current before the substorm was 3×10^{-2} A m^{-1}, the tail current must have been disrupted over a distance of about 2×10^4 km (or several earth radii). With a tail field $B_T = 20\ \gamma$ and a tail radius $R_T = 10^5$ km, the magnetic energy in a section 2×10^4 km long is

$$\frac{B_T^2}{2\mu_0} \times \pi R_T^2 \times 2 \times 10^7 = 10^{14} \text{ J}$$

The ionosphere offers resistance of about 0.1 Ω to an electrojet, and thus a current of 5×10^5 A dissipates $(i^2 r =) 2.5 \times 10^{10}$ W. Auroral precipitation dissipates a similar amount, and both hemispheres should be included, giving a total energy dissipation rate of 10^{11} W. The loss of 10^{14} J in magnetic energy from the tail can therefore support such a substorm for 10^3 s $\sim$ 20 min, which is the approximate duration of a burst of auroral activity.

There is now little doubt that the above substorm model is correct in essence, though the exact sequence of events is far from certain. One question is whether the interruption of tail current is initiated in the tail or in the ionosphere. The alternatives may be summarized as in Table 8.5.

Table 8.5   Alternative Sequences of Substorm Initiation (after Akasofu)

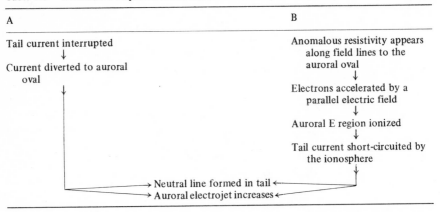

A	B
Tail current interrupted ↓ Current diverted to auroral oval	Anomalous resistivity appears along field lines to the auroral oval ↓ Electrons accelerated by a parallel electric field ↓ Auroral E region ionized ↓ Tail current short-circuited by the ionosphere

→ Neutral line formed in tail ←
→ Auroral electrojet increases ←

A critical step in option B is increasing the resistance of the auroral-oval field lines above normal so that substantial electric fields can be created along them, the normal parallel conductivity being too high for the purpose. Such an increase is called *anomalous resistivity* and will be touched on again in Chapter 9.

8.4 Further reading

Akasofu, S.-I. (1968) *Polar and Magnetospheric Substorms* (Reidel).

Akasofu, S.-I. and Chapman, S. (1972) *Solar–Terrestrial Physics* (Oxford) Chapters 6–8.

Hess, W. N. and Mead, G. D. (1968) *Introduction to Space Science* (Gordon and Breach) Chapter 5.

Matsushita, S. and Campbell, W. H. (1967) *Physics of Geomagnetic Phenomena* (Academic Press) Chapters IV-1, 2, 3, VI-1, 4, 5.
Jones, A. V. (1974) *Aurora* (Reidel).

Axford, W. I. (1969) Magnetospheric convection, *Rev. Geophys.* **7**, 421.
Nishida, A. and Kokubun, S. (1971) New polar magnetic disturbances: S_q^p, SP, DPC, and DP2, *Rev. Geophys. Space Phys.* **9**, 417.
Nishida, A. (1976) Interplanetary field effect on the magnetosphere, *Space Sci. Rev.* **17**, 353.
Burch, J. L. (1974) Observations of interactions between interplanetary and geomagnetic fields, *Rev. Geophys. Space Phys.* **12**, 363.
Kivelson, M. G. (1976) Magnetospheric electric fields and their variation with geomagnetic activity, *Rev. Geophys. Space Phys.* **14**, 189.
Hill, T. W. (1974) Origin of the plasma sheet, *Rev. Geophys. Space Phys.* **12**, 379.
Mozer, F. S. (1973) Electric fields and plasma convection in the plasmasphere, *Rev. Geophys. Space Phys.* **11**, 755.
Carpenter, D. L. and Park, C. G. (1973) On what ionospheric workers should know about the plasmapause–plasmasphere, *Rev. Geophys. Space Phys.* **11**, 133.
Chappell, C. R. (1972) Recent satellite measurements of the morphology and dynamics of the plasmasphere, *Rev. Geophys. Space Phys.* **10**, 951.
Eather, R. H. (1973) The auroral oval – a re-evaluation, *Rev. Geophys. Space Phys.* **11**, 155.
Hultqvist, B. (1975) The aurora, *Space Sci. Rev.* **17**, 787.
Beer, T. (1971) Particle precipitation at auroral heights: ground based observations, *Contemp. Phys.* **12**, 299.
Akasofu, S.-I. (1974) A study of auroral displays photographed from the DMSP-2 satellite and from the Alaska meridian chain of stations, *Space Sci. Rev.* **16**, 617.
Hargreaves, J. K. (1969) Auroral absorption of HF radio waves in the ionosphere: a review of the results from the first decade of riometry, *Proc. Inst. elect. electronics Engr.* **57**, 1348.
Barcus, J. R. (1972) Conjugate features of magnetospheric electron dynamics observed at balloon altitudes, *Space Sci. Rev.* **13**, 295.
Rostoker, G. (1972) Geomagnetic indices, *Rev. Geophys. Space Phys.* **10**, 935.
Kane, R. P. (1976) Geomagnetic field variations, *Space Sci. Rev.* **18**, 413.
Wilson, C. R. (1975) Infrasonic waves generated by aurora, *J. atmos. terr. Phys.* **37**, 973.
Paulikas, G. A. (1971) The patterns and sources of high-latitude particle precipitation, *Rev. Geophys. Space Phys.* **9**, 659.
Wilcox, J. M. (1972) Inferring the interplanetary magnetic field by observing the polar geomagnetic field, *Rev. Geophys. Space Phys.* **10**, 1003.
Feldstein, Y. I. (1976) Magnetic field variations in the polar region during magnetically quiet periods and interplanetary magnetic fields, *Space Sci. Rev.* **18**, 777.
Rees, M. H. (1969–70) Auroral electrons, *Space Sci. Rev.* **10**, 413.
Boyd, J. S. (1975) Rocket-borne measurements of auroral electrons, *Rev. Geophys. Space Phys.* **13**, 735.
Rostoker, G. (1972) Polar magnetic substorms, *Rev. Geophys. Space Phys.* **10**, 157.

Hultqvist, B. (1969) Auroras and polar substorms: observations and theory, *Rev. Geophys.* **7**, 129.

Arnoldy, R. L. (1974) Auroral particle precipitation and Birkeland currents. *Rev. Geophys. Space Phys.* **12**, 217.

Cloutier, P. A. and Anderson, H. R. (1976) Observations of Birkeland currents, *Space Sci. Rev.* **17**, 563.

Fukushima, N. and Kamide, Y. (1973) Partial ring current models for worldwide geomagnetic disturbances, *Rev. Geophys. Space Phys.* **11**, 795.

208

9
Waves in the Magnetosphere

9.1 Wave propagation in the magnetosphere

The magnetospheric plasma, by definition, is permeated by the geomagnetic field and thus can support various kinds of waves that are at home in a magnetoplasma. The source of the waves can be outside the magnetosphere, as in the case of a whistler, or the generation process can be one that operates within the magnetosphere. Whistlers were discussed in Section 7.5.1. Here we shall be concerned mainly with magnetospheric generation, with observations of the waves, and with their effects. In the magnetosphere, interactions are possible between waves in the 'cold' magnetoplasma and energetic, or 'hot', charged particles, energy being transferred from wave to particle or vice versa according to circumstances. These wave–particle interactions are thought to be important processes in the physics of the magnetosphere, and hence they merit an honourable mention in this chapter although a detailed treatment would be too advanced for an introductory text.

9.1.1 VLF waves – the whistler mode

The whistler propagation mode of an electromagnetic wave was introduced in Section 3.7.5. This mode exists for radio frequencies below the electron gyrofrequency, $f < f_B$, where $f_B = Be/2\pi m_e$ (Equation 2.18). The refractive index (n) is obtained from the Appleton–Hartree equation (2.37) by putting $Z = 0$ (no collisions), $Y_T \ll Y_L$ (QL approximation), and taking the negative sign (E-wave):

$$n^2 = 1 - X/(Y_L - 1) \tag{9.1}$$

If, also, $X \gg (Y_L - 1)$,

$$n^2 = X(Y_L - 1) = f_N^2/f(f_L - f) \tag{3.26} \tag{9.2}$$

since $Y_L = f_L/f$ and $X = f_N^2/f^2$, f_N being the electron plasma frequency, $f_N^2 = Ne^2/(2\pi)^2 \epsilon_0 m_e$ (Equation 2.20). The mode propagates in the extraordinary mode only, which is circularly polarized, rotating clockwise as seen by an observer looking along the magnetic field (Fig. 9.2). The wave is strongly dispersed, with infinite refractive index (zero phase velocity) at $f = f_L$. A pulse travels at the group velocity, given by the group refractive index (Equation 3.27).

9.1.2 Ion-cyclotron waves

If the ion motion is taken into account as well as the electron motion, the refractive index formula in the QL approximation, collisions being neglected, becomes

$$n^2 = 1 - \frac{f_N^2}{f(f \pm f_L)} - \frac{F_N^2}{f(f \mp F_L)} \tag{9.3}$$

where F_N and F_L are respectively the plasma frequency and the gyrofrequency for the ions. Both these frequencies are much lower than the corresponding electron frequencies because of the larger particle mass. The ion motions are important for electromagnetic waves in the ULF range $(f < 3\ \text{Hz})$. For frequencies $f < F_L$, and taking the upper signs (corresponding to an O-wave),

$$n^2 = 1 - \frac{f_N^2}{ff_L} + \frac{F_N^2}{f(F_L - f)} \tag{9.4}$$

since $f \ll f_L$. This wave rotates anticlockwise as seen by an observer looking along the geomagnetic field, and may be called an *ion-cyclotron wave*. At radio frequencies below f_L the extraordinary wave becomes

$$n^2 = 1 + \frac{f_N^2}{f(f_L - f)} - \frac{F_N^2}{f(f + F_L)} \tag{9.5}$$

when the ion term is included. This wave rotates clockwise. These two modes are known by various names:

O-wave
    Ion-cyclotron wave
    Pure Alfvén mode
    Slow mode
    L-wave
E-wave
    Modified Alfvén mode
    Fast mode
    R-wave

The waves are illustrated in Fig. 9.1, which shows the square of the refractive index as a function of wave frequency if $f_N = 1/4$ MHz. Waves for which n^2 is negative are evanescent.

9.1.3 Alfvén waves

At the lowest frequencies the refractive index tends to the same value for both modes. This low-frequency limit is the Alfvén wave, in which the plasma and the magnetic field move together. Section 2.4.6 derived the velocity of such a wave $(v_A = B/\sqrt{\mu_0 \rho})$ by analogy with a vibrating string. It may also be derived as the low-frequency limit of electromagnetic waves.

If we put $f \ll f_L, F_L$ in Equation 9.3 and realize that $f_N^2 F_L = F_N^2 f_L$ (from Equations 2.18 and 2.20)

$$n^2 = 1 + (f_N^2 + F_N^2)/f_L F_L$$

But also $f_N^2 \gg F_N^2$ and $F_N^2 \gg F_L^2$; hence

$$n^2 \approx 1 + f_N^2/f_L F_L = 1 + F_N^2/F_L^2$$

$$n \approx F_N/F_L = \left(\frac{Nm_i}{\epsilon_0 B^2}\right)^{1/2}$$

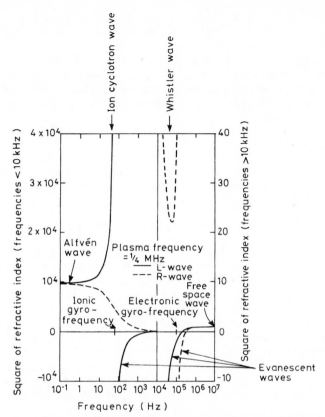

Figure 9.1 Square of refractive index (n^2) at low radio frequencies when both electron and ion motions are included, showing the whistler and ion-cyclotron waves. (After H.G. Booker, *J. geophys. Res.* **67**, 4135, 1962, copyrighted by American Geophysical Union.)

However, $Nm_i = \rho$, the density of the plasma, and $c = 1/\sqrt{\epsilon_0 \mu_0}$. Therefore

$$v_A = c/n = \left(\frac{B^2}{\mu_0 \rho} \right)^{1/2} \qquad \text{in SI units} \tag{9.6}$$

$$= B/\sqrt{4\pi\rho} \qquad \text{in cgs}$$

In terms of the Alfvén velocity, the refractive indices of the L- and R-waves discussed in the previous section may be written

$$n_L = \frac{c}{v_A} (1 - f/F_L)^{-1/2}$$

$$n_R = \frac{c}{v_A} (1 + f/F_L)^{-1/2} \tag{9.7}$$

9.2 Wave generation by magnetospheric particles

An oscillating electron radiates an electromagnetic wave at its frequency of oscillation; a familiar example is the radio transmitting antenna. Similarly, if a charged

211

particle gyrates about a fixed magnetic field it radiates a circularly polarized wave at the gyration frequency. Energy is thereby transferred from the particle to the wave, a *particle-wave interaction*.

If the energy transfer between a specific particle and a specific wave is to be efficient there must be a condition of resonance, meaning that the particle and the wave must maintain the same phase relation for a sufficient period of time. The force (**F**) on a charged particle due to the electric (**E**) and magnetic (**B**) fields of an electromagnetic wave is

$$\mathbf{F} = q(\mathbf{E} + \mathbf{v} \times \mathbf{B})$$

where q is the charge on the particle and **v** its velocity. (Compare with Sections 2.3.2 and 4.4.3.) The change of energy during the interaction is $\Delta W = \mathbf{F} \cdot \Delta \mathbf{S}$ where $\Delta \mathbf{S}$ is the distance travelled. Hence,

$$\Delta W = \mathbf{F} \cdot \Delta \mathbf{S} = \mathbf{F} \cdot \mathbf{v}\Delta t = q(\mathbf{E} + \mathbf{v} \times \mathbf{B}) \cdot \mathbf{v}\Delta t \approx q\mathbf{E} \cdot \mathbf{v}\Delta t$$

since $\mathbf{v} \times \mathbf{B} \ll \mathbf{E}$ for an electromagnetic wave. Hence the energy transfer is greater if the wave and the particle remain in a constant phase relation for a longer time (Δt). If this matching is not achieved, energy transferred to the wave during one half-cycle ($\mathbf{E} \cdot \mathbf{v}$ positive) will only be returned to the particle during the next half-cycle ($\mathbf{E} \cdot \mathbf{v}$ negative) and there will be no net gain either way.

There are several possible wave–particle resonances in the magnetosphere, but we will illustrate the point with a transverse resonance, involving the component of particle velocity ($\mathbf{v}_\perp$) transverse to the geomagnetic field. The wave with which the trapped particle resonates is in the whistler mode, and a Doppler shift has to be invoked to achieve the required matching of particle and wave velocities. This particular resonance may therefore be called the *Doppler-shifted cyclotron resonance*.

If $\mathbf{E} \cdot \mathbf{v}_\perp$ is to be constant, the wave and the particle must rotate in the same direction around the geomagnetic field. This condition is satisfied for electrons and whistler waves, both of which rotate clockwise to a stationary observer looking along the field (Fig. 9.2). The second condition is that wave and particle must preserve a constant relationship as they move along the geomagnetic field; an equivalent condition is that an observer moving with the particle must find the wave frequency to be the same as the particle's gyration frequency. The electron gyrates at f_B, but the whistler frequency may be any $f < f_B$ (Section 9.1.1), both as seen by a stationary observer. However, if the wave and particle velocities along the field are v_p and $v_\parallel$ respectively, the wave frequency as seen from the moving particle is Doppler shifted in the usual way, and

$$\frac{f_B - f}{f} = \frac{|v_\parallel|}{|v_p|} \tag{9.8}$$

is the condition for the Doppler shift ($f_B - f$) needed to bring the wave to resonance with the particle. Since the wave frequency has to be increased to achieve resonance, the wave and the particle must be travelling in opposite directions.

Taking propagation directly along the magnetic field (and thus writing f_B for f_L)

$$v_p^2 = c^2/n^2 = c^2 f(f_B - f)/f_N^2 \qquad \text{from Equation 9.2}$$

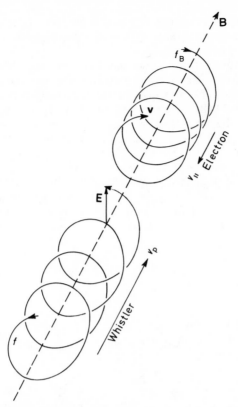

Figure 9.2 Spiral motions of electron and whistler about the geomagnetic field. For effective interaction the electric vector of the whistler must be maintained parallel to the velocity of the electron.

The parallel energy of a particle is

$$W_\parallel = m v_\parallel^2 /2 \; = \tfrac{1}{2} m v_p^2 \, (f_B - f)^2 / f^2 \qquad \text{from Equation 9.8}$$

$$= \tfrac{1}{2} m c^2 \; (f_B - f)^3 \, / f_N^2 f \qquad \text{from Equation 9.2}$$

$$= \tfrac{1}{2} m c^2 \cdot \frac{f_B^2}{f_N} \cdot \frac{f_B}{f} \, (1 - f/f_B)^3 \qquad (9.9)$$

But

$$\frac{f_B^2}{f_N^2} = \frac{B^2 e^2}{m^2} \times \frac{\epsilon_0 m}{N e^2} = \frac{B^2 \epsilon_0}{mN} \qquad \text{from Equations 2.18 and 2.20}$$

where e and m are the electronic charge and mass, N is the electron density (of the cold plasma) and ϵ_0 is the permittivity of free space. Also, $c^2 = 1/\epsilon_0 \mu_0$. Therefore, by substituting in Equation 9.9,

$$W_\parallel = \frac{B^2}{2\mu_0 N} \cdot \frac{f_B}{f} \left(1 - \frac{f}{f_B}\right)^3 \qquad (9.10)$$

213

This expresses the parallel energy of resonant electrons in terms of the ratio of wave frequency to gyrofrequency (f/f_B) and the magnetic energy density per electron of the cold plasma ($B^2/2\mu_0 N$).

The Doppler-shifted cyclotron resonance between energetic electrons and VLF electromagnetic waves in the whistler mode is of some importance in magnetospheric physics, being a likely mechanisn for the precipitation of electrons from the trapped radiation belts and from the plasma sheet to produce the aurora. In the magnetosphere the flux of particles tends to increase at lower energies. For a given emission frequency (f), and along a given field line, we therefore expect the greatest contribution to come from the electrons of lowest energy that can satisfy Equation 9.10. The term $B^2/2\mu_0 N$ is smallest in the region where the field line crosses the equatorial plane, and therefore it is usually assumed that the equatorial region is the main source of emission produced by this resonance. That being so, the appropriate value of f_B for a given magnetic latitude is that for the nose of the field line. Hence f/f_B can be calculated as a function of L-value and emission

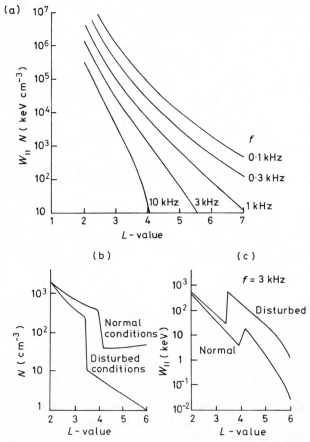

Figure 9.3 (a) Variation of $W_{\parallel}N$ with f and L for the condition of transverse resonance at the equator. Realistic models for the distribution of N (b) lead to estimates of $W_{\parallel}$ for a selected frequency (c). (After M.J. Rycroft, *Planet. Space Sci.* **21**, 239, 1973; and M.J. Rycroft, *NATO Advanced Study Institute on ELF/VLF Radio Wave Propagation*, Reidel, 1974.)

frequency; Fig. 9.3a shows the variation of $W_{\parallel}N$ as a function of L and f. (The resonance energy varies inversely with the electron density in the equatorial plane if other quantities are constant.) By taking models of N as a function of L (Fig. 9.3b) it is possible to deduce the energy of the electrons producing most of the emissions at a given frequency (Fig. 9.3c).

9.3 Observations of magnetospheric waves

9.3.1 VLF emissions

Naturally occurring VLF emissions have been observed for many years from the ground, and during the last decade from satellites. In ground-based reception, a large loop antenna feeds the signals to an audio amplifier, from which they are recorded on magnetic tape. The received signals display a variety of frequency— time characteristics. Some emissions are 'discrete' showing a well defined frequency that changes with time. Other are more or less unstructured, resembling noise. These are generally known as *hiss* or *chorus*. The frequency range is 1—10 kHz or more.

Observations from satellites are particularly useful for mapping global distributions, and they avoid one known difficulty of ground-based work — the absorption of VLF signals passing through the ionosphere. Fig. 9.4 shows bands of emission which maximize at 50—65° magnetic latitude and are generally known as the *mid-latitude* emissions. They are thought to be generated by the Doppler-shifted cyclotron resonance, outlined above. Emissions are also observed at high latitudes, and they are distinguished from the mid-latitude type by their different spectral characteristics. Since they maximize at 73—79° magnetic latitude it seems clear that they arise on open field lines and so the transverse cyclotron resonance cannot be the cause. They might be produced by Cerenkov radiation, involving a longitudinal resonance with soft electrons (< 100 eV), though this is not certain.

Another interesting result of the satellite work is the existence of co-rotating zones of enhanced emission. Normally one expects the occurrence of trapped particle phenomena to be related to the position of the Sun. However, these enhanced regions rotate with the Earth, having a life of 12 or even 24 h. They are probably due to a local increase of electron density in the plasmasphere, reducing the local value of the quantity $B^2/2\mu_0 N$. As noted in Section 9.2, this would lower the threshold energy for electrons causing the emission and so increase the emission intensity.

A similar mechanism is thought to be the cause of *plasmaspheric hiss*, a band of noise in the ELF band (3—3000 Hz) with maximum intensity at a few hundred hertz. The emission is confined to a limited range of L-values just inside the plasmapause, and the zone ends sharply at the plasmapause. Although it occurs at all local times, plasmaspheric hiss is most intense in the afternoon sector. This is where energetic electrons following nearly circular orbits around the Earth will encounter the plasmasphere bulge (Section 8.1.5). $B^2/2\mu_0 N$ will be reduced and the emission intensity will be increased.

9.3.2 Micropulsations

Another natural variation observed at the Earth's surface or in space is the magnetic micropulsation, a small fluctuation amounting to less than 10^{-4} of the geomagnetic

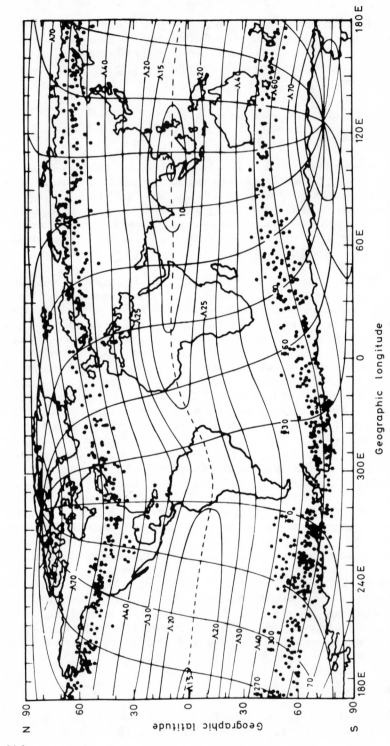

Figure 9.4 Global distribution of maxima of 3.2 kHz emissions observed from an orbiting satellite. The contours are invariant latitude and longitude. (From K. Bullough *et al.*, *Magnetospheric Physics* (Ed. McCormac), Reidel, 1974.)

216

field and having a relatively short period measured in minutes or seconds. The instruments used to detect micropulsations at the ground fall into three groups, giving:

(i) a direct recording of the magnetic elements at high sensitivity and high time resolution, i.e. a sensitive magnetometer;
(ii) a recording of the rate of change of the magnetic field using a large induction loop or a loop with a magnetic core;
(iii) a recording of the electric currents induced in the ground by variations of the geomagnetic field.

In recent years they have also been observed with magnetometers carried on satellites in geostationary or other orbits.

The range of phenomena is indicated by the international scheme of classification for micropulsations (Table 9.1). The irregular pulsations are associated with magnetic bays during substorms, and form their microstructure. Examples are shown in Figs. 9.5–9.7.

Table 9.1   Micropulsation Classification

Regular, continuous pulsations		Irregular pulsations	
	Period (s)		Period (s)
Pc 1	0.2–5	Pi 1	1–40
Pc 2	5–10	Pi 2	40–150
Pc 3	10–45		
Pc 4	45–150		
Pc 5	150–600		

Many of these pulsations are similar at magnetically conjugate points, indicating that the magnetosphere is a major influence. Pc oscillations are though to be generated at the surface of the magnetosphere, or within the magnetosphere, and to be propagated in a hydromagnetic mode. Pc 1 is attributed to bunches of particles, probably protons, oscillating between conjugate points. The period characteristic of a given latitude is related to the ion gyrofrequency at the equator along that field line, and a resonance between protons and ion cyclotron waves (which rotate about the field in the same sense) is probably involved. The interaction is analogous to that between electrons and the whistler mode discussed previously. The modulation seen in Fig. 9.5, which invoked the name *pearls*, is related to the bounce period between hemispheres, there being a bunching of particles in the troughs of the wave.

Pc 2–5 are usually explained as various oscillation modes within the magnetosphere. Pc 5 is thought to be a toroidal oscillation of the whole field line between conjugate points. The oscillations show opposite polarizations in the morning and the evening, as if arising from waves travelling on the surface of the magnetosphere (Fig. 9.8). The polarization also reverses between conjugate points. It is an interesting, but unexplained, fact that particle precipitation can be modulated at the same frequency during Pc 5 events. There is evidence that some waves in this range are generated in the ring current (Section 8.3.4).

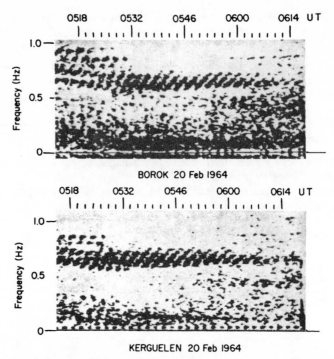

0518 0532 0546 0600 0614 UT

1.0 —

Frequency (Hz)

0.5 —

0 —

BOROK 20 Feb 1964

0518 0532 0546 0600 0614 UT

1.0 —

Frequency (Hz)

0.5 —

0 —

KERGUELEN 20 Feb 1964

Figure 9.5 Frequency-time display of Pc 1 micropulsations observed at the conjugate stations Borok and Kerguelen. The fine structure of rising tones is characteristic of Pc 1. (After V.A. Troitskaya, *Solar-Terrestrial Physics* (Ed. King and Newman) Academic Press, 1967.)

Micropulsations in the Pc 3—4 range can be used as diagnostic tools for the magnetosphere, since the period is found to depend on the electron density. The effect of the plasmapause can be seen, the period received at a given ground station being longer when the relevant field lines are within the plasmasphere. Pc 4 tend to arise inside the plasmapause, but Pc 3 outside, both in narrow latitude ranges. Current evidence suggests that these waves are field-line resonances, possibly standing Alfvén waves, which are stimulated by an event beyond the plasmapause, such as rapid movement of the magnetopause.

9.4 Wave-particle interactions

9.4.1 Particle precipitation by VLF waves

The interaction between particles and waves works both ways. When a gyrating electron feeds energy into a VLF wave by transverse resonance its velocity transverse to the geomagnetic field ($v_\perp$) is reduced while the parallel velocity ($v_\parallel$) is unaltered. Thus the pitch angle ($\alpha = \tan^{-1} v_\perp/v_\parallel$) is reduced and the particle moves nearer to the loss cone. A particle that receives transverse energy has its pitch angle increased. A spectrum of VLF waves therefore alters the pitch-angle distribution of an assembly of trapped particles, some of which move into the loss cone to be precipitated into the atmosphere at the next bounce. An interesting possibility

218

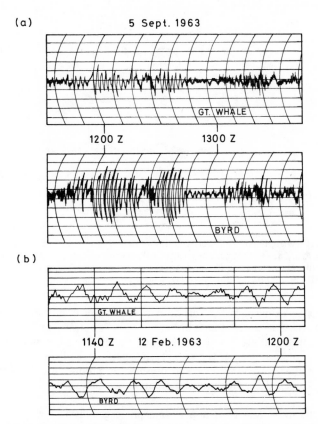

Figure 9.6 (a) Pc 4 and (b) Pc 5 pulsations at the auroral conjugate stations Great Whale River and Byrd. The conjugate similarity rules out local sources. (After J.A. Jacobs, *Physics and Chemistry in Space, Vol. 1, Geomagnetic Micropulsations*, Springer-Verlag, 1970.)

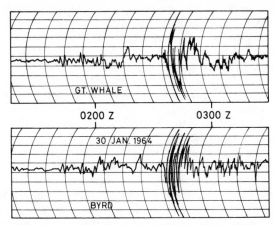

Figure 9.7 Irregular pulsations at Great Whale River and Byrd. Such events are usually sub-storm associated, and there is a degree of similarity between conjugate points, though not as much as for Pc events. (After J.A. Jacobs, *Physics and Chemistry in Space, Vol. 1, Geomagnetic Micropulsations,* Springer-Verlag, 1970.)

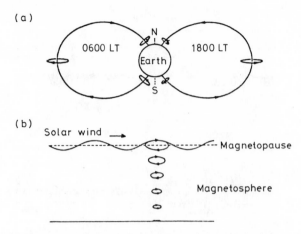

(a)

0600 LT   Earth   1800 LT

(b)

Solar wind →

Magnetopause

Magnetosphere

Figure 9.8 (a) The sense of Pc 5 polarization, as viewed from the Sun. (After T. Nagata *et al.*, *J. geophys. Res.* **68**, 4621, 1963, copyrighted by American Geophysical Union.) (b) Possible generation at the magnetopause. (After J.A. Jacobs, *Physics and Chemistry in Space, Vol. 1, Geomagnetic Micropulsations,* Springer-Verlag, 1970.)

raised by this mechanism is that electron precipitation events might be initiated by lightning flashes. Since the whistler and the electrons must be moving in opposite directions along the line of force to achieve resonance, the precipitation event should appear in the same hemisphere as the lightning that triggered it, though a higher-order whistler could presumably produce an effect in the other hemisphere. The electron precipitation event shown in Fig. 9.9 could well be such a case.

Contrary to most aspects of upper atmosphere geophysics, which are based on the 'passive' sensing of the events and phenomena that nature cares to provide, wave–particle interactions is a subject suitable for 'active' experiments in which the upper atmosphere is temporarily modified in a controlled manner. For instance, attempts have been made to stimulate precipitation events by injecting a cloud of plasma, such as ionized barium, into the magnetosphere. The idea is to induce a wave–particle resonance by increasing the local plasma density, as discussed in Section 9.3.1, which should lead to the precipitation of some trapped particles and also lead to an increase of VLF activity detectable at the ground.

An alternative approach to the controlled magnetospheric experiment is to inject VLF waves. This is readily done from the ground with a suitable radio transmitter, from which some energy will propagate into the magnetosphere in the whistler mode. At least one VLF transmitter has been constructed at a high latitude site specially for such experiments. (The Antarctic offers a bonus for VLF work, in that the thick ice cap raises the antenna far above the effective ground and thereby improves the radiation efficiency.) Commercial VLF senders have, of course, been unwittingly radiating into the magnetosphere since they first went into service in the 1920s. There are now many observations showing that these man-made waves can stimulate VLF emissions from the magnetoplasma. Power transmission lines are another source of low frequency radio emission, and it has been suggested that the unusual intensity of magnetospheric VLF emissions at longitudes corresponding to the eastern part of North America may be due to the large number of power lines in that region.

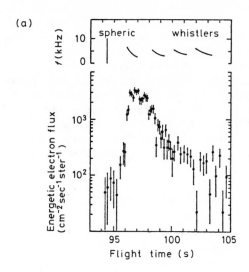

(a)

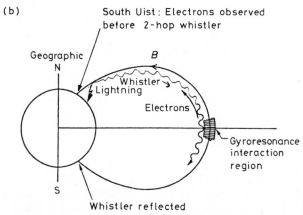

(b)

Figure 9.9 (a) An electron (> 45 keV) precipitation event observed on a rocket just after a spheric due to a lightning stroke, but before the arrival of the whistlers. (b) A possible explanation, in which the electron event is produced as the first-order whistler reaches the equatorial plane. (After M.J. Rycroft, *Planet. Space Sci.* **21**, 239, 1973.)

9.4.2 Cyclotron resonance instability

The interaction between VLF whistler waves and trapped electrons means that a non-isotropic distribution of electron pitch angles is unstable. The particles generate waves, and the waves then alter the pitch angles of the particles. Thus the pitch angles spontaneously redistribute themselves in the direction of an isotropic distribution. The rate at which this redistribution occurs depends on the initial anisotropy, which for this purpose is quantified by a parameter

$$A = (T_\perp/T_\parallel) - 1 \qquad (9.11)$$

where $T_\perp$ and $T_\parallel$ are respectively the particle temperatures perpendicular and parallel to the geomagnetic field. ($T_\perp$ and $T_\parallel$ represent the perpendicular and parallel energies of the assembly of particles. These are key quantities in theories of wave—particle interactions.) A is zero for an isotropic distribution, and positive if the plasma is hotter across the field than along it.

The practical significance of spontaneous pitch angle redistribution comes about because the atmosphere imposes a *loss cone*. Particles that move almost along the geomagnetic field while over the equator have mirror points low in the atmosphere (see Equation 7.22). The equatorial loss cone is a narrow cone (usually of a few degrees only) about the field line, from which particles will be lost to the atmosphere at the next bounce. This means that the pitch angle distribution is always anisotropic to some extent, since the loss cone is constantly being emptied.

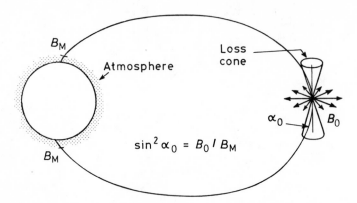

Figure 9.10   The loss cone is emptied as particles approach their mirror point and replenish by pitch-angle diffusion in the magnetosphere.

Wave—particle interactions move fresh particles into the loss cone, where they, in turn, are lost to the atmosphere (Fig. 9.10). Thus we have a mechanism whereby particles injected to the radiation belt may be dumped into the atmosphere, and which potentially satisfies two needs: first, the need for a loss process for the outer radiation belt and, second, a link in the chain that deposits magnetospheric energy into the atmosphere to give auroral phenomena.

The theory of the cyclotron resonance instability is not easy, but we shall here outline some of the key points. The mechanism is based on the generation of waves by the particles. This wave energy can be lost in various ways, and the instability can only develop if the generation rate exceeds the loss rate. The theory shows that the condition for instability growth is that the anisotropy, A must exceed a critical value

$$A_c = \omega/(\omega_B - \omega) \tag{9.12}$$

where ω is the radio frequency of the resonating waves. Reference to Equation 9.8 shows that this is like putting a condition on the parallel velocity of the particles in terms of the phase velocity of the waves. The instability grows exponentially $(e^{\gamma t})$, in which the growth rate

$$\gamma \approx \pi \omega_B \eta (A - A_c) \tag{9.13}$$

At the critical anisotropy, $\omega/\omega_B = A/(1 + A)$, from Equation 9.12. In the magnetosphere A is typically about unity, so that the wave growth is limited to waves of frequency below about half the gyrofrequency. If A is very small (distribution nearly isotropic) only waves of small ω can grow. If A is very large, $\omega \to \omega_B$. Only waves of frequency less than the gyrofrequency can take part.

There is also an energy limit on the particles involved. Rearrangement of Equation 9.10 using Equation 9.12 gives

$$W_{\parallel} = \frac{B^2}{2\mu_0 N} A_c^{-1} (1 + A_c)^{-2} \tag{9.14}$$

which states the parallel energy of the resonant electrons when the anisotropy is just critical. For a distribution of anisotropy A, particles can only be involved in the instability if their critical anisotropy is less than A. Equation 9.14 therefore gives the minimum energy of particles which can be precipitated by this mechanism. In practice the second part of Equation 9.14 is of order unity, and the criterion becomes

$$W \gtrsim B^2/2\mu_0 N \tag{9.15}$$

Fig. 9.11 is based on an electron density profile in the equatorial plane, and shows the lowest energy of particles subject to spontaneous precipitation by the cyclotron resonance instability. Particles of energy > 100 keV are subject to instability throughout the radiation belts. On the other hand there is a zone of relative stability for particles a few tens of keV in energy from just beyond the plasmapause to 7 or 8 earth radii. (Fig. 9.11 applies also to protons, which are precipitated through the resonance with ion-cyclotron waves.) This stable region may be identified with the location of the ring current (Section 8.3.4).

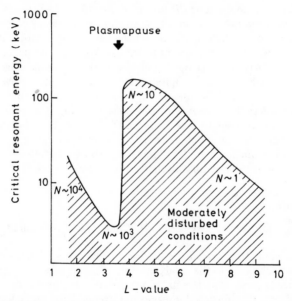

Figure 9.11   Minimum energy of particles subject to precipitation by the cyclotron resonance instability, assuming an equatorial electron density (N) profile. The plasmapause has a major effect. (From R.M. Thorne, *Critical Problems of Magnetospheric Physics*, 1972, with permission of the National Academy of Sciences, Washington, DC.)

When the instability is operating the resulting change in the pitch angle distribution may be regarded as diffusion (Section 2.2.4). It can be shown that the diffusion coefficient is

$$D_{\alpha} \sim \omega_B^2 (\delta B/B)^2 \Delta t \tag{9.16}$$

where δB is the amplitude of the magnetic vector of the VLF noise, B is the geomagnetic field, and Δt is the time for which the electrons and the VLF waves remain in resonance.

Irrespective of its physical cause, pitch angle diffusion may be described by two limiting cases according to the speed of the diffusion. The point is that the particles undergoing diffusion are mirroring back and forth along the geomagnetic field. Particles are lost near the mirror points whereas the redistribution of pitch angle occurs near the equatorial plane. If the diffusion is relatively slow, so that it takes at least several bounces for a typical particle to move into the loss cone, we have *weak diffusion* which can be likened to heat conduction along a rod. Imagine that heat (particles) is supplied to one end of the rod (pitch angle 90°), and removed (particle precipitation) at the other end (pitch angle 0°) which is held at constant temperature (loss cone). The rate of heat transfer (particle precipitation) depends on the heating rate (rate of fresh particle injection to the trapping region) and on the thermal conductivity of the material of the rod (diffusion coefficient). In the particle case, the diffusion coefficient is controlled by the intensity of the inter-acting waves (e.g. Equation 9.16). It can be seen from this analogy that the form of the pitch angle distribution does not depend on the intensity of the particle source or the actual value of the diffusion coefficient. There will be relatively few particles in the loss cone.

The other limiting case, *strong diffusion*, occurs when diffusion is so rapid that particles move into the loss cone in a time much shorter than one bounce period. In this situation the pitch angle distribution becomes almost isotropic in the vicinity of the equatorial plane because the hole left by losses at the previous bounce is rapidly filled in. It may be assumed that all the particles that have just moved in to fill the loss cone again will be precipitated at the next bounce. Now the loss rate depends on the size of the loss cone and the bounce period, but not on the value of the diffusion coefficient (provided it is large enough). The magnitude of the diffusion coefficient needed to give strong diffusion can be estimated by saying that a particle must diffuse in pitch angle by more than α_0 (the loss cone) in a time less than the bounce time, T_B. Thus, writing the diffusion coefficient as a (characteristic length)^2 divided by a (characteristic time) — see Section 2.2.4 — we obtain

$$D_\alpha \gtrsim \alpha_0^2/T_B$$

Putting in reasonable values gives $D_\alpha \gtrsim 10^{-3}$ rad^2 s^{-1} as the condition for strong diffusion.

It will be obvious from the above discussion that measurements of pitch angle distribution are relevant to the study of precipitation processes. Spacecraft observations, indeed, show that the distribution tends to the isotropic during precipitation producing the most intense auroral events. This immediately gives the theorist a valuable piece of information for deciding whether a candidate diffusion mechanism can be retained.

The cyclotron resonance instability provides a way of limiting the flux of trapped particles in the magnetosphere. An increase in the trapped flux increases the intensity of the wave turbulence, which in turn increases the diffusion coefficient. The precipitation rate therefore increases disproportionately, and eventually the number of particles held in the trapping region is limited, though the actual levels are subject to the resonance condition already discussed (Fig. 9.11). There is satellite evidence for the limitation of trapped particles; the flux increases during geomagnetic disturbances as more particles are injected, presumably from the magnetotail, but during the most intense storms the flux comes up to a constant level which is not exceeded.

224

9.4.3 Electrostatic waves

We have dealt at some length with the transverse cyclotron resonance between VLF whistler waves and trapped electrons, but it should not be thought that this is the only wave–particle interaction of significance in the magnetosphere. We have mentioned in passing the resonance between ion-cyclotron waves and trapped protons, which is rather similar in principle to the electron–whistler interaction. A different sort of wave affecting trapped particles is electrostatic rather than electromagnetic. The resonance spikes seen on a topside ionogram (Section 3.7.1) are due to electrostatic waves set up in the vicinity of the sounder. They exist locally and do not propagate to great distances. The frequencies are the electron plasma frequency (f_N), the gyrofrequency (f_B) and various combinations and harmonics. We gave an elementary treatment of f_N and f_B in Sections 2.3.4 and 2.3.2 in terms of the motion of electrons with respect to the relatively heavy protons. In both cases the restoring force is electrostatic.

The theory of electrostatic waves, as first developed for laboratory plasmas, shows that when energetic particles are present in a plasma instabilities can develop if $T_\perp > 2T_\parallel$, i.e. the plasma must be hotter transverse to the field than along it. If the anisotropy is relatively small these instabilities will be near half-harmonics of the gyrofrequency ($f_B/2$, $3f_B/2$, etc.).

Direct observations in the magnetosphere have found that electrostatic waves, particularly at $3f_B/2$, can be very intense between $L = 4$ and $L = 10$. The amplitude may exceed 10 mV m^{-1}, well above the steady electric field generated across the magnetosphere by the solar wind. The electrostatic wave amplitude increases with electron precipitation during substorms, and it has been suggested that it might be these waves, rather than VLF electromagnetic waves, which are responsible for auroral electron precipitation at the lower energies (< 10 keV). The cyclotron resonance instability is less effective at lower energy.

9.4.4 Anomalous resistivity

Because collisions between electrons and protons are infrequent in the topside ionosphere and the magnetosphere, it is usually assumed that the electric conductivity along the geomagnetic field (σ_0) is very large (Section 4.4.4). This is in some ways a useful assumption since it makes possible a mapping of electric fields between ionosphere and magnetosphere. On the other hand, the observational evidence for field-aligned particle fluxes during some auroral events (Section 8.2.3) calls the assumption into question because a substantial parallel electric field is not consistent with a negligible parallel resistivity.

To explain the appearance of parallel resistivity it has been conjectured that it is due to wave turbulence (probably electrostatic) generated by field-aligned currents. Since the waves would remove energy from the electric current there would be the appearance of a finite resistivity, and then a finite potential drop along the magnetic field becomes conceivable. This is known as *anomalous resistivity*. Though no adequate theory is available for anomalous resistivity in a collisionless plasma, observational evidence, such as that in Fig. 9.12, for an association between increased electrostatic noise and field-aligned currents provides some support for its reality.

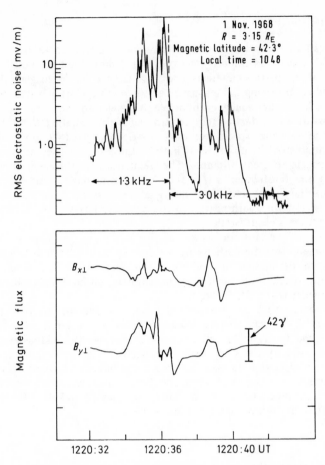

Figure 9.12 Simultaneous increase of electrostatic noise and field-aligned current at 3.15 R_E. The electrostatic frequency channel was changed from 1.3 to 3.0 kHz at the dashed line. Field-aligned currents are indicated by the deviation of the components $B_{x\perp}$ and $B_{y\parallel}$, both perpendicular to the geomagnetic field. $B_\parallel$ was relatively unperturbed. (After R.W. Fredricks, *Space Sci. Rev.* **17**, 449, 1975.)

9.5 Further reading

Helliwell, R. A. (1976) *Whistlers and Related Ionospheric Phenomena* (Stanford).

Jacobs, J. A. (1970) *Geomagnetic Micropulsations* (Springer-Verlag).

Matsushita, S. and Campbell, W. H. (1967) *Physics of Geomagnetic Phenomena* (Academic Press) Chapters IV-4, V-1.

Lanzerotti, L. J. and Fukunishi, H. (1974) Modes of magnetohydrodynamic waves in the magnetosphere, *Rev. Geophys. Space Phys.* **12**, 724.

Helliwell, R. A. (1969) Low-frequency waves in the magnetosphere, *Rev. Geophys.* **7**, 281.

McPherron, R. L., Russell, C. T. and Coleman, P. J. (1971–72) Fluctuating magnetic fields in the magnetosphere I: ELF and VLF fluctuations, *Space Sci. Rev.* **12**, 810. II: ULF waves, *Space Sci. Rev.* **13**, 411.

Orr, D. (1973) Magnetic pulsations within the magnetosphere: a review, *J. atmos. terr. Phys.* **35**, 1.

Special Issue (1974) IAGA symposium on micropulsations theory and new experimental results, *Space Sci. Rev.* **16**, 331.

Raspopov, O. M. and Lanzerotti, L. J. (1976) Investigation of Pc 3 frequency geomagnetic pulsations in conjugate areas around $L = 4$: a review of some recent USSR and US results, *Rev. Geophys. Space Phys.* **14**, 577.

Kennel, C. F. (1969) Consequences of a magnetospheric plasma, *Rev. Geophys.* **7**, 379.

Thorne, R. M. (1975) Wave–particle interactions in the magnetosphere and ionosphere, *Rev. Geophys. Space Phys.* **13**, 291.

Fredricks, R. W. (1975) Wave–particle interactions in the outer magnetosphere: a review, *Space Sci. Rev.* **17**, 741.

Fredricks, R. W. (1976) Wave–particle interactions and their relevance to substorms, *Space Sci. Rev.* **17**, 449.

Special Issue (1974) Workshop on controlled magnetospheric experiments, *Space Sci. Rev.* **15**, 751.

10
Solar Flares and Solar Protons

10.1 Solar flares

The solar flare was introduced briefly in Section 7.2.2. Not only is it a useful indicator of solar activity in general terms, but it is also the source of specific effects in the Earth's atmosphere. In this chapter we will consider two such phenomena. Physically these are quite separate from each other; what they have in common is that in each case specific effects in the atmosphere of the Earth can be related to a specific event at the Sun. In this they differ from the auroral phenomena, which relate in a statistical way to solar activity but not usually to particular solar events.

At this point it will be helpful to consider the nature of a solar flare. Observations show that a flare region contains both hot and cool plasma. The optical flare is relatively cool, about 10^4 °K, the main part being in the lower 5000 km of the chromosphere though there can be considerable upward extension into the low corona. The brightening seen at visible wavelengths is due to a local increase of plasma density from 10^{10} to 10^{13} electrons cm^{-3}. At the same time high energy plasma is produced, with a temperature of $10^7 - 10^9$ °K; the electron density becomes about 10^{10} cm^{-3}, having been 10^8 cm^{-3} before the flare. The optical flare is only about 20 km thick, and it might cover one thousandth of the visible solar disc. It seems clear that the matter involved in the flare has been greatly compressed, and it is estimated that before compression it filled a volume of some 10^{14} km^3.

A flare which releases 10^{32} erg of electromagnetic radiation probably emits another 3×10^{32} erg as kinetic energy of particles. It is generally believed that this energy is released by the annihilation of magnetic fields. Knowing that solar magnetic fields are hundreds of gauss near sunspots, and the volume of the field and plasma involved in the flare is 10^{14} km^3, it can readily be shown that the total energy in the region (at $B^2/8\pi$ cm^{-3}) comfortably exceeds 4×10^{32} erg. The magnetic field is therefore a plausible source from an energetic point of view.

Most of the energy is released in a few hundred seconds, but it takes about a day for the energy to build up sufficiently for another flare to occur in the same region. The release of energy presumably has to be 'triggered' in some way, and it is generally agreed that the magnetic fields in the active region must have the right configuration before this can occur. The form of that configuration is not generally agreed, but Fig. 10.1 shows one suggestion, which includes possible sources of some of the emissions of concern in this chapter. In this model it is thought that the conversion of magnetic energy begins at a neutral point formed above a pair of sunspots of opposite polarity. As the field collapses, particles are energized and move away from the neutral point along the magnetic field. Those protons that

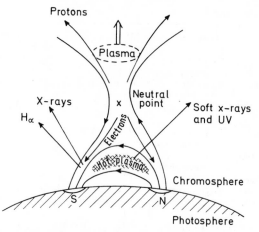

Figure 10.1 A model of the solar flare, showing possible sources for some of the known products. (After J.H. Piddington, *Cosmic Electrodynamics*, copyright © 1969. Reprinted by permission of John Wiley & Sons, Inc.)

travel outward produce *solar proton events* at the Earth (Section 10.3). Some of the electrons that travel inward are stopped in the denser gas of the chromosphere and produce bremsstrahlung X-rays which, when they reach the Earth, cause sudden disturbances of the lower ionosphere (Section 10.2). In the same chromospheric region hydrogen is ionized by the collisions of energetic electrons, and on recombination the gas emits H_α which can be observed at the Earth as the H_α *flare*. The corona becomes hot in the flare region and emits soft X-rays and ultraviolet light, both having consequences in the terrestrial ionosphere. Plasma is ejected from the Sun above the neutral point, enhancing the solar wind. In addition, many of the solar radio bursts can also be accounted for by such a model.

While it must be emphasized that the model sketched in Fig. 10.1 is by no means the only one that has been suggested, it is important to note the similarity between the model and current ideas about the substorm in the terrestrial magnetosphere (Chapter 8). The reader might like to identify the points of correspondence between the respective solar and terrestrial phenomena.

10.2 Ionospheric effects of electromagnetic radiation from a flare

The first kind of solar flare effect to be considered is characterized by the time coincidence between the ionospheric effect and the visible flare. Clearly, the cause of such an effect must have travelled from the flare at the speed of light and so must be electromagnetic in nature. In 1937 J. H. Dellinger identified *fadeouts* in high-frequency radio circuits as due to abnormal ionospheric absorption associated with solar flares. The onset of the fadeout was usually rapid, and the duration was typically tens of minutes, like that of the visible flare. Because of the sudden onset the immediate effects of solar flares are known collectively as *sudden ionospheric disturbances* (SID). The phenomenon discovered by Dellinger and known for many years as the Dellinger fade, is now usually called a *shortwave fadeout* (SWF). Since the SWF is due to abnormal absorption of the radio signal, the responsible ionization

229

Table 10.1   SID Phenomena

		Technique	Effect	Region	Radiation
SWF	Shortwave fadeout	HF radio propagation	Absorption		
SCNA	Sudden cosmic noise absorption	Riometer	Absorption	D	Hard X-rays 0.5–8 Å
SPA	Sudden phase anomaly	VLF radio propagation	Reflection height reduced		
SEA	Sudden enhance-ment of atmospherics	VLF atmospherics	Intensity enhanced		
SFE	(Magnetic) solar flare effect	Magnetometer	Enhanced ionospheric conductivity	E	EUV and soft X-rays
SFD	Sudden frequency deviation	HF Doppler	Reflection height reduced	E + F	EUV
—	Electron content enhancement	Faraday effect	Content enhanced	F	EUV

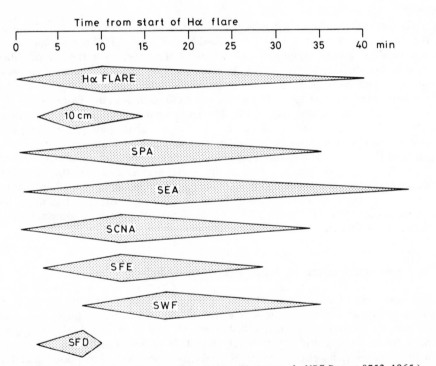

Figure 10.2   Relative timing of various flare effects. (V. Agy *et al.*, *NBS Report* 8752, 1965.)

can be placed in the D region and thus the enhancement of the electromagnetic spectrum is most likely to be in Lyman-α or X-rays, which are the major sources of D-region ion production under normal circumstances. It was known that Lyman-α was enhanced by a few percent during a flare, and for many years the SWF was

230

attributed to Lyman-α. However, when hard X-rays were observed from rockets they were seen to intensify by several powers of ten during a flare, and the X-ray enhancement is now recognized as the major cause of D-region ionization during a flare.

Many kinds of sudden ionospheric disturbance are now recognized, and these are summarized in Table 10.1. The largest effects are produced in the D region, but E and F effects can certainly be detected. The electron content, which depends mainly on the F region, is enhanced by a few percent. The various effects do not appear and vanish at quite the same time, and Fig. 10.2 illustrates the relative timing. The EUV and the hardest X-rays tend to be enhanced early in the flare, with the 10 cm radio noise, whereas the softer X-rays are of longer duration and correspond with the optical flare.

An important aspect of SID studies is the comparison between observed and computed effects of a flare on the ionosphere. The computations take the X-ray and EUV fluxes measured on a satellite, include an atmospheric model and assume the photochemistry. Fig. 10.3 shows the increases of electron density attributed to stated increments of flux in six wavelength bands. Note that most of the F-region enhancement comes from the middle bands, 260–796 Å, whereas radiations of shorter and longer wavelength contribute more at lower altitudes (see Sections 4.3.2 and 4.3.4, and Fig. 4.9). The overall profile might change as in Fig. 10.4

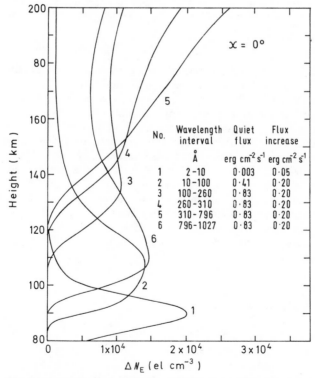

No.	Wavelength interval Å	Quiet flux erg cm^{-2} s^{-1}	Flux increase erg cm^{-2} s^{-1}
1	2–10	0·003	0·05
2	10–100	0·41	0·20
3	100–260	0·83	0·20
4	260–310	0·83	0·20
5	310–796	0·83	0·20
6	796–1027	0·83	0·20

$\chi = 0°$

Figure 10.3 The computed increase of electron density due to flux increases in stated wavelength intervals. (From A.D. Richmond, personal communication, 1970.)

231

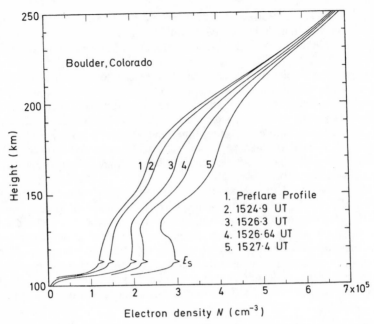

Figure 10.4 Change of ionospheric profile during a flare, deduced from SFD observations. (From R.F. Donnelly, *Report ERL 92 – SDL6*, Environmental Research Labs, 1968.)

during an actual flare. The F-region change is by no means negligible, although smaller than the D- and E-region effects in percentage terms.

All SID phenomena affect the whole sunlit hemisphere of the Earth and they show little spatial variation apart from a dependence on the solar zenith angle. Even if a solar flare is not detected optically, its direct effects can be recognized as SIDs by their spatial uniformity and time characteristics, and so the occurrence of the flare can be established with a high degree of certainty from radio observations. The detailed characteristics of the various SIDs can be used to deduce properties of the flare emissions even when they could not be directly observed. SID studies therefore contribute both to upper-atmosphere physics and, by inference, to knowledge of solar flares as such. For practical reasons that will be obvious, it has also been necessary in recent times to learn to distinguish the short-lived effects of a solar flare from those that might be produced by a nuclear explosion in the atmosphere.

10.3 Solar proton events

On 23 February 1956 a large (3+) flare occurred on the Sun, and as a result the solar proton event was discovered. The flare was followed by radio blackouts that lasted for several days, and by a large increase in the readings of ground-based cosmic ray monitors. By studying the effects on VHF forward-scatter radio links, D. K. Bailey showed that the blackout conditions were due to additional ionization

232

in the ionospheric D region. The additional ionization was produced by the arrival of energetic protons, which had presumably been released from the Sun at the time of the flare. Most of the early studies of proton events were made using riometers, a device which measures ionospheric radio absorption by means of the reduction of the intensity of the cosmic radio noise (Section 3.7.3). Riometer studies soon showed that the radio absorption is confined to high magnetic latitudes. Because the activity covers the polar caps only, solar proton events are often called *polar cap absorption events* or PCA.

On the average several events per annum are detected by riometers. The most energetic events are in addition detected at the ground by cosmic ray counters. The first recorded *ground-level event*, GLE, occurred on 28 February 1942. (The associated flare has another claim to fame, as the source of the first solar radio noise to be observed on Earth.) Fifteen GLE events were observed in the 25 years between 1942 and 1967, though this could be a slight underestimate because of the less sensitive measurement techniques in use before 1949.

Since the early 1960s it has been possible to study solar protons from space-craft, and for many years now the monitoring of energetic protons in space has proceeded almost continuously. The satellite detectors find many more events than are seen by ground-based methods. Most solar flares emit particles at the lower energies, that is up to 10 MeV, and at energies of tens of MeV the number of particles reaching the vicinity of the Earth from the Sun far exceeds the number of galactic cosmic rays. (At the highest energies, greater than about 1 GeV, galactic particles predominate.) Given that the production and release of energetic protons is a normal consequence of all solar flares, it is now thought that those producing PCA and GLE differ in degree rather than in intrinsic character. Protons must have energy of at least 30 MeV to produce radio absorption at 50 km, and 500 MeV to produce a GLE. In addition to the radio absorption in the atmosphere, affecting ionosondes, riometers and forward scatter links, solar protons affect the propagation of VLF radio waves and this provides a rather sensitive monitor for the upper-atmosphere ionization during a PCA. In spite of the name, *solar proton event*, it should be appreciated that α-particles and heavier nuclei are also emitted, the proportions representing the composition of the solar atmosphere. High-energy electrons may also be produced.

Compared to some other phenomena of solar—terrestrial physics, the solar proton event appears at first sight to be a relatively simple occurrence, in which particles leave the Sun during a flare, travel to the Earth, and there produce ionization in the atmosphere. However, consideration of the details will introduce us to a number of important physical ideas. The events are scientifically important in three respects:

(a) occurrence and relation to other solar phenomena;
(b) propagation from Sun to Earth and entry into the atmosphere;
(c) ionization effects in the atmosphere.

In addition, proton events are not without practical importance. Radio links, which may be the only form of communication available in remote areas, are seriously disrupted in polar regions. Equally, aircraft communications are affected. The protons also present a potential hazard in space travel. Practical consequences, and the monitoring and forecasting of proton events, are considered in Section 12.4.

10.3.1 Occurrence and production of solar proton events

The occurrence of PCA depends on the phase of the sunspot cycle. More than ten events can occur in an active year, or there might be none at all in a quiet one. Fig. 10.5 shows the occurrence of PCA events detected by ground-based methods over two solar cycles. Generally speaking, PCA occurrence correlates with the sunspot number, but there is some tendency for the events to avoid the peak of the cycle and to be enhanced somewhat in the declining phase, so that on average the PCA cycle lags the sunspot cycle by a year or two. As the years go by there is a tendency for more events to be reported because of the improving sensitivity of techniques and a greater awareness of the phenomena; thus some caution is needed when comparing the statistics from one cycle to the next.

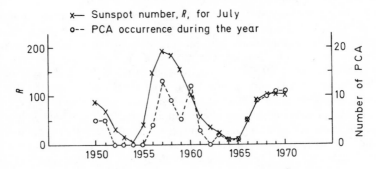

Figure 10.5   Occurrence of PCA during the sunspot cycle.

The mechanism by which a flare produces energetic particles is not altogether clear, though the energy almost certainly comes from the annihilation of magnetic field. One clue is that flares that produce proton events are also likely to generate radio noise of Type IV — a burst of long duration covering a wide frequency band. The spectrum of Type IV bursts associated with proton emission has a characteristic form, with relatively high intensity at metre and centimetre wavelengths but weaker intensity in the intermediate decametre band. The Type IV radio burst is explained by synchrotron emission from energetic electrons trapped in a magnetic cloud ejected into the solar corona. The energetic protons may come from the same energetic cloud.

The association between Type IV bursts and PCA is reasonably close and improves with the intensity of the event. Assuming that the burst and the proton cloud are released at the same time, the time of onset of the burst may be taken as the time when the protons left the Sun. It is observed that the delay between the radio burst and the beginning of the associated PCA event is shorter for more intense events: about 1 and 6 h for strong and weak events respectively. Most PCA events can be traced back to some particular flare, though with the use of more sensitive detection techniques, such as the VLF radio method, some exceptions are found. For example, of 52 PCA events detected by VLF between 1966 and 1971, 15 could not be related to a specific flare.

234

10.3.2 *Propagation from Sun to Earth*

The propagation of a cloud of protons will be considered in two parts: first the journey from the Sun through interplanetary space to the terrestrial magnetosphere; then passage through the geomagnetic field and into the atmosphere.

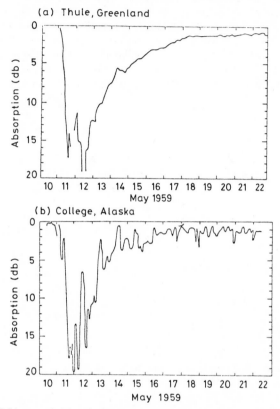

Figure 10.6 A PCA recorded by riometers at (a) Thule, Greenland and (b) College, Alaska. (From G.C. Reid, *Physics of Geomagnetic Phenomena* (Ed. Matsushita and Campbell) Academic Press, 1967.)

A typical solar flare lasts for tens of minutes, but a typical polar cap absorption event goes on for several days (Fig. 10.6). A proton of energy 10 MeV, which would be one of the slower particles of a proton event, travels at 4×10^4 km s^{-1} and would reach the Earth in 1 h if it came in a straight line. More energetic particles should arrive sooner. In fact, most PCA events are delayed several hours after the relevant flare. It seems, therefore, that both at the beginning and at the end of an event, propagation from Sun to Earth is not direct.

In looking for the cause, three hypotheses may be considered:

(i) particle emission continues after the optical and radio flares have subsided;
(ii) there is storage in interplanetary space;
(iii) particles become trapped in the geomagnetic field from which they are gradually deposited into the atmosphere.

235

Hypotheses (i) and (iii) are readily dismissed on observational grounds: (i) because a PCA may continue even though the associated flare region has moved out of sight beyond the limb of the Sun; (iii) because satellites find that the duration of the proton event in space is like that of the PCA in the atmosphere. Important clues in favour of hypothesis (ii) come from the properties of events as a function of the heliographic longitude of the associated flare. As a rule, flares towards the eastern limb of the Sun (which is on the left-hand side as seen by an observer in the terrestrial northern hemisphere) do not produce PCA. Second, the time delay between flare and PCA increases with the eastern longitude of the flare. These facts suggest that protons starting at a western heliographic longitude have better access to the Earth, and the obvious explanation is that the proton cloud tends to be guided by the interplanetary magnetic field, whose spiral form would produce just the effects observed.

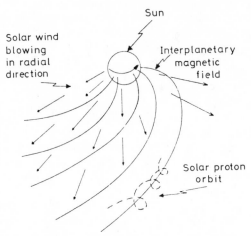

Figure 10.7   Influence of the IMF on solar protons. (From K.G. McCracken, *Solar Flares and Space Research* (Ed. De Jager and Svestka) North-Holland, 1969.)

If the gyro-radii of energetic protons is worked out for a typical IMF of 5 gamma (using Equation 2.17), it is readily shown that even for energy as high as 1 GeV the distance from Sun to Earth amounts to over 100 gyro-radii. Protons of lower energy gyrate in smaller loops and so are sensitive not only to the general shape of the IMF but also to any local irregularities and kinks (see Fig. 10.7). The propagation of protons through interplanetary space therefore becomes more like a diffusion process, in which the bulk velocity of the cloud is much less than that of individual particles within the cloud. By this means it appears possible to account for the delay and duration of the proton event and for the observed isotropy of the particles near the Earth. Since the IMF is carried out by the solar wind as indicated in Section 7.3.4, it would be expected to be more irregular in form at high solar activity and then to have a greater effect on solar protons. It is indeed observed that the delay between the solar flare and the terrestrial proton event is greatest at times of high solar activity.

On reaching the magnetosphere the energetic protons must pass through the outer geomagnetic field before they can enter the atmosphere. The theory of charged particle trajectories in magnetic fields was first worked out by Störmer for his studies of the aurora. Since the Störmer theory is valid only for single particles it is not in fact applicable to the aurora, which is dominated by collaborative phenomena. However, the theory has been successfully applied to the particles in solar proton events. For the reasons given in Section 7.6.1, a charged particle tends to follow a spiral path around a magnetic field line. The radius of curvature $(r_B = mv/Be)$ is proportional to the velocity and inversely proportional to the magnetic field. Because of the relatively high energy of solar protons, the magnetic field changes significantly over one gyration and therefore the simplification of an almost uniform magnetic field, as used for the Van Allen particles, cannot be made here. However, it is still true that particles travelling most nearly along the magnetic field are deviated least. Consequently the polar regions are the most accessible to solar protons. To reach the equator a proton has to cross field lines right down to the atmosphere. A particle of sufficient energy could do this, but in practice the equatorial regions are forbidden to the protons of solar events.

Since the radius of gyration in a given field depends on the momentum per unit charge, it is convenient to discuss particle orbits in terms of the parameter *rigidity*:

$$R = Pc/z|e| \qquad (10.1)$$

where P is the momentum, c the velocity of light, z the atomic number, and $|e|$ is the absolute value of the electronic charge. All particles with the same rigidity follow the same trajectory through the magnetic field.

Störmer investigated the special case of the dipole field, and found that the space around the dipole (representing the Earth) could be divided into 'allowed' and 'forbidden' regions, which a particle from infinity respectively could and could not reach. In general, particles can only reach magnetic latitude λ_c if their rigidity exceeds a value R_c known as the cut-off rigidity, where

$$R_c = 14.9 \cos^4 \lambda_c \qquad (10.2)$$

R_c being given in gigavolts (10^9 V). Conversely, particles of rigidity R_c may reach latitudes down to λ_c. Fig. 10.8 shows cut-off latitudes and the distance reached in

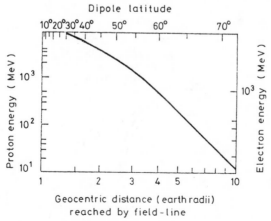

Figure 10.8 The Störmer cut-off latitude for protons and electrons. (After T. Obayashi, *Rep. Ionosphere Space Res. Japan*, **13**, 201, 1959.)

the equatorial plane by dipole field lines leaving those latitudes, for a range of proton and electron energies. The extent of the forbidden regions from the Earth is of the order of the *Störmer unit*, defined by

$$C_{st} = (M/Br_B)^{1/2} = (M|e|/mv)^{1/2} \qquad (10.3)$$

where M is the Earth's magnetic moment, and this may be used to indicate the range of energies affected by the Earth. For example, if a solar wind proton has energy 1 keV, C_{st} is greater than 100 earth radii, and thus solar wind protons cannot be expected to reach the Earth's surface. For electrons up to 1 MeV, C_{st} is always at least several hundred earth radii.

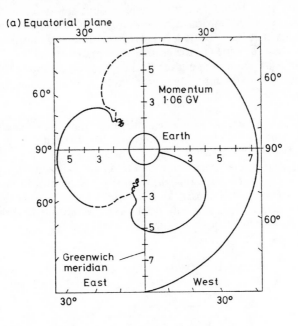

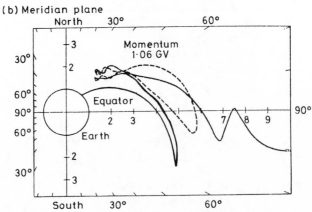

Figure 10.9 An example of a proton trajectory in the geomagnetic field. (After K.G. Mc-Cracken *et al.*, *J. geophys. Res.* **67**, 447, 1962, copyrighted by American Geophysical Union.)

To calculate the path of a proton through the geomagnetic field, whether of simple or more realistic form, the procedure is to imagine that a proton with negative charge is projected upwards from the point of impact of the real proton, since the trajectories of these two particles are exactly reversed. The complexity of the trajectories is illustrated by the example shown in Fig. 10.9. From a set of such trajectories it is possible to work out the directions in space from which protons reaching a given observing station must have come. Observations show that the more energetic protons, those with energies exceeding 1 GeV that can be monitored by ground-based cosmic-ray detectors, tend to originate more than 50° west of the Earth—Sun direction. This is further evidence for the influence of the interplanetary magnetic field on proton trajectories.

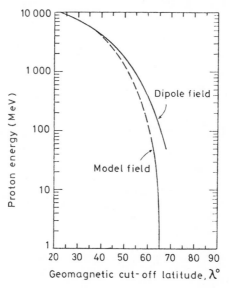

Figure 10.10 Cut-off latitudes for dipole and realistic geomagnetic fields. (From G.C. Reid and H.H. Sauer, *J. geophys. Res.* **72**, 197, 1967, copyrighted by American Geophysical Union.)

Observations of PCA with riometers show that during the main part of an event the distribution of absorption is essentially uniform and symmetrical over the polar caps, extending down to about 60° geomagnetic latitude. Störmer theory indicates that the protons causing the radio absorption should have energies exceeding 400 MeV. However, direct observations of the particles show that the cut-off rigidity at the edge of the polar cap is significantly less than the Störmer value. Two phenomena that can reduce the cut-off are a ring current (Section 8.3.4) and the distortion of the magnetosphere from the dipole form. Fig. 10.10 shows computed cut-offs for a dipole and a realistic field. If a geomagnetic storm, involving an enhancement of the solar wind and the creation of a ring current, occurs while a PCA is already in progress, it is observed that the cut-off is reduced still further.

The spatial distribution of PCA is not always uniform during its early and late phases. The absorption usually develops first near the geomagnetic poles, extending to the polar caps some hours later. Towards the end of the event it is likely that more of the absorption will be due to auroral electrons and therefore will be concentrated in the auroral zones.

Superimposed on the main features of PCA, themselves fairly well understood, are a number of effects whose explanation is less clear. Some PCA show a reduction of absorption for several hours near local noon, known as the *midday recovery*. The recoveries are localized near the boundary of the polar cap and are probably due to a local change of cut-off. A strange but statistically significant phenomenon is a seasonal variation in the occurrence of proton events. Fewer PCA occur in northern-hemisphere winter than at other times of the year (Fig. 10.11a), and events that do occur then are weaker and have longer delay times (Fig. 10.11b). There is no evidence that solar activity depends on the terrestrial seasons, and the effect cannot be explained by any discrepancy between PCA observations in northern and southern hemispheres. Presumably the relative orientations of Sun and Earth must govern the effect, but a detailed explanation is lacking.

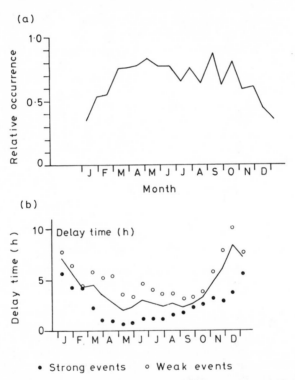

Figure 10.11   Seasonal effects in PCA: (a) fraction of flares with Type IV radio bursts that also produce PCA; (b) seasonal variation of delay time for PCA events, for strong and weak events. (After B. Hultqvist, *Solar Flares and Space Research* (Ed. De Jager and Svestka) North-Holland, 1969.)

10.3.3 *Atmospheric effects of solar protons*

When an energetic proton enters the atmosphere it loses energy in collisions with the atmospheric gas and leaves behind an ionized trail. The rate of energy loss from

240

a proton travelling through air is well known from laboratory measurements, and a graph of loss rate against distance travelled is known as a *Bragg curve*. For protons in the energy range of concern to us the energy loss rate increases as the proton slows down because it spends a longer time in the vicinity of a given air molecule. Over the range 10–200 MeV, the loss rate is almost inversely proportional to the energy, and a typical value at 100 MeV is 0.8 MeV m^{-1} of path in air at standard temperature and pressure. The energy may be assumed to be used entirely in the creation of ion pairs, each requiring about 35 eV. For a proton that enters the atmosphere from space the air density also increases along the path, and therefore the ionization is very concentrated towards the end of the path in PCA events. A proton that enters the atmosphere vertically with initial energy 50 MeV loses only 10% of its energy down to 56 km altitude, but the last 10% is lost in only 100 m between 42.1 and 42.0 km. Half the initial energy is deposited in the final 2.5 km.

If a proton event contains particles of 1–100 MeV we expect to see atmospheric effects in the altitude range 90–35 km (Fig. 10.12). The flux of particles falls off at the higher energies, however, and the recombination rate is greater at lower altitudes, so that protons of higher energy are less effective than might have been thought. Nevertheless, in some events substantial amounts of ionization are produced down to 50 km. Solar protons therefore ionize a region of the atmosphere somewhat below the normal ionosphere. Detailed computations would follow the principles outlined in Section 4.3.4, with the production rate (q) worked out as above. But since the D-region photochemistry is not adequately understood the process may usefully be turned around, measurements of the flux and energy of incident protons from spacecraft being combined with the atmospheric effects observed by radio methods to study the chemistry of the mesosphere.

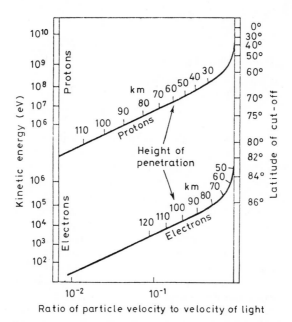

Figure 10.12 Penetration of protons and electrons, energy, and cut-off latitude as functions of velocity. (After D.K. Bailey, *Proc. Inst. Rad. Engr.* **47**, 255, 1959.)

The most striking chemical effect on PCA is the diurnal variation. The radio absorption measured by riometers is typically 4 or 5 times larger by day than by night. Fig. 10.13 shows a PCA observed at two stations at high latitude. The stations were at similar geomagnetic latitudes but in opposite hemispheres. The ionosphere over one station was continuously illuminated by the Sun during the event, whereas the other alternated between day and night. The proton flux should be similar over the two stations, and indeed the radio absorption was virtually the same when both stations were sunlit. However the absorption was greatly reduced by night. A day—night modulation can also be seen in Fig. 10.6b.

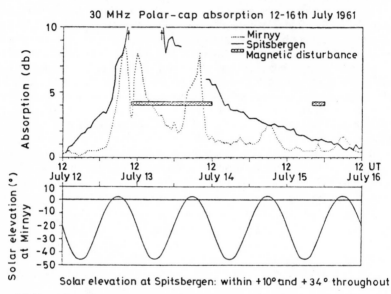

Figure 10.13   Polar cap absorption at two high-latitude stations. (From C.S. Gillmor, *J. atmos. terr. Phys.* **25**, 263, 1963, by permission of Pergamon Press.)

Without doubt the cause of the day—night modulation is a variation of the electron/ion ratio, λ, defined in Equation 4.62. Only the ionospheric electrons contribute to the radio absorption, and thus, as shown by Equation 4.64, a variation of λ is effective even though q remains constant. The behaviour over twilight is particularly instructive, showing that the atmospheric ozone layer acts as a screening layer for the solar radiation that causes electrons to be detached from negative ions. It follows that the effective radiation must be in the ultraviolet rather than in the visible part of the spectrum. This point was discussed in Section 4.3.4.

An important effect of the arrival of energetic particles that was appreciated only recently is that they may alter the chemical composition of the atmosphere. During a solar proton event, the primary protons produce secondary electrons with energies of 10s and 100s eV. These secondaries ionize and dissociate molecular nitrogen to give N_2^+, N^+ and N; and the N^+ and N may then react with O_2 or O_3 to give nitric oxide (NO). These processes go on continuously with the arrival of galactic cosmic rays, but it has been shown that the total NO production during one major PCA can exceed the annual production by galactic cosmic rays. Polar cap events are therefore an important source of atmospheric nitric oxide at latitudes above $60°$. One consequence of enhanced nitric oxide concentration is a reduction of ozone in the stratosphere by the reactions

$$NO + O_3 \rightarrow NO_2 + O_2$$
$$\underline{NO_2 + O \rightarrow NO + O_2}$$
$$O_3 + O \rightarrow 2O_2$$

10.4 Further reading

De Jager, C. and Svestka, Z. (1969) *Solar Flares and Space Research* (North-Holland) particularly the articles by De Jager and Hultqvist.

LeGalley, D. P. and Rosen, A. (1964) *Space Physics* (Wiley) Chapter 16.

McIntosh, P. S. and Dryer, M. (1972) *Solar Activity Observations and Predictions* (MIT Press) particularly the articles by Svestka and Sturrock.

De Feiter, L. D. (1976) Chromospheric flares or chromospheric aurorae?, *Space Sci. Rev.* **17**, 181.

Obayashi, T. (1976) Energy build-up and release mechanisms in solar and auroral flares, *Space Sci. Rev.* **17**, 195.

Nagata, T. (1976) Auroral flares and solar flares, *Space Sci. Rev.* **17**, 205.

Kavanagh, L. D., Schardt, A. W. and Roelof, E. C. (1970) Solar wind and solar energetic particles: properties and interactions, *Rev. Geophys. Space Phys.* **8**, 389.

Lanzerotti, L. J. (1972) Solar energetic particles and the configuration of the magnetosphere, *Rev. Geophys. Space Phys.* **10**, 379.

Scholer, M. (1976) Transport of energetic solar particles on closed magnetospheric field lines, *Space Sci. Rev.* **17**, 3.

Zmuda, A. J. and Potemra, T. A. (1972) Bombardment of the polar-cap ionosphere by solar cosmic rays, *Rev. Geophys. Space Phys.* **10**, 981.

Mitra, A. P. (1975) D-region in disturbed conditions, including flares and energetic particles, *J. atmos. terr. Phys.* **37**, 895.

Hoffmann, D. J. and Sauer, H. H. (1968) Magnetospheric cosmic-ray cutoffs and their variations, *Space Sci. Rev.* **8**, 750.

11

Storms and Other Disturbance Phenomena

11.1 Introduction

In Chapter 8 we discussed the substorm, a concept introduced fairly recently to describe the elementary unit of geophysical activity originating in the terrestrial magnetosphere. The idea of a storm is much older and less well defined, and its use is not confined to the sum of a sequence of substorms. The name plainly suggests an analogy with weather storms in the troposphere, a drastic perturbation from normal behaviour lasting, usually, from one to several days. The magnetic storm, for instance, clearly fits into this category and as a phenomenon has been known, if not by that name, since the 18th century.

There are several phenomena of the upper atmosphere whose characteristics invite the description 'storm', and it is convenient to discuss them together in the present chapter. It is likely that some of these storm phenomena, and conceivably all of them, have inter-associations that are physical as well as linguistic. But it should be remembered that this is not necessarily so. In the storm phenomena we encounter topics where the observations, as far as they go, may be on fairly sure ground, but where understanding ranges from incomplete to fragmentary. Although some of the possibilities are intriguing, the evidence in some cases will have to be considerably strengthened before it can be adjudged satisfactory.

Magnetic storms having been dealt with (Sections 8.2.2 and 8.3.4) in terms of ionospheric and magnetospheric current systems, we will look first at storms in the F region and the D region of the ionosphere. Then a major anomaly of the D region, the *winter anomaly*, will be discussed. The winter anomaly has associations with the meteorology of the stratosphere and possibly also with geomagnetic storms. Finally, we examine a controversial question: whether the influence of solar activity may extend as low as the troposphere and there influence the weather.

11.2 The F-region ionospheric storm

11.2.1 Observations

The characteristic behaviour of the ionospheric storm has been established by means of ionosondes, giving $N_m(F2)$ and $h_m(F2)$, and by trans-ionospheric radio propagation, giving electron content (I). The ratio of electron content to maximum electron density gives the slab thickness of the ionosphere

$$\tau = I/N_m \tag{11.1}$$

Like the magnetic storm, the typical ionospheric storm goes through several phases.

244

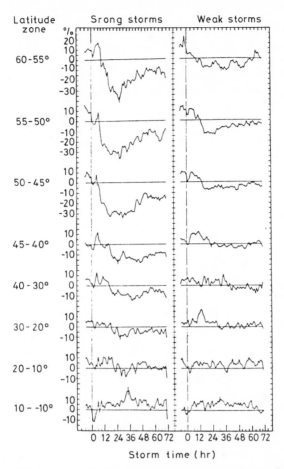

Figure 11.1 Storm-time (D_{st}) variations of maximum electron density (N_m(F2)), in eight zones of geomagnetic latitude. The ordinate is the percentage deviation from quiet-day behaviour. (After S. Matsushita, *J. geophys. Res.* **64**, 305, 1959, copyrighted by American Geophysical Union.)

During the first few hours the electron density and the electron content may be enhanced. Subsequently both quantities are depressed relative to their normal values, and there is then a gradual return to normal during the next few days. The variations are illustrated in Fig. 11.1. The various stages may be called the 'initial', 'main' and 'recovery' phases by analogy with magnetic storms, or alternatively the 'positive' and 'negative' phases. The beginning of the storm may be sudden or gradual. Ionosonde measurements at middle latitudes during the main phase show that the reduction of N_m(F2) is accompanied by a large increase in the apparent height of the maximum, h'(F2). However, real-height analysis of the ionograms shows that most of this is due to group retardation of the radio wave rather than to a genuine lifting of the region. The slab thickness (τ) does increase (Fig. 11.2), showing that the region broadens during the negative phase. A broadening is consistent with increased group retardation because a radio wave near the penetration

245

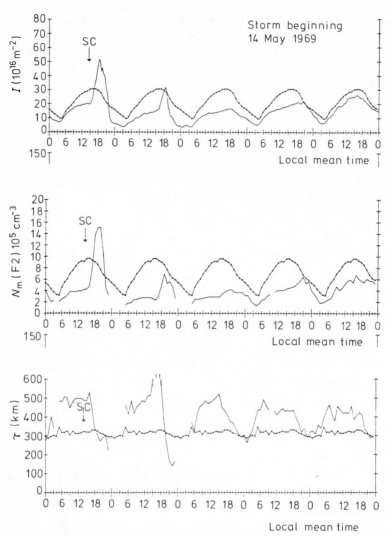

Figure 11.2   Electron content, electron density and slab thickness at a mid-latitude station during an F-region storm. SC marks the sudden commencement. The 7-day mean is also shown to indicate the normal behaviour. (From M. Mendillo and J.A. Klobuchar, *Report AFCRL–TR–74–0065*, US Air Force, 1974.)

frequency then spends a longer time in a region where the group refractive index is high and thus the pulse travels more slowly.

   To some extent it is misleading to present storm data only as a function of time, i.e. as a D_{st} variation. This may be demonstrated by a fact that appears curious at first sight, but whose explanation is really very simple. The fact is that ionospheric storms beginning at certain times of day – most notably during the night hours – often do not show any positive phase. The reason is that the positive phase appears only in regions that were on the day side of the Earth during the first few hours.

Thus, whatever the physical cause, it is clear that to make sense of the data one must recognize that the observing stations are turning with the Earth, and that the storm effects depend on the local time as well as on the storm time. The existence of longitude effects is demonstrated by Fig. 11.3, which shows that for the storm of 8 March 1970 only the more easterly stations showed enhancements; at the more westerly stations the electron content was depleted through the storm. The pattern of the ionospheric storm shows a seasonal modulation, in which the negative phase is relatively stronger, and the positive phase relatively weaker, in the summer hemisphere.

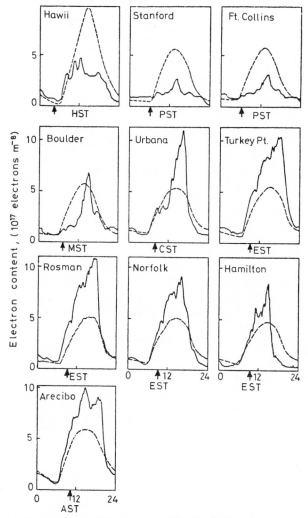

Figure 11.3 Longitudinal variation in the diurnal electron-content pattern for the storm of 8 March 1970. The Sc is indicated by the arrow, and the dashed lines show the normal pattern. The time zones are: HST = UT − 10; PST = UT − 8; MST = UT − 7; CST = UT − 6; EST = UT − 5; AST = UT − 4. (From J.A. Klobuchar *et al.*, *J. geophys. Res.* **76**, 6202, 1971, copyrighted by American Geophysical Union.)

11.2.2 Theories

Fig. 11.4 shows that if the changes of electron content, equatorial D_{st} and K_p index are averaged over many storms, characteristic patterns can be seen. Because of the outward similarities between the forms of magnetic and ionospheric storms, and because they occur together, there is a temptation to think that they must have a close physical connection. This might be the case, but such a connection has not been proved. The initial and main phases of a magnetic storm are well explained by solar-wind pressure and the ring current (Section 8.3.4). It is not at all obvious that the same causes could produce ionospheric storms.

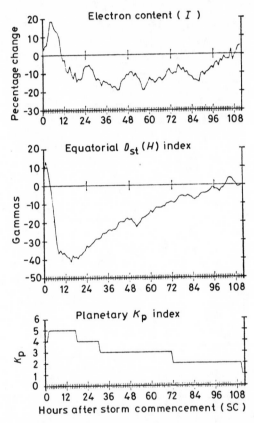

Figure 11.4   Average storm-time variations of electron content, equatorial magnetic index, and K_p. (From M. Mendillo, *Planet. Space Sci.* **21**, 349, 1973.)

Another cause that can be dismissed is a change in the flux of ionizing electromagnetic radiation. There is no direct evidence for variations of solar EUV emissions during ionospheric storms (except those associated with solar flares). Moreover, it is observed that the marked changes that occur in the F2 region do not extend to the F1 region. This, also, suggests that the ionizing radiation remains constant.

248

The basic principles of F2-region behaviour, as far as they are understood, were introduced in Sections 4.3.3 and 5.2.4. Presumably the ionospheric storm should be explicable by a modification or modifications of these mechanisms, and so one should look at production by energetic particles; changes in recombination rate due to a change of chemical composition or of temperature; movements of plasma along the geomagnetic field lines between ionosphere and protonosphere; and effects of neutral-air winds or electric fields.

The question of ionization by energetic particles is really one for the high-latitude ionosphere (Section 5.4.2). During ionospheric storms the F region is enhanced in the auroral zone, which is not surprising since particle precipitation is a major source of ionization in those regions most of the time. The enhancement of K_p and other indices (Fig. 11.4) shows that auroral-zone activity is high during storms at middle latitudes.

We saw in Section 8.1.5 that the plasmasphere is depleted during a storm, and that a gradual refilling occurs during geomagnetically quiet periods. It is not known with certainty where the plasma goes to. Possibly it moves down to the F region, where it might produce the positive phase of the storm in the evening sector. On the other hand it might be detached from the co-rotating plasmasphere and be dispersed by the general magnetospheric circulation. What is clear is that the plasmapause moves inward and the mid-latitude ionospheric trough moves to lower latitude. Thus, places within the range of movement of the trough (generally between L-values 6 and 3) will experience a depletion of electron density. The beginning of the negative phase of some storms has been attributed to such a movement of the trough.

In addition, it is believed that changes in the wind pattern are important. The idea that the circulation of the thermosphere is different during storms is based on the heating of the upper atmosphere at auroral latitudes by the auroral electrojet and particle precipitation. Electrojets can dissipate as much as 0.5 W m^{-2} in the E region during severe disturbances, and auroral particles contribute a smaller amount. The tidal winds normally present in the thermosphere are driven by the absorption of a much smaller flux, some 0.5 mW m^{-2} of solar EUV at altitudes above 120 km, which produces a poleward flow by day and an equatorward flow by night (Section 6.3.3). These flows are typically between 100 and 300 m s^{-1}. During a storm the day-time wind is reversed, now flowing towards the equator at speeds up to 500 m s^{-1}, with the return flow probably near the 100 km level (Fig. 11.5).

Consider a parcel of air of unit volume, which rises at velocity W and receives energy E per second from an external source. The energy input (a) heats the gas and (b) supplies the mechanical work done by the gas on its surroundings as it expands in moving to regions of lower pressure. The heating rate

$$\frac{dQ}{dt} = \rho C_p \frac{dT}{dt} = \rho C_p \frac{dT}{dh}\frac{dh}{dt} = \rho C_p \frac{dT}{dh} W$$

where ρ is the air density, C_p is the specific heat per unit mass at constant pressure, T is the temperature, h the height and W the vertical wind speed. If the pressure is P, mechanical work is done by the air parcel at the rate

$$-\frac{dP}{dt} = -\frac{dP}{dh}\frac{dh}{dt} = \rho g W$$

since $dP/dh = -\rho g$ from the barometric equation, where g is the acceleration due to gravity.

249

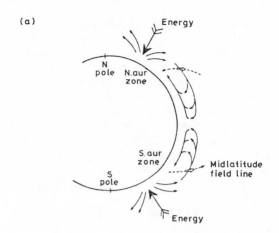

(a)

Neutral-air wind → Wind-produced ion drift ⇒

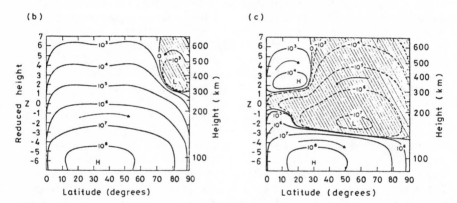

(b)   (c)

Figure 11.5 (a) A sketch of the storm circulation in the upper atmosphere. (From H. Rishbeth, *Rad. Sci.* 9, 183, 1974. Crown copyright, Appleton Laboratory, Slough.) (b, c) Computed circulation of the upper atmosphere (b) under solar heating and (c) with the addition of Joule heating by electrojets. (After R.E. Dickinson *et al., J. atmos. Sci.* **32**, 1737, 1975.) In (b) solar heating produces a Hadley circulation, with air rising over the equator and sinking over the poles. There is a small contrary cell at high latitude and high altitude, due to heat flowing in from the magnetosphere outside the plasmapause. The addition of an auroral heat source in (c) reverses the circulation at F-region heights, leaving a poleward flow at lower levels. At 300 km the equatorward wind is about 50 m s^{-1}.

Thus

$$E = \rho \left(C_p \frac{dT}{dh} + g \right) W$$

$$W = E \Big/ \rho \left(C_p \frac{dT}{dh} + g \right)$$

(11.2)

In an isothermal atmosphere $dT/dh = 0$, $W = E/\rho g$.

The first results of the equatorward wind is to lift the ionization by the mechanism discussed in Section 5.2.4. Though the amount of lifting is not large by day,

250

it acts to reduce the recombination rate and thus to increase $N_m(F2)$ and the electron content for the positive phase. The bulk motion of the neutral air is also important because it alters the composition of the upper atmosphere at middle latitudes. In general, because of diffusive separation, the ratio of atomic to molecular gases increases with altitude. The storm circulation also carries up air from the lower levels and thereby enriches the molecular species at the higher levels, and this enriched air arrives at middle latitudes after a time delay of several hours. As shown in Section 4.3.3, this means that the ratio q/β is reduced, and thus the electron density is reduced to give the negative phase of the storm. Although the details are probably more complex than in this simple picture, the foregoing ideas are thought to be a promising start to understanding the mechanisms of the ionospheric storm.

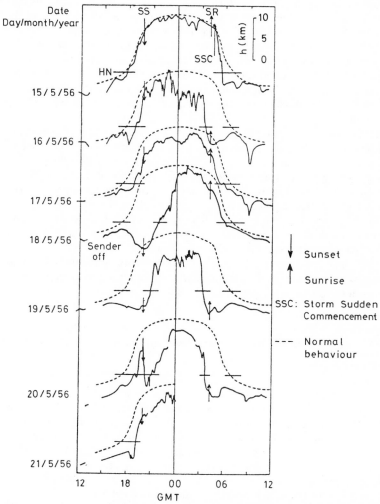

Figure 11.6 D-region storm as seen by VLF radio: reflection height of 16 kHz waves over the path Rugby–Cambridge, UK, during a magnetic storm in 1956. (From J.S. Belrose, *AGARD Report 29*, 1968.)

251

11.3 D-region storm effects

Here again we are concerned mainly with effects in the mid-latitude ionosphere. Disturbances of the high-latitude D-region are directly caused by the precipitation of energetic particles, and have been dealt with in Section 8.2. The mid-latitude D region, however, undergoes longer term changes that are more puzzling.

Most studies of the D-region ionospheric storm have been based on radio propagation, in particular of LF (30–300 kHz) and VLF (3–30 kHz) waves. The effect on VLF signals is illustrated by Fig. 11.6. As we saw in Section 3.7.2, VLF waves are reflected as at a sharp boundary whose height is very sensitive to the electron density. During the storm, the reflection height is reduced, indicating an increase in the local electron density at heights near 70 km. This has been called the storm *after-effect*, since it typically follows a magnetic storm and continues for several days after the magnetic data have returned to normal. During the VLF after-effect the absorption of LF radio waves is increased, which again indicates that the lower part of the D region has been enhanced. LF effects have been detected for up to 8 days after a magnetic storm. The after-effects have been ascribed to the precipitation of energetic electrons (100s keV or more), which had been built up in the inner part of the magnetosphere during the magnetic storm. This view is supported by the observation that the effects seem to be confined to latitudes poleward of about $50°$.

Another interesting possibility is a change in the chemistry. It is a curious fact about the D region that its chemistry, complex at middle latitudes under quiet geomagnetic conditions, becomes simpler during some disturbances and at high latitudes. The water cluster ions (Section 4.3.4) disappear at altitudes above 70–75 km during polar cap absorption events, so that the dominant positive ions become O_2^+ and NO^+ and the negative ions revert to the simple species O_2^- and O^- throughout most of the D region. The relative concentration of the larger ions decreases as the activity increases, so that the effective electron recombination rate becomes dependent on the production rate.

Nitric oxide is produced during precipitation events by the mechanism outlined in Section 10.3.3, and its enhancement will increase the production rate of ionospheric electrons by solar Lyman-α radiation (Section 4.3.4).

11.4 The winter anomaly of ionospheric radio absorption

11.4.1 Observations

The winter anomaly of ionospheric radio absorption is one of the classical anomalies of the ionosphere. It has been intensively studied and the observational facts are beyond question. Yet it has proved difficult to understand. One important point is that the phenomenon cannot be dealt with in isolation but, on good evidence, it involves relationships between the mesosphere and the stratosphere. This introduces an interesting aspect of present-day atmospheric science, which seeks to investigate how the different levels of the atmosphere influence each other. It remains to be seen how far such ideas can be extended.

The winter anomaly shows up in radio absorption measurements, particularly those made by the pulse-reflection method (Section 3.7.1) and by the reception of continuous low-frequency radio transmissions after reflection from the ionosphere. It is possible that the anomaly actually comprises several components, but in essence it amounts to two facts:

(a) on the whole, radio absorption is greater in winter (by about a factor of two) than would be predicted by a simple extrapolation from measurements made in the summer;
(b) the absorption is much more variable in winter than in summer, and groups of adjacent days show absorption levels that are abnormally high even with respect to the general enhancement (a); on the other hand, the absorption on some days is as low as would be expected of summer conditions.

Fig. 11.7 shows some examples that include both facets of the anomaly. These observations show the absorption either at one time of day (noon in Fig. 11.7a), or for one particular value of the solar zenith angle ($\cos \chi = 0.1$ in Fig. 11.7b). According to simple theory, the absorption in decibels should vary with zenith angle as $\cos^n \chi$ (Fig. 11.7a plots $\cos \chi$ and $\cos^{1/2} \chi$); whatever the law, one would expect that the absorption would be the same for a given value of χ at all times of year (dashed lines of Fig. 11.7b). In the northern hemisphere this is true for most of the year, typically from March to October. The behaviour tends to be anomalous in the winter months of November to February, the largest anomalies being in January and February. The winter anomaly also occurs in the southern hemisphere but with some differences. The effects are smaller and the major events occur very late in the winter – usually in November, which would be equivalent to May in the northern hemisphere. In neither hemisphere does the pattern repeat exactly from one year to the next.

Probably, more remains to be discovered about the morphology of the anomaly, but it is generally considered to be essentially a mid-latitude phenomenon, in which the magnitude and duration of periods of high absorption decrease with decreasing latitude. In the northern hemisphere, for which the best observations are available, it does not occur south of about $40°$ N. Above that latitude the steady part of the anomaly (a) occurs over the whole of the winter hemisphere. By contrast, the variable part (b) is of limited extent on any one day, and the correlation coefficient between pairs of observing sites falls off significantly over 200 km. This is illustrated in Fig. 11.8.

The immediate cause of the winter anomaly is quite clear. Rocket measurements have shown that on anomalous days the electron density between 70 and 80 km is enhanced by as much as a factor of ten (Fig. 11.9). Since there is no enhancement of ionizing radiation that can explain the major effects – a qualification regarding minor effects is indicated in Section 11.4.3 – one must look for a change of chemistry. The three favourite mechanisms are (1) increased ion production rate due to an increase of NO concentration, (2) decreased loss rate due to removal of water-cluster ions, and (3) effect of temperature changes on the chemistry. (1) and (2) require large-scale transport of species from the auroral zones. In Section 4.3.4 we pointed out the ledge in the electron density profile between 80 and 90 km, which is ascribed to the hydration of ions below the ledge. In explanation 2 the ions below 85 km revert to the non-hydrated form on anomalous days. There is also direct evidence for increased NO concentration (explanation 1), and for temperature changes (explanation 3) associated with wind towards, not away from, the auroral zone! Clearly, the cause of the changed electron-density profile is still debatable.

253

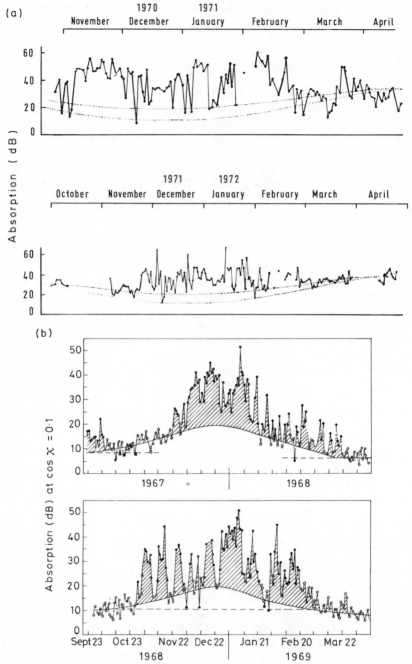

Figure 11.7 Winter anomaly of ionospheric radio absorption. (a) Absorption at noon for 1970–72. (After J.K. Hargreaves, *J. atmos. terr. Phys.* **35**, 291, 1973, by permission of Pergamon Press.) (b) Absorption constant zenith angle for 1967–69. (From H. Schwentek, *J. atmos. terr. Phys.* **33**, 1647, 1971, by permission of Pergamon Press.) The trends from summer are shown dotted.

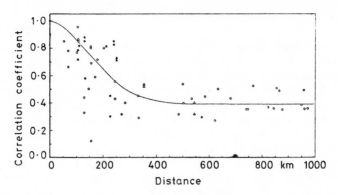

Figure 11.8   Correlation coefficient of absorption with distance for 1970–72. (After H. Schwentek, *Methods of Measurement and Results of Lower Ionosphere Structure*, Akademie–Verlag, 1974.)

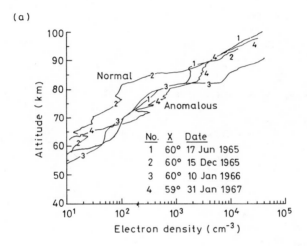

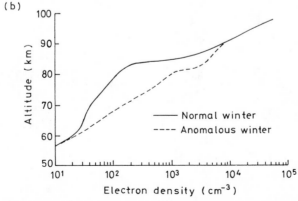

Figure 11.9   Electron density profiles measured from rockets: (a) individual measurements; (b) average profiles for normal and anomalous winter conditions. (From M.A. Geller and C.F. Sechrist, *J. atmos. terr. Phys.* **33**, 1027, 1971, by permission of Pergamon Press.)

255

11.4.2 Relation to the stratospheric warming

It is not known with certainty why the ion composition of the mesosphere alters on some winter days, but an important clue came to light in the mid-1960s. It has been known for many years that at the 10 mbar level (about 30 km altitude) the atmosphere has a warm region over the pole in summer and a cold region over the pole for most of the winter. But on certain days of the winter the polar atmosphere becomes warmer than usual, by 10–30 °K, there being a change of circulation pattern as well. (Changes of nearly 100 °K were observed at 45 km in early 1971.) Such warm periods, generally known as *stratospheric warmings*, or *stratwarms* for short, were first observed during the 1950s by rocket techniques. More recently, satellite remote-sensing methods have delineated the temperature distribution on the global scale, and shown that during the stratwarm a hot region develops to

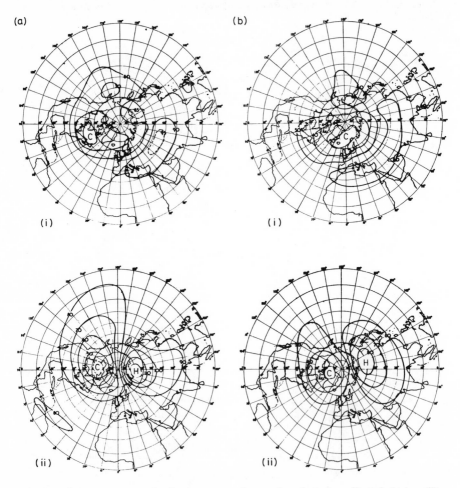

Figure 11.10 Temperature distributions over the north pole before (i) and during (ii) a stratwarm in early 1971, at the (a) 2-mbar and (b) 20-mbar levels. C: cold, H: hot. (After J.J. Barnett *et al.*, *Nature*, **230**, 47, 1971.)

256

one side of the pole, as shown in Fig. 11.10. Measuring around the pole at constant latitude, the temperature shows one maximum and one minimum, and this pattern is said to be 'wave number 1'. Wave number 2 and higher orders can appear at times in the stratosphere, and in the troposphere the polar front typically exhibits a high number such as 5 or 6. At stratospheric levels, however, the characteristic of a major stratwarm, which happens only in winter, is that the temperature distribution becomes dominated by wave number 1. Major warmings may occur once or twice during a winter, but in some years there are none at all.

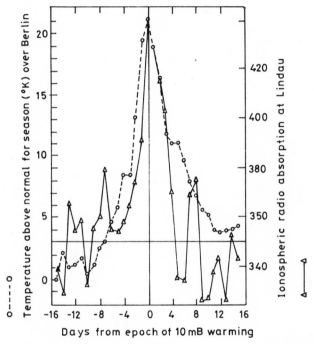

Figure 11.11 Superimposed epoch analysis comparing 10-mbar temperatures over Berlin and radio absorption over Lindau. (After A.H. Shapley and W.J.G. Beynon, *Nature*, **206**, 1242, 1965.)

The key discovery of the 1960s was that a major stratwarm is always accompanied by a period of anomalous radio absorption. This is illustrated by the superimposed-epoch analysis shown in Fig. 11.11. On the other hand, it is debatable whether there is a correlation in general between the stratospheric temperature and the D-region radio absorption above a given station — some investigators have produced evidence for, and others against, this proposition. Nevertheless, if attention is confined to the major events, established by global observations of each pheno-menon, the association seems to be indisputable. Thus we have a clear relation between stratospheric parameters such as temperature and wind, and the chemistry of the D region. For energetic reasons, one assumes that the first influences the second, rather than the reverse, since a planetary wave established in the lower part of the atmosphere would be expected to propagate upward.

257

11.4.3 Relation to geophysical activity

The possibility of an influence originating at a higher level enters the problem when the role of geophysical activity is considered. Whether or not auroral and magnetic disturbances had any connection with the winter anomaly was for many years a debatable issue. Studies published up to the mid-1960s generally concluded that magnetic activity (usually quantified by the index K_p) and mid-latitude radio absorption in the ionosphere were not associated. Then it was found that there was a correlation with energetic particle precipitation into the mid-latitude ionosphere, as measured on an orbiting satellite. In itself this was not too surprising, since energetic particle precipitation causes abnormal radio absorption at high latitudes. The mid-latitude effect was therefore in essence an equatorward extension of the phenomenon of auroral radio absorption (Section 8.2). The difficulty with this explanation is that the amount of precipitation is not sufficient to account for the amount of absorption during a major anomaly. But with attention thus redirected to the question, further studies, again based on superimposed-epoch analysis, showed that enhanced mid-latitude absorption and stratwarms tended to follow high-latitude magnetic disturbances by 3–5 days. More recent studies have tended to confirm that relations between high- and mid-latitude phenomena are found more easily if a time delay of several days is allowed for. This applies to the stratwarm as well as to the D-region winter anomaly. For example, the event shown in Fig. 11.10 came to a maximum several days after a period of enhanced geophysical activity, when auroral-zone precipitation was intense and magnetometers were disturbed; be it noted, though, that close inspection of the observations shows that the stratospheric warming had already begun to develop before the geophysical activity increased. The intense solar proton event of August 1972 produced effects like a winter anomaly in the sub-Antarctic region.

Two kinds of hypothesis are put forward to explain these associations:

(a) The stratospheric warming is an instability to which the winter atmosphere is prone, and it may be triggered by an influx of energetic particles at high latitude. Calculations (Section 8.2.4) show that the energy deposition by energetic particles is considerable, and could lead to significant local heating even at 80–90 km. It is not at all clear, though, how any perturbations produced at that level could penetrate down to the stratosphere.

(b) The chemical composition of the upper atmosphere is altered by particle precipitation, particularly at high latitude. The general effect is to reduce the effective electron loss rate and thereby increase the equilibrium electron density. These species have relatively long lifetimes, and if they are produced by auroral activity at a time when a stratwarm is in progress, they could be carried by the altered circulation to middle latitudes and there produce the winter anomaly (as in Section 11.4.1).

Explanation (a) says that one phenomenon causes another; (b) relies on a chance coincidence of two phenomena to produce a third. Both explanations have their attractions, and it is too soon to say which, if either, is correct.

A moral to be drawn from the study of mesosphere–stratosphere relations is that in the complexity of the middle atmosphere no phenomenon can be properly understood in isolation. It is a topic where some of the ideas which are now accepted as plausible theories or as tolerable hypotheses would have been thought highly improbable at one time, and a reminder that one should be cautiously open-minded when considering the more intractable problems of the atmosphere.

258

11.5 Solar activity and tropospheric weather

11.5.1 Introduction

We conclude our discussions of storm phenomena by introducing a topic that is speculative, controversial and intriguing — the possibility that the consequences of solar activity might extend below the upper atmosphere (where its influence is unquestionably a major one) to affect weather patterns in the troposphere. Perhaps this topic is misplaced among storm phenomena, but it is included at this point because there are superficial similarities, such as of duration and of association with geomagnetic effects, with other storms that we have discussed; because it appears to be a problem in which different levels of the atmosphere must be considered together, indeed, it is possibly the extreme case of such interactions; and because it provides us with a good example of an area of atmospheric science in embryo form. In its partly developed state the topic of Sun—weather relationships is frowned on by some, but grasped with excitement by others; a topic where some of the observations have gained respectability, but the theory is still in disarray.

The question of sunspots affecting the weather is an old one, dating back more than three centuries. Over the years many attempts have been made to correlate the sunspot numbers with parameters related to the weather, such as rainfall, temperature, crop yields, length of growing season, and so on. Such is the influence of the weather in human affairs that if there is some aspect of solar activity that affects the weather, even if only slightly, it is important that it should be known. Unfortunately, some of the Sun—weather relationships that have been claimed have not been such as to enhance the reputation of the topic. Either the result claimed has lacked statistical significance, or the correlation has appeared ridiculous for lack of any conceivable mechanism. As a result, the topic had been in disrepute since the 19th century. The recent resurgence of interest has come mainly from students of space and the upper atmosphere, who had new kinds of data to hand and who asked what atmospheric behaviour went with it. No doubt the lack of good mechanisms contributed largely to the subject's former disrepute. And while some of the new evidence is convincing in a statistical sense, the mechanisms remain obscure. The fact is, however, that the evidence for such effects is now so strong that it can no longer be ignored. The development of adequate explanations is one of the main tasks of further investigations.

In this section we will summarize some of the evidence related to Sun—weather relationships, as an indication of the kinds of facts that have come, and are coming, to light. Whether all of them are relevant to whatever final picture eventually emerges, remains to be seen. We conclude with some points relevant to the search for mechanisms.

11.5.2 Solar cycle and weather

The kind of analysis that suggests a connection between the solar cycle and the weather is illustrated in Fig. 11.12. Here the rainfall in three zones between 50° and 70° N has been plotted against the year. The rainfall figures are taken over a number of stations and are smoothed. The sunspot numbers shown have also been smoothed, and they have also been normalized so that each cycle goes between zero at sunspot minimum and 100% at sunspot maximum, a procedure that removes

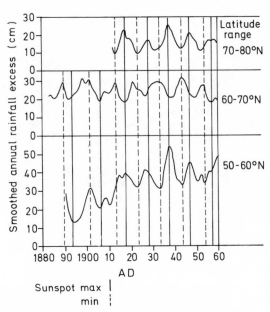

Figure 11.12 Correlation between rainfall and the maxima and minima of sunspot cycles. (After J.W. King, personal communication, data from Xanthakis, 1973.)

the variation of sunspot numbers from one cycle to the next. What is most striking in Fig. 11.12 is the regular variation of the annual rainfall at 60–70° N, and in particular the fact that its periodicity is just the same as the sunspot cycle, the respective variations being close to antiphase throughout the run of data. At the higher latitude, 70–80° N, over a shorter data series, the variations are in phase. In the lower latitude zone of 50–60° N, the author evidently had greater difficulty, for there the data were restricted to the American sector and even then a change of phase appears about 1910. The oscillation in the rainfall that is being shown here amounts to some 10 cm per annum.

Similar results have been obtained for the southern hemisphere, in that the 11-year component of Australian rainfall, amounting to 1 or 2 in per annum, varies in phase with the 11-year sunspot cycle at 43° S but in antiphase with it at 17° S. Other relations of this kind have been found, and some of the oscillations can be as large as 50% of the average annual rainfall: a large variation indeed where agriculture is concerned. That the phase of the 11-year rainfall variation depends on the latitude and in some cases also on the longitude, suggests that the main effect is to shift the rain belts rather than to alter their intensity. If that is so, one would expect to find locations – and perhaps a majority of locations – where the solar cycle has no apparent effect.

At some latitudes the significant variations seem to take place over 22 years rather than 11. The 22-year period is sometimes called the *double sunspot cycle*, and its physical significance may be in the fact that the polarity of the Sun's magnetic field reverses with each new sunspot cycle. Rainfall at certain places in Brazil and South Africa appears to be correlated with the double sunspot cycle in such a way that the rainfall passes through a maximum at one sunspot cycle and a minimum at the next. Again the modulation is a significant fraction of the average rainfall.

Similar results have appeared at Adelaide in Australia. The July temperatures in central England oscillated by nearly 1 °C in phase with the double sunspot cycle between 1750 and 1880. In the USA also, the 22-year cycle appears to exert an influence, with a tendency for droughts at every second sunspot minimum. Fig. 11.13 shows the drought periods in the High Plains of the Central USA, in the 500 or 600 km to the east of the Rocky Mountains. There are also 22-year variations of temperature at certain seasons and in certain areas. Recent analysis of the ratio of deuterium to hydrogen in tree rings, which is taken to indicate the temperature when the wood was formed, shows evidence of a 22-year cycle in the climate (in this case at 10 000 ft altitude in California). The result shown in Fig. 11.14 covers 1000 years of tree growth and suggests that the 22-year cycle revealed in the records of relatively modern times (Fig. 11.13) may actually have been at work for many centuries. (However, see the next section for further comments on the long-term behaviour of the Sun.)

Not only do the periods typical of the sunspot cycle show up in meteorological variables like rainfall and temperature, but they also appear in related quantities. The incidence of lightning in Great Britain correlates remarkably well with the

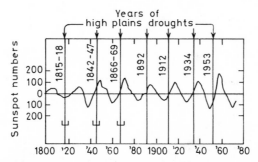

Figure 11.13 Sunspot numbers and droughts in the High Plains. Vertical lines show the central years of drought periods, the first and last years are given for the earlier ones which were less accurately recorded. Every drought except the first occurred near a sunspot minimum. (From W.O. Roberts, *NASA Report SP-366*, 1975.)

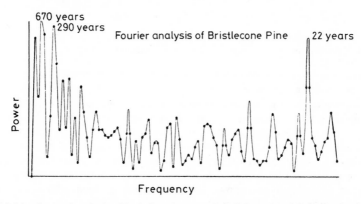

Figure 11.14 Spectral analysis of deviations from normal of the ratio of deuterium to hydrogen for bristlecone pines over 1000 years. (From L. Svalgaard, *Physics of Solar Planetary Environments, Proceedings of STP Symposium,* 1976, copyrighted by American Geophysical Union.)

261

11-year cycle (Fig. 11.15). Averaging has here suppressed the short- and medium-term variations. According to Fig. 11.15 the incidence of lightning strikes varies typically by a factor of 1.7 from the minimum to the maximum of the sunspot cycle. The correlation between lightning or thunderstorms and sunspots is often surprisingly good, correlation coefficients of 0.9 having been reported in some comparisons.

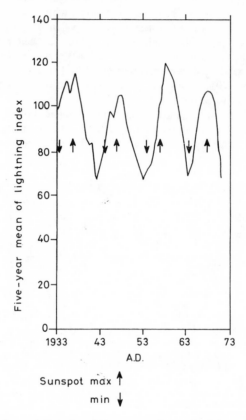

Figure 11.15   Lightning variations in relation to maxima and minima of the sunspot cycle. (After J.W. King, personal communication, data from Stringfellow, 1974.)

An even more serious consequence of weather variations is the state of agriculture, and with a decline in world food stocks over the last few years the variations of production due to the weather are large enough to give cause for concern. Grain production, for instance, varies about 10% with the weather. Various analyses that have been made suggest strongly that wheat production varies with the sunspot cycle. Some of the evidence is summarized in Table 11.1. The indications are that more wheat is produced in the northern hemisphere at sunspot maximum, and more in the southern hemisphere at sunspot minimum. Overall, world production does better at sunspot maximum. These variations of food production may, of course, carry over into world economics.

In 1972, drought in the Moscow area led to a poor wheat crop in the USSR. Large purchases were made from the USA, and these purchases, coinciding with

262

Table 11.1 Notes on Wheat Production in Relation to the Solar Cycle*

Years of sunspot maximum		Late 1957	1968
Years of sunspot minimum	1954	1964	1974
World-wide production		Greater for 1958 than for next 5 years	Greater for 1968 than average of next 4 years
Northern hemisphere production			
China		22% greater in 1956–58 than in 1960–65	
USSR		Greater in 1956–58 than in 1960–65: for 1958, 54% greater than for 1963	Progressively lower output from 1970 to 1972: very poor in 1975
Canada			27% greater in 1967–69 than in 1970–73
Southern hemisphere production			
Argentina	Maximum in 1954		Maxima in 1963–64: average of 1954, 1963, and 1964, 50% greater than 1965–73
Africa	Greater in 1954 than in any of next 8 years		
Australia	Average of preceding 8 years was 83% more than 1957 production		

*Data from J.W. King, 1974.

new demands from elsewhere, used up the surpluses. International grain prices, and subsequently meat prices, soared. Drought also struck India and the sub-Sahara region, where millions of people went hungry because local supplies dried up at the same time that world reserves were running out. The general increase in basic food prices that began around this time played a part in triggering the current (1975–77) economic recession. 1975 was another year when the USSR had a very poor wheat harvest, coinciding with sunspot minimum. The wheat crop of the USA was faced with disaster in 1976 due to drought in the early part of the year, but rain came later on, just in time to save the crop. By contrast 1977 has seen good wheat harvests, that in the USA being one of the best ever. (It is tempting to remark here that during 1977 solar activity has been recovering strongly from its recent prolonged minimum!) An earlier period of severe drought occurred in the 1930s, when dust bowl conditions in the American mid-West exacerbated the Great Depression.

The scientific attitude to facts like these should be one of caution, since what seem like associations can so easily prove to be no more than coincidences that do not stand the test of time. It is when a large number of coincidences, all of the same kind, come to light, that people begin to wonder if there might not be some underlying cause after all, and that is about the current situation in a topic like

263

sunspots and agriculture. To become science, however, the observations need to satisfy objective statistical tests, and plausible theories must emerge — both seemingly difficult in this area.

11.5.3 Long-term climatic change

One of the more obvious ways in which solar activity might affect the weather is through the entry of energetic particles, whose flux is greatest in the auroral zones. Some weight is lent to this idea by analyses showing that the lower-atmosphere effects that appear to be associated with the sunspot cycle also maximize at auroral latitudes. Presumably, the auroral-zone precipitation would have to be able to alter the atmospheric circulation if it was to exert any influence over weather patterns. If this did happen it is clear that the geographical source of the modifications would be controlled by the geomagnetic field, since that is what determines the location of the auroral zones. It is a strange fact that there are marked similarities between certain properties of the troposphere and the intensity of the geomagnetic field at ground level. Temperature and humidity tend to be low, and surface pressure high, where the magnetic intensity is greatest. The height of the 500 mbar level also shows a very good correlation (correlation coefficient -0.96) with magnetic intensity around a zone at $60°$ N. Now it is quite possible that these agreements are no more than coincidence, since the distribution of land masses is a more obvious cause for longitudinal variations of the mean troposphere. Nevertheless, they do lead to the intriguing speculation that long-term climatic changes might be associated with the slow secular variation of the geomagnetic field. Some evidence in favour of this hypothesis has been produced: for instance in the variation of temperature and magnetic intensity over nearly 5×10^5 years deduced from a deep-sea core (Fig. 11.16). The correlation coefficient between low temperature and high magnetic intensity is here 0.67. The probability that a correlation coefficient of this magnitude could arise by chance in that set of data is only 0.1%.

In recent history the occurrence of sunspots has been well documented, and the 11-year sunspot cycle is quite obvious in the data. However not all cycles are of

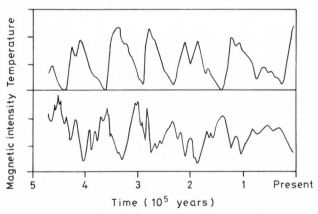

Figure 11.16   Temperature deduced from oxygen isotope data, and magnetic intensity, from a deep-sea core. (After J.W. King, personal communication, data from Wallis *et al.*, 1974.)

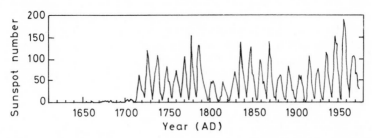

the same amplitude and modulations of longer period appear (Fig. 11.17). What must be kept in mind is that these solar observations cover a minute span in the life of the Sun and in the history of the Earth whose climate is of interest to us. That the Sun has behaved regularly in recent history is no proof that it has always been so well behaved, nor guarantee that it will continue so in the future. Attempts to extend the sunspot record back before the 18th century have brought the surprising result illustrated in Fig. 11.17, that between 1645 and 1715 very few sunspots occurred. The dearth of sunspot reports is supported by a lack of reports of aurorae, and by the weakness of the solar corona as seen during eclipses. The Maunder minimum, as it is known, corresponded with an exceptionally cold period in Europe.

The Maunder minimum also shows up in carbon-14 analyses of tree rings. The carbon-14 content indicates the cosmic ray intensity at the time the wood was formed, and since the cosmic ray flux at the Earth is modulated by solar activity it is thought that carbon-14 analysis indicates solar activity. Such analyses go back for some 7000 years and show several fluctuations suggesting times of high or low solar activity. Some of the less ancient ones find supporting evidence in documents of the time, though the records are not as good as those for the Maunder minimum: the Sun appears to have been inactive between AD 1400 and 1510 (the Sporer minimum), and between AD 1120 and 1280 (the Medieval maximum); reports of aurorae and of sunspots seen with the naked eye were more frequent than during the preceding or the following centuries. These two anomalies corresponded with exceptionally low and high temperatures on the Earth respectively, and there are correspondences between earlier periods as well. There is some evidence, therefore, that in the long term the activity of the Sun and the terrestrial climate vary together.

11.5.4 The solar wind, geomagnetic storms and trough development

The sunspot numbers indicate the general state of activity of the Sun in a statistical way, but of course the effects of solar activity at the Earth are seen as discrete events such as flares, proton events and auroral substorms. The main consequence of high solar activity is that these same kinds of events occur more often and with greater intensity. If solar activity appears to affect the weather in the long-term statistics, one might also expect to see meteorological responses to some of the discrete events.

265

Various studies have been made along these lines. These show that in the 24 h after a solar flare, the altitude of the 500 mbar pressure level increases at latitudes 45–65° and decreases at high latitude (above 70°). The results are symmetrical with respect to the geomagnetic rather than the geographic pole. An increase of global thunderstorm occurrence has been reported on the third and fourth days after an H_α flare.

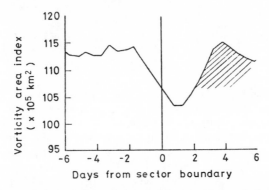

Figure 11.18   Characteristic variation of the vorticity area index during the passage of a sector boundary in the solar wind. (After L. Svalgaard, *Correlated Interplanetary and Magnetospheric Observations* (Ed. Page) Reidel, 1974.)

Some of the most striking results appear to relate tropospheric effects to the sector structure of the solar wind (Section 7.3.4). One such investigation uses a *vorticity area index* to quantify the circulation of the air at the 300 mbar level (altitude about 9 km) in the northern hemisphere poleward of 20° N. The vorticity of a rotating rigid disc is just twice the angular velocity, and low pressure systems in the atmosphere have increased vorticity. The area index was defined as the area in km^2 over which the vorticity exceeded some stated threshold value. Fig. 11.18 gives the result of a superimposed epoch analysis, and shows that the vorticity decreases 1 day after a solar-wind sector boundary crosses the Earth, and passes through a maximum 3½ days after the boundary crossing. The analysis includes 54 sector boundaries, and a later study using 131 crossings has given the same result. (Similar effects are seen throughout the troposphere down to the 850 mbar level, but they do not appear in the stratosphere above 100 mbar.) Although the physical reasons for this effect are still obscure, the result itself has been subjected to detailed criticism and scrutiny by various workers and its statistical significance is beyond question. An important point regarding this result is that the solar wind structure unquestionably had its origin in the Sun, and there is no conceivable way that meteorological phenomena on the Earth could affect it.

Since cyclones are low-pressure systems it might be expected that atmospheric pressure would also vary with the solar-wind sectors, and such is the case. On the average a constant-pressure level is lower over the poles than over middle latitudes. Three or 4 days after a boundary crossing, the pressure tends to increase over the poles but to decrease at middle latitude. The consequence is increased storminess at middle latitude and stronger anticyclones at high latitude.

Fig. 11.19 illustrates another effect associated with sector crossings, this time in the vertical electric field and the air-earth current through the atmosphere

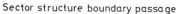
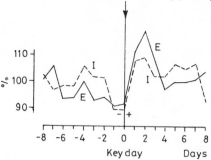

Figure 11.19 Superimposed epoch analysis of the change of vertical electric field (*E*) and air-earth current (*I*) around the passage of a sector boundary. (After Reiter, *J. atmos. terr. Phys.* **39**, 95, 1977, by permission of Pergamon Press.)

measured on a mountain 3 km above sea level. According to this analysis, which used observations over a period of 4 years, the electric field increases more than 20% in the 2 days after a solar-wind sector boundary has swept past the Earth. In the circumstances of the measurements it can be said that these changes in the electric field will indicate corresponding changes in the potential of the ionosphere with respect to the ground as it appears across the low-conductivity troposphere. It is generally thought that global thunderstorm activity generates most of the potential difference, but it would be premature to speculate further what mechanism allows the solar wind to exert an influence on the global electric circuit.

Several analyses have reported 27-day periodicities in winds or rainfall, which might be related to the solar rotation period. A number of studies have found associations between tropospheric phenomena and geomagnetic activity, but significant though they might be in a statistical sense, such results are open to the objection that geomagnetic activity could conceivably be influenced by meteorological disturbances propagating into the upper atmosphere. Special caution is needed when considering such results.

Most of the results discussed in the present section appear only or most strongly in the winter. This fact must be important in the consideration of mechanisms. Either, it seems, the solar effects couple more effectively into the winter hemisphere, or the winter atmosphere is for some reason more readily affected by them.

11.5.5 Mechanisms

While many scientists feel convinced by the statistical evidence in favour of a solar-activity influence on weather and climate, most will admit that finding the mechanism or mechanisms seems a daunting task. That the mechanism is not known does not, of course, disprove the existence of the phenomenon. Often quoted, and worth repeating in this context, is the story of Lord Kelvin and the magnetic storm. In his presidential address to the Royal Society of London in 1892, Lord Kelvin considered the hypothesis that terrestrial magnetic storms were due to magnetic waves emanating from the Sun and, on the basis of some

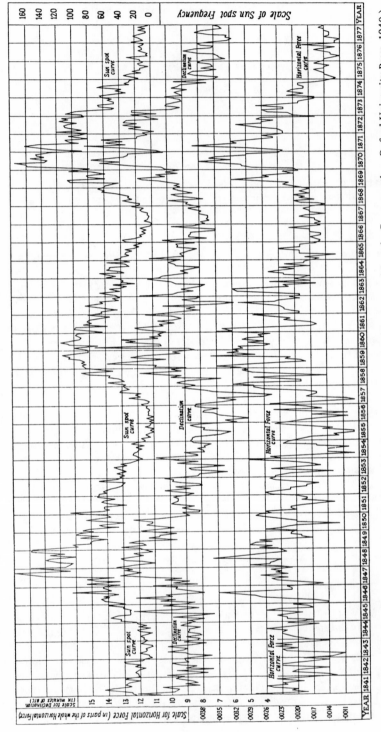

Figure 11.20   Three cycles of sunspot numbers and magnetic variations. (After Chapman and Bartels, *Geomagnetism*, Oxford University Press, 1940.)

calculations of the energy required, concluded 'against the supposition that terrestrial magnetic storms are due to magnetic action of the Sun, or to any kind of dynamical action taking place within the Sun, or in connection with hurricanes in his atmosphere, or anywhere near the Sun outside'. He continued 'It seems as if we may also be forced to conclude that the supposed connection between magnetic storms and sunspots is unreal, and that the seeming agreement between the periods has been a mere coincidence.' What led the great physicist to this monumental error was his assumption that he knew all the relevant facts. But he did not, because he did not know, and in 1892 could not have known, about the solar wind which introduces mechanisms quite different from the one Kelvin had assumed. One should beware of too readily dismissing an observational result just because the mechanism has not come to light.

There is another point from the history of geomagnetic storms that may be instructive in regard to weather effects. When Sabine announced the correlation between sunspot numbers and geomagnetic activity in 1852 it was on the basis of only two cycles of data. (The idea suffered a set-back with Kelvin's attack of 1892 — but who doubts it now!) Fig. 11.20 shows three cycles, in which the correlation looks no better than some of those now being offered between solar activity and weather phenomena. The moral is that correlations tend to be noisy in geophysics, and should not be scoffed at on those grounds alone; the underlying influence might still be important. On the other hand, the scientific community expects to see stronger evidence emerging if the correlation is to gain general acceptance, and to see reasonable theories appearing in due course.

In such a speculative field as this it would not be appropriate to go into the proposed mechanisms in great detail, but some points should be made to illustrate the inherent difficulties. The first concerns the energy budget. The energy that drives the troposphere comes mainly from sunlight absorbed at the Earth's surface. The amount absorbed is

$$U_{EM} = \pi R_E^2 F(1 - A) \tag{11.3}$$

where R_E is the earth radius, F the solar constant, and A the albedo (~ 0.5). Putting in numbers, $U_{EM} = 8.9 \times 10^{16}$, $W = 8.9 \times 10^4$ TW (1 TW = 10^{12} W). Most of this ultimately returns to space, but on the way it is involved in the weather system, giving temperature gradients and evaporating water to drive convective systems and produce instabilities. The total energy of the solar wind striking the magnetosphere is

$$U_s = \pi R_M^2 (\rho v^2/2 + B_s^2/2\mu_0) v \tag{11.4}$$

where R_M is the radius of the magnetosphere as seen from the Sun, ρ the density of the solar wind and B_s is the interplanetary magnetic field. The first term of Equation 11.4 is the kinetic energy per unit volume of the solar wind, and the second term is the energy density of the IMF (Section 7.3.4). Less than 1% of U_s is thought to enter the magnetosphere. Assuming 1%, and taking $R_M = 12 R_E$, $\rho = 8 \times 10^{-2}$ kg m^{-3}, $v = 400$ km s^{-1} and $B_s = 10 \gamma$ (= 10 nanotesla) gives for the corpuscular and magnetic energy entering the magnetosphere, $U_c = 5 \times 10^{-2}$ TW. Thus, $U_c/U_{EM} = 6 \times 10^{-7}$. The energy from the solar wind thus looks minute compared with that reaching the troposphere by electromagnetic radiation.

To increase the ratio one might look for extreme conditions. In the winter hemisphere U_{EM} might be as low as 6×10^3 TW due to reduced illumination and increased cloud and snow. In an intense magnetic storm, 10^{18} J might be injected from the magnetotail to the closed field region of the magnetosphere (causing aurorae, electrojets and ring currents) in 10^4 s ($\sim$ 3 h). U_c might then rise as high as 10^2 TW. The greatest ratio U_c/U_{EM} that appears likely to occur is therefore 1.7×10^{-2}. This still looks too small to have much effect, and a difficulty to be added is that there would also have to be a mechanism for feeding the magnetospheric energy into the troposphere; even if such a mechanism were known we would be lucky if it turned out to have a high efficiency.

Because of the inadequacy of a straight approach via the energetics, attention has turned to so-called *trigger mechanisms*, whereby it is supposed that a small amount of energy from outside might switch the troposphere from one quasi-stable mode to another. This is not necessarily a far-fetched idea, since weather systems are largely unstable and, once started, tend to develop from their own internal energy. A small amount of energy, correctly applied, might therefore have a large effect. (Weather-modification experiments embody the same assumption.)

We may gain further insight into the mechanism problem by considering an attempt to explain one observed phenomenon — the change of vorticity in the North Atlantic region at the passage of a solar-wind sector — rather than looking at the energy requirements on the global scale. The observations indicate that the effect is equivalent to increasing the angular velocity of a disc of air 500 km in radius from 4×10^{-5} to 6×10^{-5} rad s^{-1}. The moment of inertia of such a disc is $\pi r^4 \rho/2$ for radius r and air density ρ. Taking the air above the 300 mbar level, $\rho = 3 \times 10^3$ kg m^2. Thus $I = 2.9 \times 10^{26}$ kg m^2. The energy of the rotating disc is $E = I\Omega^2/2 = 5.3 \times 10^{17}$ J if $\Omega = 6 \times 10^{-5}$ rad s^{-1}, which is similar to the energy of a magnetic storm. The power required to increase Ω from 4×10^{-5} to 6×10^{-5} rad s^{-1} in 1 day is

$$dE/dt = I\Omega \cdot d\Omega/dt = 2.7 \text{ TW}$$

The energy of rotation increases by 2.3×10^{17} J. Since the largest value of U_c was 10^2 TW over 3 h (or 12.5 TW over a day), it appears that sufficient energy is available.

The difficulty arises when one looks for a mechanism for coupling magnetospheric energy into rotational energy of the troposphere, since in most respects the two regions are remarkably well shielded from each other. Hardly any energetic particles or X-rays penetrate as deep as the tropopause (at about 16 km altitude). Although there can be marked temperature changes above 120 km at auroral latitudes, the energy flux of a very intense auroral beam (perhaps 1 W m^{-2}) is still only 10^{-3} of that due to sunlight. Again, the atmosphere has two temperature minima, the tropopause and the mesopause, which effectively insulate the troposphere from temperature changes in the thermosphere. One possible coupling mechanism involves ionospheric currents, which can reach $J = 10^6$ A during disturbances. In the presence of the geomagnetic field (of induction B), the force on the neutral air communicated from the ionization is $J \times B = 60\ N$ m^{-1} if $B = 0.6$ G ($= 6 \times 10^{-5}$ T). Over the diameter of a 500 km disc of air, this force would give angular acceleration $d\Omega/dt = 2JBR^2/I = 10^{-13}$ rad s^{-2}. However, the requirement for a change of 2×10^{-5} rad s^{-1} in one day would need the acceleration to be 2×10^{-10} rad s^{-2}. Therefore the available torque is not nearly sufficient, even if it were possible to communicate horizontal neutral motions from the electrojet region (105 km) to the troposphere.

The foregoing discussion will illustrate the difficulties of dynamical approaches, but many other ideas are abroad as well, all coming into the 'trigger' category. Weather and climate are very sensitive to the amount of solar radiation reaching the Earth's surface. Various estimates suggest that over a solar cycle the solar constant might change by as much as 2% or as little as < 0.1%. Strange as it may seem, this fundamental quantity has not yet been measured with precision, but studies of the history of solar activity suggest that in ancient times solar patterns might have been quite different from those we see today, and there is no reason to assume that the solar constant has always had its present value. Cloudiness affects the amount of solar radiation reaching the surface, so mechanisms that would alter cloudiness have to be considered as well. Cloud cover produces a greenhouse effect and is therefore an excellent candidate as a trigger mechanism. Another group of theories concentrates on changes of chemical composition, particularly ozone. One such mechanism was indicated in Section 9.3.3, and operates during solar proton events. In addition to causing climatic changes, it has been suggested that ancient solar proton events may have influenced the evolution of life on Earth by hastening the extinction of certain species. The proton flux increases the nitric oxide concentration, which in turn destroys ozone; with the ozone level reduced the atmosphere transmits more ultraviolet, which is harmful to life.

Whatever mechanisms eventually become acceptable, it seems clear that they will be subtle, probably including several processes acting as links in a chain of cause and effect. (For an excellent example of a long cause-and-effect chain consider how the Sun produces a luminous aurora (Chapters 7 and 8).) To take an example from Sun–weather relations, we will consider the droughts in the High Plains area of the Central United States (Fig. 11.13). It is thought that the immediate cause of the loss of precipitation in drought years may be that less of the warm, moist air from the Gulf of Mexico reaches the lee of the Rocky Mountains; the deflection of the moist air might be due to a strengthening of the jet-stream westerlies over Colorado; as a cause of this circulation change, the formation of high cirrus clouds following polar aurora has been suggested (an occurrence for which there is some slender evidence); the cloud would alter the circulation through a modification of the radiation budget; and so on! Obviously, it could be many years before all the links in such a sequence of events were proved beyond reasonable doubt.

It should be clear, now, why the topic of Sun–weather relationships, fascinating and potentially important though it may be, is not yet on the same level of scientific certainty as most of the material in the present text. Nevertheless it is at least possible now to formulate questions that require an answer*.

How many of the seemingly random meteorological variations are really responses to a variable solar forcing function?

To what extent do short-term Sun–weather effects add up to produce climatic effects?

How can mathematical models of these effects be formulated?

What are the mechanisms by which Sun–weather effects are caused?

What are the properties of the double sunspot cycle that are associated with Sun–weather effects?

Are any of the effects subject to accidental modification or deliberate control by man?

* Set due to J. W. King.

One side benefit of the controversies has been some rethinking of the nature of scientific evidence and particularly the value of correlations in geophysics. How clear-cut should evidence be before a phenomenon can be believed, in a field where controlled experiments are not possible? While opinions differ, most people at least agree that these are important questions for the conduct of geophysical science, and it is conducive to the good health of the science that they should be given an airing from time to time.

11.6 Further reading

Svalgaard, L. (1974) Solar activity and weather, in *Correlated Interplanetary and Magnetospheric Observations* (Ed. Page) (Reidel).

Murata, H. (1974) Wave motions in the atmosphere and related ionospheric phenomena, *Space Sci. Rev.* **16**, 461.

Matuura, N. (1972) Theoretical models of ionospheric storms, *Space Sci. Rev.* **13**, 124.

Rishbeth, H. (1975) F-region storms and thermospheric circulation, *J. atmos. terr. Phys.* **37**, 1055.

Mitra, A. P. (1975) D-region in disturbed conditions, including flares and energetic particles, *J. atmos. terr. Phys.* **37**, 895.

Roberts, W. O. and Olsen, R. H. (1973) New evidence for effects of variable solar corpuscular emission on the weather, *Rev. Geophys. Space Phys.* **11**, 731.

Willis, D. M. (1976) The energetics of Sun—weather relationships: magnetospheric processes, *J. atmos. terr. Phys.* **38**, 685.

12
Some Practical Aspects of
Upper-atmosphere Studies

12.1 Science and engineering of the upper atmosphere

Most people would agree that in a philosophical sense science and engineering are different kinds of human activity. Science is a method of gaining knowledge, and as such its achievements are primarily intellectual. While recognizing that knowledge cannot be absolutely certain, the objective of scientific investigation is to generate knowledge with the maximum degree of objective certainty. When a scientist does practical work in the course of his investigation it is as a step in the process of getting to understand something rather than as an object in itself; the primary achievements of scientific activity occur not in a laboratory but in a person's mind. Engineering, nowadays often called technology, is the other way round. Its main purpose is developing devices to perform certain specified tasks. The device might be a system rather than a tangible machine, and thus it might exist on paper only, never to emerge as a solid object. Nevertheless the engineer's output is about doing rather than just understanding.

Although engineering and science have different objectives in the general sense, it is natural that there should be strong links between them, with each drawing from and feeding into the other. In studies of the terrestrial upper atmosphere and of solar–terrestrial relations the links are traditionally strong – indeed, a little thought will show that they are essential to the progress of the subject. First, as will by now be obvious, nearly all aspects of upper-atmosphere science are engineering based. Not only do experiments and observations depend on technology, much of it of an advanced nature, but also the present day theorist relies heavily on computer techniques. Second, the information and the understanding of phenomena that come from scientific investigations feed across to the engineers who are concerned with various practical endeavours, these being mainly aerospace activities and communications. These are the aspects to be considered in the present chapter. Some of the applications are long term, in the sense that the nature and the properties of the upper atmosphere determine once and for all the systems that can be supported. We shall be more concerned here with the effects of variations, particularly those originating with solar activity. Effects on radio communications and navigation will be considered, then the status of services for monitoring solar–geophysical activity, to minimize its disruptive effects, will be reviewed. Last but not least, we outline the international data services on which so much upper-atmosphere work depends.

12.2 Predictions for radio propagation

The science of the upper atmosphere has seen two major periods of growth, the first during the 1930s when radio communications were developing, and the second during the late 1950s and early 1960s when space activities were first becoming practical. Overall, the field of upper-atmosphere studies has grown rapidly; if judged by the output of written reports and papers, upper-atmosphere and space physics has shown one of the most rapid growth rates in all 20th century science (with output doubling every 5–6 years). A major stimulus to this growth has been the information requirement of new technologies, and it is probably true to say that the initial boost to ionospheric research arose because reliable predictions of ionospheric conditions were needed for high-frequency (HF) radio communications.

Today, because of the development of satellite communications (using signals of such high frequency that the ionosphere has little effect on them − though see Section 12.3) and improvements to submarine cables, the proportion of traffic that goes by HF radio is smaller than hitherto. The inference that ionospheric radio propagation is therefore less important would, however, be false. The total, as against proportionate, use of HF radio circuits is actually greater now than ever before, and a substantial research effort is still being put into the improvement of ionospheric predictions.

There are many aspects of ionospheric behaviour which affect radio propagation, but the first and most basic task of a prediction scheme is to estimate the *maximum usable frequency* or MUF. For every circuit there is a maximum radio frequency above which the signal will penetrate the ionosphere and be lost to space. For a vertically incident signal the MUF is the quantity $f_E F2$ (see Equation 3.8), which

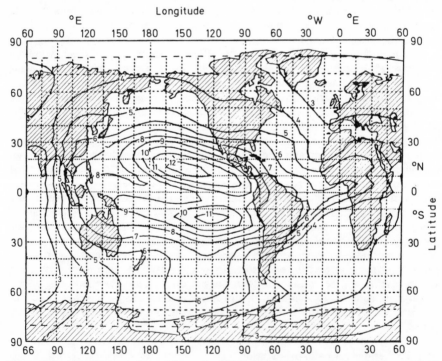

Figure 12.1   World map of monthly median MUF (Zero)F2 − identical to $f_E F2$ − at 00 UT in April when the sunspot number is 10. (From W. Roberts and R.W. Rosich, *Report OT/TRER 13*, Institute for Telecommunication Sciences, Boulder, Colorado, 1971.)

is simply related to the maximum electron density of the ionosphere above the transmitter. When transmitter and receiver are separated, the MUF exceeds $f_E F2$ because the signals are incident obliquely on the ionosphere. As a rule the received signals are strongest at the highest frequencies because the absorption increases with decreasing frequency (Equation 3.12); thus, propagation conditions are usually best close to, but just below, the MUF.

In the absence of ionospheric disturbances the MUF depends in a known way on station separation, latitude, longitude, the time of day, and the sunspot number. For many years MUFs were predicted by hand for each individual circuit, using the observed behaviour of the ionosphere and the trend of solar activity. The experience built up over the years has now been condensed into a standard set of world maps showing the MUF for F2 reflection at zero distance (= $f_E F2$), for F2-region propagation over 4000 km, and for E-region propagation over 2000 km. The contours show monthly median values, and there is a set of maps for each 2 h of the UT day, for every month of the year, and for three sunspot numbers. Nomograms are used for interpolation. As an example, Fig. 12.1 shows the map of monthly median MUF(Zero)F2 for 00 UT, April, and sunspot number 10. Computer programs have also been developed to give the same information and also signal strength, reliability, and other factors needed by the communications engineer. Although it is nearly 20 years since computer techniques began to assist in the prediction of radio propagation, it is worth noting that the present state of consolidation and refinement has been reached only during the last few years (since about 1970).

The MUF data are essentially long term. They are used for planning new radio circuits and for deciding in advance the best operating frequencies for existing circuits, all of which takes place within the restrictive framework of international and national frequency allocations*. An overseas broadcasting organization, for example, will have been assigned frequencies in several of the HF bands designated for broadcasting. Since the programmes and the frequencies will be published in advance and cannot be altered at short notice, an accurate prediction of which frequency to use at each time of day is clearly important to the efficiency of the operation.

In principle, then, the procedure might seem quite straightforward. The greater problems arise from the short-term variability of the ionosphere: the several anomalies, storm phenomena and irregularities discussed in previous chapters. Table 12.1 summarizes some of the effects. These disturbances are of solar origin, and to minimize disruption of communication circuits the practice is to make short-term propagation forecasts based on the recent experience of radio conditions and the trend of solar–geophysical activity. (The prediction of solar–geophysical activity is covered in Section 12.4.) Such forecasts differ in detail between countries, but the general style is illustrated by the example in Appendix 12.1. The short-term forecast contains the expected propagation conditions for the next 7 days, the latest prediction of sunspot numbers for the next year for use in preparing MUF values, and revisions to MUF predictions for a 2-week period in the near future.

*The electromagnetic spectrum is, and has been for many years, an oversubscribed natural resource. The uses of the various bands are rigidly specified by international agreements which are revised every few years to meet changing requirements. Within a given country the allocations made to individual users are monitored by a government agency and infringements are subject to penalty at law. An interesting account of events before and during the development of international frequency regulation is given in *The Birth of Broadcasting* by Asa Briggs, Vol. 1, Sections II.4 and V.4.

Table 12.1   Disturbances to ionospheric radio propagation

Disturbance	Ionospheric effect	Propagation effect	Typical duration
SID	Increased D-region electron density by day	Increased absorption; loss of signal	¾ h
Storm	Increased D-region electron density at high latitude	Loss of signal on polar routes	Several hours at a time, over several days
	Decreased F-region electron density, world-wide*	Loss of signals near MUF; narrowing of useful radio spectrum	1 day
PCA	High D-region electron density over polar caps	Loss of polar communications	Several days

*The effect increases with latitude and may exceed 30% at latitudes greater than 45° (see Fig. 11.1).

The forecast also includes a *quality figure*, defined from 'useless' to 'excellent' on a scale of 0–9 (5 means 'fair' and 7 is 'good'). The likelihood that ionospheric storms (Chapter 11), solar flare effects or PCA (Chapter 10) will occur are also predicted.

Besides the regular forecasts that are sent out by mail to the users, a comprehensive service will also provide rapid warnings to specific users if particular kinds of event, e.g. solar flare, occur. The rapid warning will be sent by telephone or telegram, or may be incorporated into a service broadcast such as a standard frequency and time transmission. Such warnings will often be specific to particular circuits, such as the important north-Atlantic links between Europe and North America.

Irregularities such as sporadic-E, scintillations, spread-F and travelling disturbances also degrade the performance of HF communication circuits by focusing or de-focusing the signal, by allowing propagation over more than one path so that two or more signals that may mutually interfere are received simultaneously, and by causing fading (such as rapid 'flutter fading' at 10–100 Hz on paths crossing the equator). A good example is the effect of travelling disturbances (Section 6.5) on high-frequency direction finders, one of whose main applications is to search-and-rescue operations following ship or aircraft disasters at sea. A modern direction finder may be able to measure the direction of arrival of a received signal with an angular error of only 0.1° in theory. In practice, however, the errors tend to be much larger, typically several degrees, due to irregularities in the ionospheric reflecting layer. It appears difficult to predict these errors in detail, but something can be done to correct them as they occur by simultaneous monitoring of gravity waves, particularly by the HF Doppler technique (Section 3.7.1).

12.3 Global navigation systems and international timekeeping

In general the position of a point on or above the Earth's surface can be determined by measuring its relative distance from four satellites at known positions in space. Suppose that radio pulses are transmitted simultaneously from each satellite and received back at the point whose position is to be fixed. Comparing the arrival times of two of the signals defines a surface related to the positions of the sources. (In the special case when the signals arrive together this surface is just the plane bisecting and normal to the line joining the sources.) A comparison of three signals locates the receiver on a line, and if the altitude of the receiver above the Earth is known there is now enough information to fix the position in latitude and longitude. If the height is not known a fourth signal is needed. For engineering reasons pulsed signals may be inconvenient, and in practical systems it is more usual to achieve the same result by transmitting a continuous wave carrying a sinusoidal modulation. The receiver then measures phase rather than time difference, but for our purposes the two are equivalent and it is convenient to discuss all systems in terms of time measurement.

The principle of position fixing by means of radio ranging was first applied during the Second World War, and for many years now ship and aircraft navigation has been aided by radio navigational systems in which three transmitters are located on the Earth's surface. Over short distances such systems are very accurate, but to reach beyond a few hundred kilometres ionospheric radio propagation is involved and the variability of the ionosphere imposes a limit to the accuracy. The most accurate ground-based systems for long-range navigation operate in the VLF (3–30 kHz) band. Being reflected from the D region, VLF signals show greater stability than those of higher frequency that are returned from the F layer, but nevertheless there is a large day–night change (Section 5.2.3) which alters the time delay over a given propagation path, and there can be perturbations during solar flares (Section 10.2) and aurorae (Section 8.2.2).

The *Omega* navigational aid, now under construction and in partial operation, will use eight transmitters in the VLF band 10–14 kHz to provide a world-wide position fixing system for ships and aircraft. The accuracy will be limited by ionospheric variations, some of which, such as the day–night change, are sufficiently regular that they can be compensated for, at least in part. Others are less predictable, and disturbance phenomena such as solar flare effects (Section 10.2) and polar cap events (Section 10.3) could cause positional errors of several kilometres. Solar flares affect mainly low and middle latitude propagation paths across the daylit hemisphere, while the polar cap events affect polar paths. The D-region storm after-effect (Section 11.3) will also affect middle latitudes, particularly by night. These phenomena are all related to known disturbances, essentially of solar origin, and some prediction may therefore be possible. In addition there will be ionospheric variations due to the irregularity of the reflection region — notably due to travelling disturbances (Section 6.5) — and these will cause random errors amounting perhaps to several hundred metres.

The errors introduced by variations of the ionosphere are greatly reduced in satellite navigation systems, which use radio frequencies well above the penetration frequency of the ionosphere. However the errors are not completely avoided, and whether or not they matter depends on the application and the accuracy required. At the speed of light a radio signal travels 1 km in 3.3 μs and 10 m in 33 ns (1 ns = 10^{-9} s). These very small time differences can be measured without great difficulty by modern electronic techniques. The problem is, rather, that at a frequency such as 150 MHz a typical electron content of 10^{17} electrons m^{-2} introduces a time delay of 0.5 μs, corresponding to 150 m in range. It should be remembered that the electron content often exceeds 10^{17} m^{-2}, sometimes by as much as a factor of ten. Such range errors are unacceptably large.

The first operational satellite navigation system (NNSS) was designed to work at 150 MHz, but because of ionospheric effects a second transmission, coherent with the first but at 400 MHz, was added. Using both signals a correction can be determined because the ionospheric time delay varies inversely with the square of the frequency (Section 3.7.3). Although correction is feasible, dual-frequency systems are undesirable because of their extra cost and more difficult maintenance, factors that weigh heavily with the user, since the cost of equipping a fleet of ships or aircraft with a new navigational system is far from trivial. Thus, future systems, which are being tested in the late 1970s for operational use in the mid-1980s, are expected to transmit near 1600 MHz, at which frequency an electron content of 4×10^{17} m^{-2} introduces a time error of 15 ns and a range error of only 5 m. For ship and aircraft navigation, such a system, illustrated in Fig. 12.2, should be sufficiently accurate. While the traditional requirements of navigation should be well met, new applications are arising in geodesy and surveying which call for errors as small as 1 cm for purposes such as monitoring faults in earthquake zones and studying continental drift. Such applications would call for ionospheric corrections, with the electron content known to better than 1%.

Thus the more demanding tasks required of satellite navigation systems will still need ionospheric corrections, and these are more difficult to obtain than might be thought at first sight. There are several reasons for this. Despite the long series of electron content measurements that have been made by observing geostationary

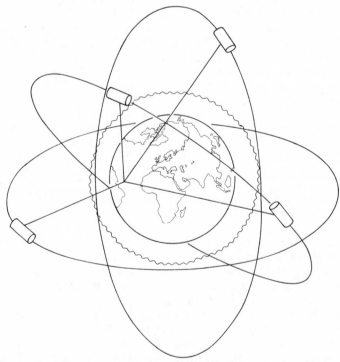

Figure 12.2   A possible navigation system using four satellites, orbiting at 19 000 km with periods of several days. The transmissions would be near 1600 MHz to minimize propagation effects. (From J.A. Klobuchar, personal communication.)

radio beacons at various terrestrial sites (Section 3.7.3), the global distribution of electron content is not sufficiently well known, particularly since the effects of perturbations like ionospheric storms should be included. Second, most of these measurements have been based on the ionospheric Faraday effect, which because of the weighting by the geomagnetic field provides only an approximate value that will often be in error by more than 20%. Finally, the electron content that is required for a correction is the slant content between satellite and ground station, not just the vertical content through the ionosphere. A measurement therefore has to be modified by means of an appropriate ionospheric model (including both the height and spatial dependence of electron density) before it can be applied to a signal coming from some other direction. Considerable effort is being put into ionospheric modelling (the task of finding a mathematical representation of the global ionosphere, including its temporal variations) at the the present time, and the current (1976) state of the art is that it would be relatively easy to correct system errors to 50% but very difficult to reduce them to 5%. The level that is likely to be achieved in practice lies somewhere between these limits.

International timekeeping presents problems that are closely related to those of global navigation. To compare the frequency standards and the clocks based on them, which form the basis of timekeeping in various countries, it is necessary to interchange time signals. One method is to carry crystal clocks physically from one country to another, so that they may be compared side by side in the laboratory. Time signals are also exchanged by radio in the VLF band, utilizing the most stable mode of ionospheric propagation available. However as time standards become more accurate — and the primary standards are now achieving errors as low as 15 ns day^{-1}, equivalent to 1 s in 180 000 years! — so more accurate intercomparisons are needed. Satellite transmissions will provide high accuracy until limited, eventually, by ionospheric uncertainties.

Some of the applications of precise time comparison are important to scientific experiments, an interesting example being wide-baseline interferometry in radio astronomy. This technique requires exact relative timing between signals received from the astronomical object (such as a quasar) by radio telescopes separated by hundreds or even thousands of kilometres. Thus, satellite timekeeping, a technique receiving an input from ionospheric science, becomes in its turn one of the tools of astronomical science.

12.4 Forecasting solar activity and geophysical disturbances

12.4.1 The customers

There are several groups of people who need to know when a geophysical disturbance occurs, and if possible would like to be warned in advance of the occurrence. One such group is the scientific community itself. There are many experiments that have to be activated at the start of an event of some particular type. A rocket launch is a clear example. In such a case local monitoring by ground-based equipment will play a large part in the launch decision, but it is also helpful to have predictions of when an event is likely. In other cases, experimenters observing continuously may wish to switch to a different mode of operation, e.g. one with greater resolution, during the event. Finally, even if the observations are not modified it is often useful to have prompt information about significant events so that those periods can be selected for detailed analysis.

Thus there is a range of geophysical experiments and operations whose good management is aided by warnings and forecasts of solar–geophysical events. However, this is only about one half of the market, for there are also some practical activities that can be affected in a serious manner by disturbances. These are summarized in Table 12.2.

Table 12.2   Practical Consequences of Disturbances

Event	Effect	Users
Geomagnetic storm	Induced earth currents Field perturbations Enhanced particle fluxes in space	Electric power companies Magnetic surveyors Satellite operators
Ionospheric storm	Ionospheric changes	Radio communicators Ship and airlines
Solar flare (X-rays)	Sudden ionospheric disturbance	Radio communicators Ship and airlines
Polar cap event (protons)	Damaging radiation	Space agencies Airlines

Geomagnetic storms accompany aurorae and an increased level of substorm activity in general. At such times strong currents flow in the high-latitude ionosphere (Section 8.3.3). These induce currents in the ground beneath and in any long conductor such as a power line. In certain regions, such as Canada, the northern United States and Scandinavia, the induced currents are sufficient at times to trip the overload protectors and thereby disrupt the distribution system. Induced currents also cause problems for long oil pipelines, in which they lead to increased corrosion. The disturbance of the geomagnetic field near the ground is a direct interference with magnetic prospecting. The technique of magnetic prospecting for minerals from aircraft requires geomagnetically quiet conditions, and therefore information about the imminence of disturbances is needed if wasted flights are to be avoided.

Field perturbations can affect the operation of satellites for communication or other purposes, because some satellites are orientated in space by reference to the geomagnetic field. The magnetic field orientation is quite well known during quiet periods, but large errors can occur if orientation manoeuvres are made during a stormy period. In a large storm the magnetopause can be closer to the Earth than a geostationary satellite (at $6.6 R_E$), and then such a satellite will find itself in the solar wind whose magnetic field is entirely different from the Earth's in both direction and magnitude. Spacecraft also accumulate electric charges in space, and during a disturbance, when the trapped particle population is enhanced, there can be electric discharges between different parts of the spacecraft. Though these might not cause any physical damage, the discharges produce intense local radio interference which can appear as spurious messages in the spacecraft command receiver and thereby disrupt operations. The operators at the control centre need to know the cause of such failures when they occur so that rapid and appropriate remedial action can be taken. Warnings of the present and pending levels of geomagnetic activity are a valuable input to such operations.

Geomagnetic storms are usually accompanied by ionospheric storms. There are many aspects to the ionospheric storm (see Chapter 11), but the most significant for communications are:

(i) changes of electron density and therefore of critical frequency;
(ii) increased radio absorption; and
(iii) the occurrence of scintillations.

Communications interruptions associated with geomagnetic activity are most serious at high latitudes and, in general, for propagation paths crossing the auroral zones. Long-range aircraft communications make considerable use of ionospheric propagation, and there are many routes crossing the polar regions. Sudden ionospheric disturbances due to solar X-ray flares (Section 10.2) are, likewise, at least an embarrassment and at most a hazard in aircraft and ship communications. These affect the whole of the Earth's sunlit hemisphere.

The only solar disturbance to present a direct health hazard is the proton event (Section 10.3). The most dramatic consequence is for men in space, for whom the proton radiation during a major event might even be fatal. While no serious effects are known to have occurred so far the risk is well appreciated, as evidenced by the intensive solar–geophysical monitoring that was operated during the Apollo flights to the Moon. To take an example, had the energetic proton events which occurred in August 1972 arrived instead during an Apollo flight the astronauts would certainly have been affected. If caught on the lunar surface, protected only by space suits, the radiation doses could have proved fatal. In practice the flight plan would have been altered so that the astronauts were recalled to the command module orbiting the Moon, but even with the greater protection afforded there some radiation sickness would have resulted. Manned lunar landings may be behind us now, but proton radiation will again have to be reckoned with during the flights of the space shuttle in the 1980s. The shuttle will fly well within the magnetosphere, and therefore solar proton fluxes can only be encountered over the solar caps (Section (10.3.2). However, the longer duration of flights will tend to increase the total dose again.

Somewhat nearer home for most of us is the effect of solar protons on passengers in aircraft at high altitudes over the polar regions. The proton fluxes are small at 30 000 ft, a typical altitude for subsonic aircraft, but supersonic aircraft fly at twice that height and there the radiation could be significant during a large event. During August 1972 the dose at 65 000 ft reached 400 mrem h^{-1}, which might be acceptable for astronauts or others in high-risk occupations, but is outside the conservative safety limits imposed for fare-paying civilians. The occurrence of a large proton event may therefore require a modified flight plan to be put into effect. Even though the hazard can readily be avoided a reliable monitoring and prediction service is obviously essential.

12.4.2 Space environment services

The forecast and warning requirements discussed above are met by a world-wide organization of Warning Centers, part of the International Ursigram and World Days Service (IUWDS). The IUWDS is a permanent service of URSI, the International Scientific Radio Union, which, via the International Council of Scientific

Unions (ICSU), comes within the ambit of the United Nations Organization. The IUWDS is comparatively new, having been formed in 1962, but it was actually built on earlier arrangements for the rapid exchange of solar and geophysical data that had been built up since about 1930. In the present arrangement there is a World Warning Agency which is located within the US National Oceanic and Atmospheric Administration in Boulder, Colorado, and there are Regional Warning Centers in Moscow, Paris, Darmstadt, Tokyo and Sydney, supplemented by a number of Associate Regional Warning Centers in some other countries. Each regional centre serves the customers in its own region with forecasts and alerts as required, and thus the balance of activities varies from place to place. However the basic data are shared daily between the centres so that the best global information is available to each. The procedure is that each regional centre first advises the WWA in Boulder of its own forecast, and the personnel at WWA make up a consensus forecast which takes account of all the information, including the opinions received from the regional centres. The consensus is then communicated to each regional centre, for use as appropriate in the advice sent to the customers.

In addition to its role as WWA the Space Environment Services Center in Boulder serves as a regional centre for the Western Hemisphere. Forecasts are issued every 8 h, to cover the following 72 h. The main forecast is issued at 2200 UT and comprises a number of points:

(i) a report of solar features and activity, and immediate geophysical effects during the preceding 24 h, and a forecast of solar activity for the next 72 h;
(ii) a report of geomagnetic activity over the past 24 h, including proton observations at satellite altitudes, and a forecast in general terms of geomagnetic activity for the next 72 h;
(iii) probabilities for the occurrence of X-ray flares and proton flares during the coming 72 h, and a forecast of polar cap absorption events for the next 24 h;
(iv) observed and predicted values for the intensity of 10 cm solar emission;
(v) observed and predicted values of geomagnetic A indices.

Supplementary forecasts at 0500 UT and 1430 UT update the main forecast as required.

These reports in themselves cover many of the needs of those wanting prompt information on solar–geophysical conditions. In addition, some users want to know when an event is in progress, and alerts are sent out to them by telephone or telegram as soon as an event within their range of interest is recognized. Since the X-rays from a flare reach the Earth at the same time as the visible emissions it might seem that such an alert must always come too late to be of any use. However the operator of a communication circuit, for example, finds it helpful to know the reason for a failure immediately. Moreover it is sometimes possible to predict the ultimate intensity of an SID from observations of its initial development; the alert may therefore contain an element of prediction, even for a disturbance lasting only tens of minutes. A solar proton event, on the other hand, develops relatively slowly and may last for several days; here the alert is plainly a practical input to the user's operational strategy.

At the kernel of a solar–geophysical service is the prediction of flare activity. Although flares are also seen in white light, they are most prominent in H_α (6563 Å) light, coming from the chromosphere, and current flare prediction methods depend heavily on features such as filaments and plages seen in H_α. It has been found that the magnetic field structure in active regions may be inferred from the appearance of these H_α regions. The forecaster inspects the solar disc (which may be displayed on a TV screen) for neutral lines where the magnetic field reverses, particular attention being paid to active regions near the lines of polarity reversal. Each region is tested against a number of criteria, such as the development of new magnetic fields, the merging of active regions, reversal of polarities from the normal, and twisting of the neutral line. Reference is made to past activity in deciding how to interpret the tests. Some ancillary information, such as solar radio fluxes, X-ray emissions, and optical activity at the solar limb, all of which refer to conditions in the corona, are taken into account. The forecast does not follow automatically but depends on the training and experience of the forecaster. An example of a solar analysis and forecast, made during the Skylab flight of 1973, is shown in Fig. 12.3.

The flare forecast includes the probability that a flare with a specified X-ray output will occur within the next 24 h. Table 12.3 indicates the classification which

Table 12.3   X-ray flare classification

Class	Energy flux in $1-8$ A measured near the Earth ($\mathrm{erg\ cm^{-2}\ s^{-1}}$)
X	$\geqslant 10^{-1}$
M	$10^{-2} - 10^{-1}$
C	$10^{-3} - 10^{-2}$

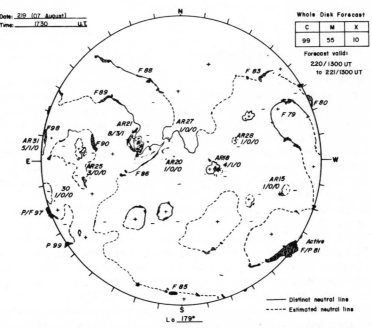

Figure 12.3   Solar analysis and flare forecast made on 7 August 1973, during a Skylab flight. The analysis shows magnetic polarities, neutral lines and active regions. The flare forecast gives the percentage chance that a flare of Class C, M, or X will occur during a 24-h period. (From J.R. Hirman, personal communication, 1974.)

has been in use since 1969. The forecast is made in seven categories, the highest being Class $X \geqslant 50\%$, Class $M \geqslant 90\%$; the lowest, Class $C < 50\%$. To assess their accuracy, the forecasts are continually compared with subsequent events. The assessment process is itself sophisticated, for it is necessary to account for the fact that a forecaster who was merely guessing would score some successes by chance. The state of the prediction art at the present (1976) time is that predictions of whether or not a given active region will flare sometime are excellent and not far below the performance of a perfect forecaster. Predictions of which day the flare will occur are not as accurate, but are still much better than the performance of a hypothetical 'no skill' forecaster. In 1974, solar X-ray data from geostationary satellites began to be readily available, providing a quantitative measure of the development of the active regions, and the accuracy of flare forecasts improved significantly at that time. Such measurements are now received in real time, and further improvements may be expected as solar monitoring becomes more detailed in the future. Some of the approaches to solar monitoring are indicated in Section 12.4.3.

Forecasts of solar radio emissions are based largely on the current levels and the observed trends. For forecasting geomagnetic activity two factors are considered. During the increasing phase of the 11-year solar cycle geomagnetic storms often follow the major flares. At such times geomagnetic forecasts depend heavily on the solar-flare observations and forecasts with a time delay inserted. But during the declining years of a sunspot cycle geomagnetic disturbances show a strong tendency to recur with the 27-day solar rotation (Section 7.2.1), and hence the 27-day history of the activity may be used as a predictor. In general, of course, both factors have to be kept in mind by the forecaster.

12.4.3 Monitoring the space environment

A forecasting and warning service can only operate effectively if fed with a stream of accurate and up-to-date information. At its most general the information should cover optical, radio and X-ray emissions from the Sun; the state of the solar wind and the arrival near the Earth of energetic protons; and those effects of solar activity which are observable in the terrestrial atmosphere or at the surface — the state of the ionosphere, aurora detected by optical, radio or radar methods, and perturbations of the geomagnetic field. Ideally the information should be provided continuously and in real time, i.e. with no delay between the data being acquired by a sensor and displayed in the forecast centre.

The nearest approach to the ideal is probably the Space Environment Laboratory Data Acquisition and Display System (SELDADS) that is being developed to serve the World Warning Agency facilities at Boulder. Fortunately the system is less ungainly than its name. Data are received in various ways. For example, H_α reports are received by teletype from six of the solar telescope sites of the Global Solar Flare Patrol network, while the image from the seventh, located in Boulder, is displayed in real time. On the other hand, readings of X-rays, protons and the geomagnetic field in space, all of which are being monitored by geostationary satellites, are received via the satellite telemetry. Among ground-based observations of the atmosphere's response to solar activity, high-latitude measurements are particularly important and the readings of magnetometers and riometers in Alaska are transmitted to the Agency four times an hour. There are plans to extend this

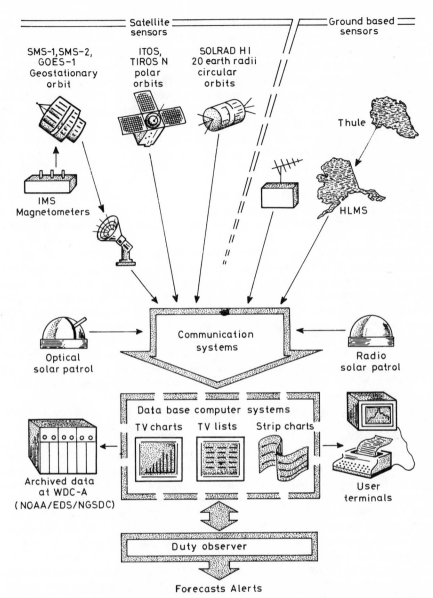

Figure 12.4 The SELDADS approach to space environment monitoring, as planned for 1978. Satellite and ground-based sensors are included (HLMS stands for High Latitude Monitoring Station), as well as inputs from optical and radio solar patrols. The output data are available to local users, the forecaster, and for archiving. (From D.J. Williams, *Report ERL 357–SEL*, Environmental Research Labs., Boulder, Colorado, 1976.)

concept by using a geostationary satellite to relay the readings from a larger ground-based array of magnetometers, thus achieving real-time coverage. Fig. 12.4 gives an idea of the scope of SELDADS in 1978.

The above outline indicates the various approaches to space monitoring that are proving practical, and also illustrates the level of technical sophistication that is

285

being applied to this practical aspect of solar–terrestrial work. Not only do the data that are thereby brought together form the basis of the warning and forecasting services, available to both the scientific community and commercial users, but they also have a more permanent value as a data bank. The policy is to retain the data in full detail for 4 days, and in an averaged form for 32 days. Within a month of the occurrence of some interesting geophysical or solar event, therefore, a great deal of information about it can be obtained quickly from the forecast centre. After a month the data are passed on to the World Data Center, whose function is to archive and to disseminate information in the longer term, as will be seen in the next section.

12.5 Solar–geophysical data services

Solar–terrestrial physics is largely an observational science, concerned not so much with what might happen but rather with what actually does happen in the solar–terrestrial environment, and there is a large synoptic element in many of the observations. The observer sets out to investigate some specific problem, but once his equipment is properly installed and working he has to take what nature sends by way of natural phenomena. Thus it is in the character of the science that many observations accrue that are not quite in line with the observer's original expectation and that will not be amenable to immediate explanation. Moreover, there are observations that cannot be interpreted alone but which begin to make sense if they are set beside others, perhaps from a different technique or a different geographical region. Data can therefore be said to have a potential value, and the principle is widely accepted within the solar–terrestrial community that reliable data should not be lightly discarded.

To facilitate the exchange of data between scientists and to preserve a reasonable body of information in geophysics as a whole, a system of World Data Centers has been established under ICSU (the International Council of Scientific Unions). The scheme was begun in 1957 at the start of the International Geophysical Year (1957–58) because it was foreseen that a large body of data was about to be generated and would need to be open to the scientific community if maximum benefit were to accrue. In retrospect we can appreciate the wisdom of that decision. The IGY arrangements proved to be very effective, and studies based on IGY data continued to appear in the scientific literature for many years after the IGY period itself had ended. Consequently the WDCs have continued to operate, and, indeed, have expanded considerably over the years to keep pace with the growth of geophysical science. The data kept in the Centers are made available at minimum cost to all *bona fide* scientists, who are expected to submit their own observations to the Centers as and when appropriate.

The World Data Centers cover the topics in Table 12.4. (We are only concerned here with the first two major headings, but the full list is given for completeness.) The scheme provides for three Centers, covering different geographical regions. WDC-A is in the USA, WDC-B in the USSR, and WDC-C is divided into two parts: C1 in various countries of Western Europe, and C2 in India and Japan. This geographical distribution is for the convenience of the users. The Centers interchange all data amongst themselves and each in fact assumes an international responsibility, opening its archives to any scientific investigator irrespective of national origin.

One valuable service of the Data Centers is the publication of data summaries,

Table 12.4   Disciplines Covered by World Data Centres

Solar–terrestrial physics
    Solar and interplanetary phenomena
    Ionospheric phenomena
    Flare-associated events
    Geomagnetic variations
    Aurorae
    Cosmic rays
    Airglow

Rockets and satellites

Meteorology
    Atmospheric chemistry
    Atmospheric radioactivity
    Solar radiation
    Ozone
    Noctilucent clouds
    Atmospheric electricity
    Surface and upper air observations

Oceanography

Glaciology

Solid-earth geophysics
    Seismology
    Tsunamis
    Gravimetry
    Earth tides
    Recent movements of the Earth's crust
    Rotation of the Earth
    Marine geology and geophysics
    Magnetic measurements
    Paleomagnetism and archeomagnetism
    Volcanology
    Geothermics

---

such as *Solar–Geophysical Data* which is published monthly by WDC-A. Much of the information contained therein is only 1 or 2 months old, recent enough to be useful in the preliminary analysis of results from on-going experiments. The topics include geomagnetic activity, solar active regions, flares, and the state of the solar wind. The Data Centers also prepare special reports from time to time, bringing together data of one kind, or perhaps a wide collection of data about some unusual geophysical event such as a severe magnetic storm. These special reports assist the progress of geophysical science not only through the facts they disseminate but also by keeping the working scientist aware of current work by other techniques and in other countries.

The collection and dissemination of data may strike some as a mundane business, but in geophysics it is one of the essentials if there is to be progress. A healthy, vigorous science draws strength from many roots, for instance, the interest of able people, and the willingness of government and other bodies to grant financial support. The tradition of data sharing and international cooperation in upper-atmosphere geophysics is without doubt one reason why this particular branch of science has blossomed so profusely in recent years.

12.6 Further reading

Davies, K. (1966) *Ionospheric Radio Propagation* (Dover) Chapters 4–9.

Briggs, A. (1961) *The Birth of Broadcasting* (Oxford University Press).

De Jager, C. and Svestka, Z. (1969) *Solar Flares and Space Research* (North-Holland) Part 5.

McIntosh, P. S. and Dryer, M. (1972) *Solar Activity Observations and Predictions* (MIT Press). Article by J. B. Smith.

Reijnem, G. C. M. (1976) Major legal developments in regard to space activities 1957–1975, *Space Sci. Rev.* **19**, 245.

Johnson, F. S. (Ed.) (1965) *Satellite Environment Handbook* (Stanford University Press).

Valley, S. L. (Ed.) (1965) *Handbook of Geophysics and Space Environments* (McGraw-Hill).

COSPAR (1972) *International Reference Atmosphere, 1972* (Akadamie-Verlag).

Solar–Geophysical Data, published monthly by World Data Center A, Boulder, Colorado.

Appendix 12.1

U.S. Department of Commerce
Office of Telecommunications
Institute for Telecommunication Sciences
TELECOMMUNICATION SERVICES CENTER
Boulder, Colorado 80302

WF- 614

11 August 1976

Weekly Radio Telecommunications Forecast

A. Forecast of HF Radio Conditions 12-18 August 1976

1. Radio conditions should be unsettled at the beginning of the period and then improve as the effects of recent geomagnetic activity diminish. Solar activity is expected to be low to very low throughout the interval. MUFs should be near to slightly above seasonal normals. Radio conditions should remain quiet until about 24 August.

The daily high latitude radio quality is expected to be: 5-6-6-6-6-6-6

2. Sporadic excess attenuation (Solar Flare-Induced SWF on daylight paths)

Relative Attenuation	Probability of Occurrence Each Day
Moderate	15%
Large	10%

3. Ionospheric storms:

Relative Importance	Period	Probability
Minor	24-27 August	50%

4. Polar Cap Absorption: Probability of occurrence-- slight

B. Review of HF telecommunications for the past week. Telecommunications were quiet through most of the review interval, but became unsettled after an increase in geomagnetic activity. 9 August. Solar Activity, though somewhat enhanced in comparison with previous weeks, was low to very low in terms of flare activity. MUFs were slightly to moderately above seasonal normals through August 8 and then became normal to slightly below normal .

High Latitude Observed Radio Quality

August	4	5	6	7	8	9	10
Whole day index	6	6	6	6	6	6	5
6-hour indices	6666	6666	6776	6666	6666	6566	5466
Geomagnetic Activity A_{FR}	5	8	5	5	5	14	7
1700Z 2800 MHz Flux (provisional)	80	82	82	81	80	81	81

C. The following effective solar activity indices (12 month moving average Zurich sunspot numbers) are for use with the new "Ionospheric Predictions" OT/TRER 13.

1976-1977	AUG	SEP	OCT	NOV	DEC	JAN	FEB	MAR	APR	MAY
	9	8	8	7	7	6	6	5	5	5

D. Semimonthly revised ionospheric predictions. The following factors for 16-31 Aug. apply to monthly median predictions for August based on the solar activity index, SS No. 8 .

UT

	00	06	12	18
U.S. East Coast	.90	1.15	.95	1.00
U.S. West Coast	1.00	1.00	1.00	1.05
Europe	1.00	.90	1.00	.90
Alaska	.90	.90	.90	.90
Central Pacific	.95	1.00	.90	.90
Central Asia	.90	.95	.95	.90
S.E. Asia	1.00	1.00	.95	.85
Japan	1.05	.95	.85	.95
M.East	.90	1.00	.95	.90

USODMM - LAL

Questions

The reader who has mastered the text will find little difficulty with these questions. On the whole they are at a level suitable for an examination, and indeed many of them could readily be adapted for that purpose. They should also be found useful as a basis for discussion in tutorial sessions, and for private study as an aid to revision.

1. Write an account of the main features of the vertical distribution of the atmosphere above 50 km with regard to pressure, temperature and composition. Assuming that the relative proportion of major constituents at 100 km is much the same as at the ground, show by numerical example how it is that atomic oxygen comes to be the dominant neutral species at 300 km.

2. A topside ionogram shows resonance spikes corresponding to the O- and E-modes at 0.95 MHz and 1.55 MHz respectively. Work out the electron density and the magnetic field at the satellite.

3. Define the optical depth, τ, for an atmosphere irradiated by monochromatic radiation, vertically incident, where the atmosphere contains n ionizable atoms per unit volume and the absorption cross-section is σ. If the atmosphere is in hydrostatic equilibrium with scale height H, show that $\tau = \sigma H n$. Comment on the significance of the condition $\tau = 1$, and hence show that

$$\tau = \exp - (h - h_m/H)$$

where h_m is the height of maximum ion production.

4. A satellite of mass 750 kg, area 5 m^2, is decaying from circular orbit. At what altitude (approximately) will it begin its last circuit of the Earth?

5. What is the shortest wavelength you would specify for an incoherent scatter radar to explore the mid-latitude ionosphere between altitudes 80 and 1000 km, and why?

6. Suppose that electrons are produced in the F region by photo-ionization at a rate q, and lost by a two-stage recombination process, where the recombination coefficient is α and the attachment coefficient is β. Show that at equilibrium $1/q = 1/\beta N + 1/\alpha N^2$, where N is the equilibrium electron density, transport being neglected.

7. Explain why the protonosphere starts higher up in a hot exosphere than in a cooler one. If the transition from heliosphere to protonosphere occurs at 800 km when $T = 700\,°K$, where would you expect it to be at $1000\,°K$, neglecting effects due to other constituents?

8. Discuss and contrast various approaches to observing and measuring the upper atmosphere, stressing the advantages and disadvantages of each group. Illustrate your points by referring to particular instruments and techniques.

290

9. What do you understand by the term 'anomaly' as applied to the behaviour of the ionosphere? Name, and give a brief account of, the major anomalies of (i) the D region, and (ii) the F region. To what extent do you consider the term 'anomaly' as obsolete in these cases?

10. Assuming that collisions and the geomagnetic field may be neglected, derive a simple formula showing that the electron content may be determined if two harmonically related radio signals are emitted from a satellite beyond the ionosphere. Anticipate some problems of this technique, and consider how they might be overcome.

11. Show that a particle with charge e and mass m gyrates in a magnetic field of induction B with angular frequency $\omega_B = Be/m$. Explain briefly how the motions of electrons and ions in the upper atmosphere depend on the relative values of ω_B and the collision frequency ν. Hence explain why the E region may be thought of as a dynamo and the F region as a motor.

12. Show that the electron density in a Bradbury layer increases indefinitely with altitude. What are the factors that determine the height of the peak of the F2 layer? Assuming a thermospheric temperature of $1000\,^\circ K$, and neglecting the geomagnetic field, estimate the height of the F2 peak if the effective attachment coefficient at 200 km is $10^{-2}\,s^{-1}$, and the plasma diffusion coefficient at the same height is $10^9\,cm^2\,s^{-1}$. Taking the geomagnetic field into account, what will be the effect of an equatorward wind of $100\,m\,s^{-1}$ at middle latitude? At what latitude will such a wind have greatest effect on the ionosphere under equinoctial conditions?

13. Starting with the Chapman production function, show that
    (a) the maximum production rate at zenith angle χ is given by $q_m(\chi) = q_m(0)$ $\cos\chi$, where $q_m(0)$ is the maximum production rate when $\chi = 0$;
    (b) the height of a level of constant electron production, well below the layer maximum, varies as $\ln(\sec\chi)$. How can this be used to determine the scale height of the neutral atmosphere?

14. Show that the separation between positive ions and electrons due to gravity is approximately $\delta h = \lambda_D^2/H$, where λ_D is the Debye length and H the scale height.

15. Which parts of the ionosphere are most likely to absorb an incident radio signal, and why?

16. Suppose the E-region critical frequency maximizes 10 min after local noon. What does this tell you about the E region and its likely behaviour after sunset?

17. What is the connection between fertilizer and the science of the upper atmosphere? (In other words, write an essay of not more than 100 pages on the ionospheric D region, succinctly indicating its scientific and practical importance, and why, though it is the lowest part of the ionosphere, it is also the most difficult part to study.)

18. Quoting specific techniques, show how the radio refractive index may be exploited to investigate the vertical profile of the ionized upper atmosphere when the vertical profile of the refractive index is (i) slowly varying, (ii) discontinuous, and (iii) irregular. What fundamental differences of technique depend on whether the critical frequency or the wave frequency is the greater?

19. A 6 MHz radio wave reflected from the ionosphere fades to a depth of 10% in amplitude. The correlation time of the fading is 20 s, and spaced receiving antennas indicated a fading distance over the ground of 50 m. Between two

antennas 50 m apart in an east–west direction, similar fades appear to be displaced in time by 1 s. What can you deduce from these observations about the irregularities producing the fading? What further observations would you make to improve your information?

20. At height 120 km an acoustic-gravity wave has a period of 6 min and a horizontal wavelength of 20 km. What is the vertical wavelength and the directions of propagation of the phase and the energy? How would your interpretation differ if the observations related to a height of 150 km?

21. If the solar wind plasma contains 5 protons and 5 electrons/cm^3, and the IMF is 5 γ, work out the Alfvén speed and Alfvén Mach number of the solar wind with respect to the Earth. (Make the calculation in both MKS and cgs units, and verify that the answer is the same.)

22. Cars are travelling along a busy highway at 50 m.p.h. If the typical driver reaction time is 0.5 s, what happens if the distance between cars is allowed to narrow to less than 36 ft? What has this to do with the magnetosphere, and with what consequences?

23. If the magnetotail field is 20 γ, what must be the temperature of the particles in a stable plasma sheet containing 10 particles cm^{-3} ?

24. Consider the gyro-radius of protons in the solar wind. Hence say what energies should be possessed by solar protons arriving isotropically at the orbit of the Earth.

25. A polar cap event produces 5 dB absorption of 30 MHz radio waves, as measured with a riometer. The cut-off is observed at 72° geomagnetic latitude. Assuming the effective recombination coefficient in the D region to be 2×10^{-6} cm^3 s^{-1}, estimate the amount of abnormal ionization, and the flux of protons producing it. (For simplicity assume a mono-energetic flux.)

26. To what extent is a knowledge of the form of the magnetosphere and the physical processes at work within it important for understanding solar–terrestrial relations? Illustrate your answer by reference to specific disturbance phenomena, emphasizing the role of the magnetosphere in each.

27. Assuming conservation of the first adiabatic invariant, show how the equatorial pitch angle of a geomagnetically trapped particle determines its mirror point. Show that the integral of parallel momentum of the particle between mirror points l_1 and l_2 depends on

$$\int_{l_1}^{l_2} (1 - B/B_{\mathrm{M}})^{1/2} \, dl$$

where B is the magnetic induction at a point on the field line, and B_{M} is that at the mirror points.

28. Discuss the role of the IMF in determining the structure and dynamics of the magnetosphere. If the cross-tail electric field is 0.9 mV m^{-1} and the plasmapause is observed to be at $L = 4$, work out an expression for the co-rotation electric field as a function of L.

29. Discuss the conditions leading to the generation of VLF waves in the magnetosphere. A station at 60° geomagnetic latitude observes emissions at 10 kHz but not at 3 kHz. Some time later emissions appear at 3 kHz. What interpretation would you put on these observations? What further information would you like to obtain and why?

292

30. Write brief notes to indicate the geophysical significance of: (a) nose whistlers; (b) the clefts; (c) the knee of the plasma distribution; (d) the tail of the magnetosphere; (e) Van Allen's belt.

31. Discuss qualitatively and quantitatively the effects of an increase in the magnetospheric electric field on the mid-latitude ionosphere. How far could an increase of this kind account for the changes observed during an ionospheric storm?

32. Identify and compare points of similarity between the solar flare and the terrestrial substorm.

33. Use an image method to work out the position of the magnetopause on the Earth–Sun line (image moment = 28 × earth moment; location of image $40\,R_E$).

34. Suppose that a field line originally going out to $10\,R_E$ collapses in to $5\,R_E$ during a substorm. Assuming as a first approximation that the field line is dipolar throughout, estimate the effect of the collapse on particles that originally were moving (a) transverse to the field, and (b) almost directly along the field.

35. Estimate the magnitude of the electric field across the magnetosphere when $K_p = 1$ and $K_p = 6$, given that the plasmapause occurs respectively at 5.3 and 3.5 earth radii. Hence, assuming a reasonable value for the tail magnetic field, calculate the speed of plasma convection from the lobes towards the neutral sheet. Compare this with the solar wind speed under the same conditions, and comment.

36. In a central heating system the water pump is driven by an induction motor, in which a magnetic field rotating at 50 Hz spins a copper disc. How large must the disc be for reasonable coupling (conductivity of copper $= 6 \times 10^5$ mho cm^{-1})? Use the same physical principles to comment on the penetration of external electric fields into the ionospheric F region.

37. Write an essay on the ionospheric storm, showing to what extent the observations may plausibly be explained by means of known mechanisms, and discussing why it is important to develop a full and correct understanding of this phenomenon.

Index

Hydrogen-alpha line, 139, 181, 229, 282–284
Hydromagnetic wave, 23–24
Hydrostatic equation, 54–55
Hydrostatic equilibrium, 54–55, 157

Impedance probe, 29
Incoherent scatter, 45–49, 134
Indirect sensing, 25
Infrasonics, 135–136
Initial phase of magnetic storm, 203
Inner belt of trapped particles, 165, 166–167
Integral invariant, 163–164
International Geophysical Year (IGY), 183, 284
International Scientific Radio Union (URSI), 279
Interplanetary magnetic field (IMF), 147–149, 172–173, 236
Inverted-V event, 191
Ion,
    hydrated, 76
    negative, 76–77
    terminal, 77
Ion drag, 123, 128
Ion production rate, 61–64
Ion-cyclotron waves, 209–210, 217
Ionization efficiency, 62, 70
Ionization potential, 68
Ionosonde, 37–39, 134
Ionosphere, 2, 4, 53, 60
Ionospheric layer,
    alpha-Chapman, 65
    beta-Chapman, 65
    Bradbury, 72
    D, 60, 74–77, 92–93
    E, 60, 70–72, 89–91
    F, 60
    F1, 70–72, 92
    F2, 72–74, 93–98
Ionospheric radio propagation, 37–50
Ionospheric storm, 244–251, 281
Irregularities in ionosphere, 112–115

Kinetic theory, 7–11
Knee of plasmasphere, 157–159
K_p index, 177, 188

L parameter, 164
Langmuir probe, 27–28
Laser, 35
Layer conductivity, 86–87
Lightning, 259–260
Lobes of magnetosphere, 159–160, 199–200
Longitudinal conductivity, 85
Low-energy plasma in magnetosphere, 158–161
Lyman-alpha line, 74

M-region, 183, 188
Magnetic bay, 188
Magnetic disturbance, 188
Magnetic storm, 203–204, 280, 284
Magnetic moment (of trapped particle), 162
Magnetometer, 32, 217
Magnetopause, 53, 152–154
Magnetosonic wave, 24
Magnetosheath, 155
Magnetosphere, 4, 53
Magnetotail, 155–157
    current, 198–200
Main phase of magnetic storm, 203–204
Mantle aurora, 184
Mass spectrometer, 30–31
Maximum usable frequency (MUF), 274–275
Maxwell–Boltzmann law, 8–9
Mean free path, 10
Merging (of magnetic fields), 174
Mesopause, 52–53
Mesosphere, 52–53
Meteor trail, 115
Micropulsations, 190, 215–219
Midday recovery (of PCA), 240
Mirror point (of trapped particle), 162
Monochromatic electron flux (in aurora), 190–191
Motor region, 126–127

Negative ion, 76–77
Nitric oxide, 58, 74, 242–243, 252
Nitrogen, atomic, 58
Noctilucent clouds, 135

Open field-lines in magnetosphere, 172–173, 197–198
Ordinary wave, 23, 39, 42, 43–44
Outer belt of trapped particles, 165–168
Ozone, 57, 77, 243, 271

Pc pulsations, 217–218
Pi pulsations, 217
Partial reflections from ionosphere, 41–42
Pearls, 217
Pedersen conductivity, 85
Penetration frequency, 38
Permeability of free space, 6–7
Permittivity of free space, 6–7
Phase screen, 108
Phase velocity, 18
Photo-electron, 48
Photochemistry, 3
Pitch angle, 162
    anisotrophy, 221–223
    diffusion, 223–224
Plage, 139
Plasma diffusion coefficient, 67, 73
Plasma drift, 126–127
Plasma frequency, 14–15, 37–39
Plasma line, 47